Springer Proceedings in Energy

More information about this series at http://www.springer.com/series/13370

Sebastian Groh · Jonas van der Straeten
Brian Edlefsen Lasch · Dimitry Gershenson
Walter Leal Filho · Daniel M. Kammen
Editors

Decentralized Solutions for Developing Economies

Addressing Energy Poverty Through Innovation

Springer

Editors
Sebastian Groh
Microenergy Systems Research Group, ZTG Sekr.
Technische Universität Berlin
Berlin
Germany

Jonas van der Straeten
Microenergy Systems Research Group, ZTG Sekr.
Technische Universität Berlin
Berlin
Germany

Brian Edlefsen Lasch
Microenergy Systems Research Group, ZTG Sekr.
Technische Universität Berlin
Berlin
Germany

Dimitry Gershenson
Berkeley Rural Energy Group
University of California, Berkeley
Berkeley, CA
USA

Walter Leal Filho
Faculty of Life Sciences
Hamburg University of Applied Science
Hamburg
Germany

Daniel M. Kammen
University of California Berkeley
Berkeley, CA
USA

ISSN 2352-2534 ISSN 2352-2542 (electronic)
Springer Proceedings in Energy
ISBN 978-3-319-15963-8 ISBN 978-3-319-15964-5 (eBook)
DOI 10.1007/978-3-319-15964-5

Library of Congress Control Number: 2015932434

Springer Cham Heidelberg New York Dordrecht London

Printed on acid-free paper

Springer International Publishing AG Switzerland is part of Springer Science+Business Media (www.springer.com)

Foreword by ADA

ADA is a nongovernmental organization approved and cofinanced by the Luxembourg Directorate for Development Cooperation and Humanitarian Affairs and is placed under the High Patronage of Her Royal Highness the Grand Duchess Maria Teresa.

Over the past 20 years, ADA has been dedicated to building and catalyzing the financial inclusion of populations excluded from conventional banking channels in developing countries. ADA empowers microfinance institutions and networks, and helps them obtain funding for sustainable growth through the Luxembourg Microfinance and Development Fund. It also supports them through specific technical assistance projects enhancing human resources and capacity building through the implementation of management tools. Convinced that the development of healthy local financial institutions is essential to the success of any sustainable development strategy, ADA offers its know-how to individual States and collaborates with them to support their inclusive finance expansion strategy. ADA makes available its expertise in areas including youth financial inclusion, access to green energy through microfinance, microinsurance, and linking remittances to savings for migrants.

Research and innovation are of the founding blocks for the definition of ADA's activities. The provision of adapted financial services to underserved populations requires research on the one hand in order to understand markets and document unknown practices, and development or innovation on the other hand to resolve problems and propose solutions responding to people with unsatisfied needs. This means targeting excluded populations, understanding their context, and helping institutions design financial products that have greater impact by enabling populations to achieve their own development objectives and improve their quality of life. ADA strongly believes financial inclusion is a means and not an end in itself. Therefore, if vulnerable populations need better energy services, which should be clean, renewable, efficient, and affordable, access to financial services providing them with the opportunity to acquire the equipment, the technology, or the technical assistance required for this purpose, should be provided. In this sense, ADA believes that it can contribute to poverty alleviation and autonomous development through financial inclusion.

Reverse innovation or trickle-up innovation are terms referring to any innovation that is adopted first in the developing world. ADA believes this is an eye-catching field for applied research. The processes for developing and marketing goods and services arising from grassroots ingenuity are subject to become more inclusive as they are relevant to markets lower down the pyramid. We applaud and thank BREG and MES for their Symposium and for choosing to focus it on Innovating Energy Access for Remote Areas and discovering untapped resources. We strongly believe this approach is critical for improving peoples' living conditions and advancing inclusive and sustainable development in the developing world.

ADA's research and development activities focus on establishing the link between scientific research and practical experiences in the financial inclusion sector. This frequently entail, bringing private and public actors together to share experiences and learn from each other. The MES and BREG Symposium in Berkeley have strongly contributed to these efforts.

We were delighted to have the opportunity to share our experience in linking access to finance with access to energy with participants and for having the opportunity to have a first-hand account to research findings. We congratulate researchers for their insightful papers and encourage microfinance experts and other financial industry providers to take into account the technical solutions presented as investment opportunities. ADA will be happy to continue its collaboration with BREG and MES in the future and will also encourage, young as well as experienced researchers to continue their good work. Their contribution is fundamental for capacity building purposes, for the professionalization of the microfinance sector and for advancing inclusive innovation.

Olivier Massart
ADA, Operations Director

Juana Ramirez
ADA, Research and Development Project Officer

ADA asbl, 39 rue Glesener
L-1631 Luxembourg
Tel.: +352 45 68 68–1, Fax +352 45686868
adainfo@microfinance.lu, www.adamicrofinance.lu
Inclusive Finance. Increasing Autonomy. Improving Lives.

Preface by Microenergy Systems Research Group

In 2001, at the Technische Universität Berlin Institute for Energy Technology, two young researchers, Noara Kebir and Daniel Philipp, traveled to Bangladesh to explore a seemingly simple question: How is it possible that in a country like Bangladesh solar energy is implemented on a sustainable basis, whereas in Germany it is considered a mere fancy dream? Intrigued by the example of *Grameen Shakti*, which since its founding in 1996 continues to successfully expand the dissemination of solar home systems in rural areas of Bangladesh, they founded a new research focus at their University, which they named *Microenergy Systems (MES)*.

Microenergy systems can be defined as *decentralized energy systems based on small energy appliances, which provide households, small businesses or farms with locally generated energy, enabling a spatial link between energy demand, supply and generation*. The systems provide solutions for single households or micro-enterprises (e.g., solar home systems, irrigation pumps, biogas plants) as well as technologies for several interconnected households or communities (e.g., mini-grids). In recent years there have been growing expectations that microenergy systems can play an important role in shifting energy policy as well as creating co-benefits where development goals (e.g., reducing poverty) are combined with climate mitigation and adaptation actions.

A systemic perspective on microenergy systems stresses the need for a holistic and rigorous interdisciplinary research methodology, including feasibility analysis, product development, manufacturing, planning, implementation, servicing, and use of decentralized energy systems, combining different perspectives stretching from end-users to policy decision-makers. Therefore, the term "Microenergy System" refers to a broader understanding of the respective technical artifact in interrelation with its natural conditions, social context, economical system, organizational structures, and political framework. The introduced research concept is demand-driven and deduced from practical needs rather than a product of pure theoretical consideration, and consequently exists outside the purview of any individual discipline.

We find that there is a strong need to consolidate the knowledge accumulated in the various overlapping fields of research focused on such solutions, through regular exchange as well as through joint exploration and education. The aim of these efforts is to enable the transfer of strategies, programs, and tools that can be tailored to the diverse myriad of local conditions. This was the original goal of the founders of MES and continues to be our driving force for our ongoing activities including the third event we have organized on decentralized energy supply: the MES-BREG symposium held at the University of California, Berkeley, Berkeley, in 2014.

Sebastian Groh
Jonas van der Straeten
Researcher, Microenergy Systems

Preface by Daniel M. Kammen

After what seems like years of slow and uneven action in the off-grid and remote energy sector, there are dramatic signs of innovation. One only needs to assess the progress of mobile phone coverage across the Global South to see new technologies, business models, and modes of commerce and governance. At the same time, the global energy sector is scattered, and lacks rigorous efforts to support innovation to build greater sustainability in the services that are made available—generally as commercial products—to billions of people. For much of the last century, the energy sector was dominated by monopolies with partial or complete ownership of the regional grids and large-scale power plants. Research has shown that in areas with these monopolies the metrics of energy sector innovation are markedly low, perhaps because such monopolies are not motivated to disseminate results, or that they patent defensively, or simply underinvest in new scientific, technical, and business models. This raises a red flag if we are to employ an innovation-based strategy to confront the major energy-related challenges facing our society today.

Thankfully the energy field overall has seen more change and innovation over the past decade than arguably at any time in the last 100 years. New large-scale energy technologies and practices are only part of that change. The rapid evolution of decentralized approaches to energy access has been dramatic. Increasingly, bottom-up business and community-based efforts are playing important roles in changing the types and quality of energy services available worldwide. Today, a genuine hope for energy inclusion has gained momentum. This new paradigm is visible in high-level United Nations and national development efforts, through a vibrant private sector, and through the expectations that a diverse provider and consumer community brings to the equation. The value of diversity and the demand for quality service has been instrumental in driving innovation.

It is in this spirit that last year's symposium had *innovation* as its focal point. *Innovating for Energy Access* set out to bring together leading experts in the field to share their experiences on innovative approaches and to jointly drive new thought. The idea that innovative approaches can work with previously underutilized or unrecognized resources is central to this symposium, as this may lead to circumstances or cues for the development of successful and sustainable energy access programs. Such untapped resources may be seen in the discovering of synergies within areas such as pre-existing service infrastructures, supply chain and value chain management, natural resource availability, financing schemes, leap frog technologies, and more.

I particularly thank Microenergy Systems from the Technische Universität Berlin and the Berkeley Rural Energy Group for their efforts in making this symposium a success. A vision of an inclusive, service-driven, and innovating energy sector will enable this moment of hope to be seen as part of a wave of change. This symposium and the resulting volume of collected papers will hopefully be one important contribution to that evolution.

With best regards, to all those who read and benefit from these case studies and assessments.

Daniel M. Kammen
Professor Energy and Resources Group
Goldman School of Public Policy
Department of Nuclear Engineering

Founding Director
Renewable and Appropriate Energy
Laboratory, University of California
Berkeley

Preface by the Organizing Team

The Western world lives by a paradigm of electrification that was formed at the end of the nineteenth and the beginning of the twentieth century. The story revolves around the extension of transmission and distribution infrastructure and the development of large-scale centralized generation. Today, the ubiquitous supply of electricity has become self-evident in much of the US and Europe, and sometimes passes for a barely noticed feature of peoples' daily lives in much of the industrial world. However, for many countries in the Global South, electricity is a luxury for the few. We do not know today what historians in 100 years will write about the path to universal access to modern energy in South Asia or Sub-Saharan Africa. Maybe they will write about the technology transfer of mobile payment and DC grids—from the Global South to the North. We cannot know the story yet—but as a research community, we all have the opportunity to make our contribution to it.

The following pages are a compilation of the research that was presented at the jointly organized *MES-BREG Symposium, Innovating Energy Access for Remote Areas*: *Discovering untapped resources*. The event was a resounding success, and it is our great pleasure to share the presented works with the general public. We hope that the publication of this knowledge will motivate the energy access research community to come together for future conferences, develop stronger collaborations, and more freely exchange knowledge and ideas.

The MES biannual conference events began in 2011, but this year marks the first collaborative symposium, which was organized in partnership with the Berkeley Rural Energy Group (BREG). MES and BREG are excited by the outcome of this collaboration, and we hope that this is just the first in a series of many smaller themed symposia that will take place in new locations around the globe, staggered year-for-year with our major Berlin events.

The MES-BREG event showed us that there remains a plethora of research questions that needs to be addressed by the international research community on energy access and decentralized energy systems. As the field is rather interdisciplinary, problem-focused, and practice-oriented, we need the support of the traditional established disciplines to tackle the remaining challenges. For this reason, the empirical problem—the lack of access to sustainable energy worldwide—needs to be introduced and emphasized in mainstream theoretical debates. At the same time, we need to maintain our close connection to the practitioners in the field, the entrepreneurs, the consultants, and to the decision makers.

The event held at Berkeley marks a few developments of which we are particularly proud. First, we are extremely grateful to *ADA*, the Luxembourg organization dedicated to inclusive finance, for its generous support of the symposium as well as for its continued dedication to helping researchers publish their work on the relationship between microfinance and the environment. Second, the symposium was also organized in coordination with Environmental Research Letters, which will publish extended versions of select papers. Finally, as in the previous years, we thank our conference organization partner *MicroEnergy International* (*MEI*), a Berlin-based consultancy active in Africa, Latin America, and Asia, aiming for the provision of fair and sustainable energy for all. Without their continuous support, these events would not be possible.

We also would like to thank the authors who submitted more than 70 papers on innovations in energy access from across the world. Our sincere gratitude also goes to our distinguished Scientific Committee that took the time to review and provide feedback on these papers. We would like to thank the *Hans-Böckler Foundation* for their continuous financial support to MES, as well as *ADA, GIZ/Bangladesh,* and *the Pakistan Poverty Alleviation Fund'* for providing scholarships for researchers from the Global South to come to Berkeley. We are also very grateful to *REEEP* for supporting a workshop during the symposium. We also thank *BERC, the Trojan Battery Company*, and *SEA-RAE* for their financial support in hosting the poster session networking event and reception. Finally we are also grateful to *energypedia,* our media partner for this event, for making sure that the main conference takeaway messages are accessible to the global energy access community.

Your organizing team

Sebastian Groh, MES
Jonas van der Straeten, MES
Brian Edlefsen Lasch, MES
Dimitry Gershenson, BREG/RAEL
Daniel M. Kammen, BREG/RAEL

Conference Summary

What is moving the off-grid world? A status quo analysis

With the turn of the twenty-first century, the global energy landscape is changing profoundly. Rapidly improving technologies for decentralized electricity generation are challenging traditional grid systems across the world. The Sustainable Energy for All initiative (SE4All) formed by the UN projects that 70 % of the currently unelectrified population will gain access to electricity by 2030 through mini-grids or small, stand-alone off-grid solutions (SE4ALL 2013). These factors have all contributed to a surge of interest in decentralized power, among academics as well as practitioners and investors.

"From the bottom up"—the decentralized track of electrification

National governments have also become increasingly aware of the limits of top-down, supply-driven electrification approaches. The World Bank recently echoed this shift by publishing "From the bottom up: How Small Power Producers and Mini-grids Can Deliver Electrification and Renewable Energy in Africa," a comprehensive report on the changing policies and practices of energy access in the African continent. The report highlights a fascinating fact that was unheard of only a few decades ago: decentralized energy development appears as a recommendation in virtually every national electrification strategy across Sub-Saharan Africa (Tenenbaum et al. 2014).

The future of rural electrification—drawing from lessons learned

The rural electrification of Bangladesh, where Solar Home Systems are installed at a rate of around 50,000–70,000 per month, is a well-known success story in the decentralized energy field. In his presentation at the MES-BREG conference, Professor Rezwan Khan illustrated that case and gave us an outlook into the future of rural electrification in Bangladesh, with implications for the greater Global South. The potential applications are diverse and promising: new uses of solar PV for irrigation and cold storage of agricultural products; the clustering of PV panels and the establishment of DC-nanogrids; and ultimately, through a "swarm electrification" approach, the connection of SHS clusters and stand-alone grids to the national grid system.

However, as profound as the changes of the supply side of rural electrification may seem, the evolution of consumer energy use deserves similar attention. Innovation and improvement in the cost and efficiency of end-use devices, above all for mobile phones and LED lamps, have changed the off-grid world significantly. For more than a 100 years, telecommunication and lighting have been a privilege for urban elites in countries of the Global South, but are now becoming accessible and affordable to large numbers of rural off-grid and low-income clients. Just in Sub-Saharan Africa, mobile phone subscriptions have grown by nearly 18 % CAGR since 2007, reaching nearly 253 million unique mobile subscribers in 2013 (GSMA 2013). Concurrently, off-grid lighting markets have been booming. The Lighting Africa initiative of the World Bank and IFC forecasts 20–28 million cumulative sales of solar lighting products in Africa by 2015 (Lighting Africa 2012). This rapid spread of these small and inexpensive mobile electrical appliances is challenging the age-old access story. One no longer needs a grid connection, a solar home system, or a diesel generator to attain two key uses of electricity: communication and lighting.

Innovative approaches for new markets

Fundamental changes in technology costs and connectivity have also led to the emergence of a wide array of innovative business models for off-grid energy. At the same time, the same factors are improving the ability of researchers to access supply chain actors, develop new technological solutions, and more closely liaise with policy makers and government officials. We at MES-BREG 2014 were able to see this first hand, and the following pages provide a short summary of the research presented.

Key Topics

New Experiences, Simulations and Visions from the Field of AC and DC Minigrids

This book is divided into three sections which outline related topics around innovation in energy access. The first section focuses on mini-grids, and the emerging role of DC power. First, Sebastian Groh and his colleagues suggest the concept of "Swarm Electrification", a bottom-up approach for building up microgrids by interconnecting existing small-scale DC generation units (Groh et al., paper "Swarm Electrification" in this book). The approach is likened to the concept of swarm intelligence, where each individual node brings independent input to create a conglomerate of value greater than the sum of its parts. In the given scheme, otherwise unused electricity from individual Solar Home Systems can be traded between households. As the systems tend to be sized to ensure the battery can also be fully loaded in the rainy season, as well more efficient DC-run appliances increasingly penetrate the market, more electricity is generated by the panel during sunny days as can be stored in the battery, a concept that is discussed by Hannes Kirchhoff in his paper (Kirchhoff, paper "Identifying hidden resources" in this book).

Rezwan Khan makes a powerful argument for DC grid technologies, challenging the current paradigm of AC infrastructure in his paper on "A Concept of DC Nano-grid for Low Cost Energy Access in Rural Bangladesh" (Khan and Brown, paper "A Concept of DC Nano-grid" in this book). In fact, according to Khan (as elaborated during his keynote address), Bangladesh will have converted its national grid to DC within the following 15–20 years. This trend is not only expected in the developing world, as Paul Savage from Nextek Power Systems suggested during his keynote speech. According to his team's research, 80 % of all electricity worldwide is used by native DC power electronics, and DC domain segments are rapidly expanding in industrialized countries.

Research suggests that mini-/ microgrids hold the promise of becoming the most cost-efficient technology for the "decentralized track" of electrification (Tenenbaum et al. 2014). Still, as we learn from two case studies presented in this context from Bangladesh (Khan and Huque, paper "Experience from First Solar Minigrid" in this book) and India (Chandran-Wadia et al., paper "Prospects for Electricity Access in Rural India" in this book), there are major challenges to overcome during the next years: the inability to attract private investors, the slow dynamics of matching the supply and demand of a particular site, and the dilemma between affordability of energy access versus a sustainable business model. Two of the papers in this section also provide important design considerations for mini-grids (Chowdhury et al., paper "Off-grid Rural Area Electrification" in this book), as well as an innovative participatory tool for design and planning of microgrids through a simulation gaming approach (Abdullah and Kennedy, paper "A Simulation Gaming Approach" in this book).

Focusing on implementation, Muchunku et al. present first-hand insights from their analysis of financial performance and model efficacy for a project running since 2012 based on an energy center model (Muchunku and Ulsrud, paper "The Energy Centre Model" in this book). In "Lessons from the Edge," Newell and colleagues investigate how the high human cost of delivering energy on the battlefield led the US Army to reexamine assumptions and risks in the provision and use of energy and thereby derive lessons for instilling energy resilience in rural communities (Newell et al., paper "Lessons from the Edge" in this book).

Innovations in Value Chains and Financing Schemes

In the first paper in this section, Henrik Beermann and colleagues propose an innovative conceptual framework for energy-based upgrading in agricultural value chains and illustrate it with the example of rice farming in the Philippines (Beermann et al., paper "Value chain thinking and energy projects" in this book). Using the case of rural India, Kumar et al. introduce methods of advanced solar irrigation scheduling for improvements to agriculture (Kumar et al., paper "Advanced Solar-Irrigation Scheduling" in this book). Tackling the end of life concerns, Alexander Batteiger proposes an analytical framework and next steps toward a sustainable waste management system for retired SHS storage systems (Batteiger, paper "Towards a Waste Management System" in this book).

While value chains are critical to improving access, universal access largely remains a financing problem. However, appropriate finance mechanisms to overcome the investment barrier for sustainable energy technologies in emerging markets are still largely underdeveloped. At MES-BREG 2014, these topics were discussed at length, with four researchers sharing their recent discoveries in this sector. In this book, Natalia Realpe presents a paper on the question of how to scale up green microfinance loans (Realpe, paper "How to Scale Up Green Microfinance" in this book); Satish Pillarisetti discusses the innovative group-based products that have been developed for rural communities in India through partnerships between retail banks and locally based NGOs (Pillarisetti, paper" Microfinancing decentralized solar energy systems" in this book); Izael Da Silva showcases the key insights from his team's work with the LUAV project in Uganda (Da Silva et al., paper "The LUAV-Light Up a Village project" in this book); and Dominique Diouf explores the barriers to impact investing in sustainable energy in West Africa (Diouf, paper "Exploring the Barriers to Impact Investing" in this book).

Mobile connectivity has grown far beyond the extent of the electricity grid in most emerging markets. As we learned in our session on ICT and Energy, there are more than 643 million people worldwide covered by mobile networks but without access to electricity, representing up to 53 % of the global off-grid population. The mobile industry is pushing the demand for decentralized electricity supply in remote areas. In parallel, it also provides a range of new channels for innovative energy business models and projects: the off-grid telecom tower infrastructure, mobile operators distribution networks, machine to machine connectivity, mobile payments and mobile services. Much more research is needed to explore the synergies between the ICT and energy sectors, but also the challenges, e.g., in regard to privacy and consumer protection. The conference session on the nexus of Energy and ICT addressed some of the questions that the sector currently faces. Here we include texts by Michael Nique et al, who present their team's work at GSMA highlighting the opportunities to employ mobile technologies to foment off-grid energy access (Nique and Smertnik, paper "The Synergies between Mobile Phone Access and Off Grid EnergySolutions" in this book) and Bruce Nordman, who discusses the possibility of using smart appliances to reduce costs in off-grid energy installations by responding to local energy prices in real time (Nordman and Bugossi, paper "Optimizing Device Operation" in this book).

The role of Big Data was also covered at MES-BREG 2014 during the panel session moderated by Peter Alstone: "Innovating at the Nexus of Big Data and Energy Access". The panelists, Kate Steel from Google, Lesley Marincola from Angaza Design, and Michael Nique from GSMA all agreed on the value of this data but raised the questions: who is allowed unrestricted access, for what purpose, and at what cost? These questions need to be answered in the coming years as part of a critical push to better understand the sector. Other burning questions that were raised included: What happens when hundreds of thousands of off-grid users become more "legible" to big companies and potentially governments? What are the dangers of the ability to remotely control the usage of energy and mobile payment services? How can user privacy be protected while allowing companies to improve their services for consumers?

Implementation and Decision-Making

The delivery of adequate energy services requires a deep understanding of consumer conditions. What is the willingness and ability to pay for sustainable energy or energy efficiency? What existing sources of energy are used and can potentially be substituted? What is the role of energy in local value creation processes? These remain the major questions that all business and project developers have to deal with. The conference provided a number of valuable contributions to the debate on the right approaches and methods to answer these questions. Two papers on demand assessments were presented, one on the financing needs for thermal insulation measures for housing in Kyrgyzstan (Bakteeva and van der Straeten, paper "Financing Energy Efficiency and Climate Adaptation Measures" in this book) and one on the demand for Solar Home Systems in Pakistan (Ajaz and Taylor, paper "Demand Assessment of Solar Electrification" in this book).

To date, access programs find it difficult to scale up and meet the high expectations of donors and beneficiary communities. Against this background, Sesan investigates a context-responsive approach that has the chance to facilitate the dissemination of locally appropriate interventions based on a case in Western Kenya (Sesan, paper "Scale vs. Substance" in this book). Going beyond the common knowledge on health risks due to indoor air pollution, Verástegui and Pilco show first evidence from Peru's rural context in the simultaneous lack of modern energy devices for lighting and cooking (Verastugi and Pilco, paper "How big is small? Enough to not breathe oil!" in this book).

A key driver of the expansion of renewable and decentralized energy generation lies in the national electrification strategies of governments and regional bodies. In her analysis of decision-making for off-grid rural electrification in Colombia, Valencia argues that costs related to electricity provision through diesel are not adequately or correctly taken into account, and argues for a planning methodology which is more comprehensive, mindful of final costs of service for the end user, and attentive to the long term sustainability of service (Valencia, paper "Analysis of decision-making for off-grid rural electrification" in this book). On the basis of semi-structured interviews, Richter and Blechinger explore the barriers to the development of renewable energies for power generation in the Caribbean and outline a strategy of how to overcome these barriers (Richter and Blechinger, paper "Barriers to the development of renewable energy" in this book). Finally, Terrapon-Pfaff, Dienst and Ortiz look into the planning processes of small-scale sustainable energy projects in developing countries with regard to their gender sensitivity (Terrapon-Pfaff et al., paper "The role of gender concerns in the planning" in this book).

Contents

About the Editors

Sebastian Groh lives and works in Bangladesh as the director of the companies ME SOLshare Ltd. and ME Fosera Ltd. aiming at energy poverty through latest innovation in rural electrification. Sebastian is further pursuing his Ph.D. on the role of energy in development processes. Having turned his back on the trading floor, he has been working for the past five year as a consultant to MicroEnergy International across the world at the intersection of energy service provision and microfinance.

Jonas van der Straeten is a Ph.D. student in the Research Group Microenergy Systems at the Technische Universität Berlin, which looks into access to modern energy in the global South from a multidisciplinary and user-focused perspective. His current research focuses on the history of the power sector in East Africa. In parallel, Jonas works for MicroEnergy International. He has project experience in South East Asia, Central Asia, and East Africa.

Brian Edlefsen Lasch is a practicing Solar Engineer based in Berlin, Germany. With the EPC company ib vogt GmbH, he designs and builds photovoltaic farms worldwide. As a Consultant with MicroEnergy International, he advised and implemented rural energy initiatives, including microfinance enabled projects in Mexico, Peru, Bangladesh, Pakistan, and the Philippines. He holds degrees in Solar Technology from Technische Universität Berlin and Electrical Engineering from Santa Clara University in California.

Dimitry Gershenson is a Ph.D. student in the Energy and Resources group at UC Berkeley. His research focuses on energy access finance, investment in emerging markets, and business model innovation for decentralized power in East Africa. Dimitry also works as a Renewable Energy and International Development Specialist for Ecoshift Consulting and supports the efforts of the Berkeley Rural Energy Group as an Advisor.

Walter Leal Filho is Senior Professor at Hamburg University of Applied Sciences (Germany) and holds the Chair of Environment and Technology at Manchester Metropolitan University, UK. Professor Leal Filho has wide international experience on energy efficiency and renewable energy, having undertaken a set of projects in Latin America, Africa, Caribbean, and South Pacific. He holds a number of doctorates for his work on environment and sustainability issues, and has in excess of 300 publications to his credit.

Dr. Daniel M. Kammen is the Class of 1935 Distinguished Professor of Energy at the University of California, Berkeley, with parallel appointments in the Energy and Resources Group, the Goldman School of Public Policy, and the Department of Nuclear Engineering. He was appointed the first Environment and Climate Partnership for the Americas (ECPA) Fellow by Secretary of State Hilary R. Clinton in April 2010. Kammen is the Founding Director of the Renewable and Appropriate Energy Laboratory (RAEL).

Part I
New Experiences, Simulations and Visions from the Field of AC and DC Minigrids

Chapter 1
Swarm Electrification: Investigating a Paradigm Shift Through the Building of Microgrids Bottom-up

Sebastian Groh, Daniel Philipp, Brian Edlefsen Lasch and Hannes Kirchhoff

Abstract The study investigates a bottom-up concept for microgrids. A financial analysis is performed through a business model approach to test for viability when replacing a researched energy expenditure baseline in Bangladesh. A case study of Bangladesh illustrates the potential for building on the existing infrastructure base of solar home systems. Opportunities are identified to improve access to reliable energy through a microgrid approach that aims at community-driven economic and infrastructure development. Network effects are generated through the inclusion of localized economies with strong producer-consumer linkages embedded within larger systems of trade and exchange. The analyzed approach involves the linkage of individual stand-alone energy systems to form a microgrid that can eventually interconnect with national or regional grids. The approach is linked to the concept of swarm intelligence, where each individual node brings independent input to create a conglomerate of value greater than the sum of its parts.

Keywords Energy access · Bottom-up · Microgrids · ICT · Bangladesh

Introduction

Across the developing world, considerable amounts of national incomes are invested in infrastructure development, such as the national electricity grid (Dobbs et al. 2013). Still they fail to cater to large shares of their (rural) populations, as 1.3 billion people lack access to basic electricity (IEA 2012). For an acceleration of economic and social development these challenges need to be addressed as they are inhibiting—or at least delaying—people's development paths (Groh 2014).

S. Groh (✉) · D. Philipp · B.E. Lasch · H. Kirchhoff
Research Focus Microenergy Systems, Technische Universität Berlin,
Center for Technology and Society, Secr. HBS, Hardenbergstr. 16–18,
10623 Berlin, Germany
e-mail: groh@ztg.tu-berlin.de

S. Groh et al. (eds.), *Decentralized Solutions for Developing Economies*,
Springer Proceedings in Energy, DOI 10.1007/978-3-319-15964-5_1

Higher and more volatile resource costs and infrastructure resilience to climate change are yet to be adequately considered. It is against this framework that McKinsey's Global Institute has set out the trillion dollar question on infrastructure productivity (Dobbs et al. 2013). The study, however, is often centered on two key players of development: the government and the private sector. This approach fails to take into account "the crucial third agent, in whose name development is carried out: people organi[z]ed as communities and collectives, people seen not as 'beneficiaries' of the state or 'consumers' of private services but as drivers of their own destiny, empowered to self-provision basic needs and to govern from below" (Kothari and Shrivastava 2013). This research investigates a new conceptual approach on rural electrification where the infrastructure of a microgrid is built through the people's own resources from the bottom-up. Experience has shown that user-centered concepts can contribute to the pursuit of sustainable and effective energy access models (Tenenbaum et al. 2014). Grid-based solutions, on the other hand, can offer great potential to provide stable and sufficient electricity supply for productive uses, which play a key role in bolstering economic development (Kaygusuz 2011). Here, discussions usually follow a dichotomous character (Tenenbaum et al. 2014). There are either centralized (e.g. national grid extension) or decentralized solutions (e.g. stand-alone SHS or isolated microgrids). Hence, the economic calculus is based on the (non-) viability of grid extension, which is measured by the distance-based cost of extension. Remote villages with low load factor and demand need to be electrified with a "second class" solution through a decentralized approach (Mandelli and Mereu 2013). Further, discussions are often reduced to on-grid and off-grid population, leaving potential solutions for an estimated one billion people out of scope (AGECC 2010). This group has been referred to as the "temporarily on/off-grid sector" and is further targeted in the step-wise electrification approach presented here (Groh 2014, p. 85). Furthermore, a consensus has been formed on the imminent need of low carbon development scenarios for developing countries in order to prevent dangerous anthropogenic climate change but without undermining their development goals (Jakob et al. 2013). Nussbaumer et al. (2012), Sovacool (2012) and Pachauri (2011) give extensive overviews evaluating different approaches to measure energy poverty. Based on these articles, Groh (2014) discusses the financial implications of people living in remote areas for the case of Arequipa (Peru). A central result is that structural handicaps in the sense of deprivation of a certain level of energy service quality, physical and economic isolation from distribution systems and infrastructural poverty are key factors keeping people in their currently poor economic states. The authors assume that in certain scenarios a paradigm shift away from a top-down, centralized, and fossil-fuel based scheme will lead to better results in terms of economic and social impacts. Furthermore, the authors hypothesize that such a paradigm shift could improve existing decentralized methods for rural energy, including stand-alone one-off Solar Home Systems (SHS) and baseline energy fuels such as kerosene. This research seeks to test this hypothesis through the analysis of a newly developed sharing-based energy infrastructure approach, based on decentralized growth incentives and resource efficiency. The concept follows the principle of a bottom-up initiative in the sense that it is a decentralized track which is generally carried out through non-governmental entities such as cooperatives, community user groups, or private entrepreneurs. However, the concept further envisions a readiness toward the actors and infrastructure of the centralized track, being the utilities and the national electricity grid.

The objective of the study is therefore to investigate the feasibility of an approach where the people themselves start building upon their present resources in order to form a balancing network and prepare themselves for an eventual grid connection. The underlying research question raised here is to what extent a grid can be built bottom-up and in an economically sustainable way thereby meeting the challenges current trends in microgrids for rural electrification face. Based on previous research by Kampwirth (2009), Sarker et al. (2012) and Unger and Kazerani (2012) the authors elaborate on a bottom-up concept drawn from an approach that follows the basic principles of swarm intelligence in distributed information and communications technologies networks and test for its viability. In this scheme, each individual node brings independent input to create a conglomerate of value ostensibly greater than the sum of its parts. In the way that each node in a swarm intelligence network shares information with its neighbors to achieve a compounding network effect, individual stand-alone household energy systems could share electrical power—in that they are linked together to form a microgrid—to achieve a networked grid effect. Upon applying frameworks to evaluate the concept, a microgrid developed in this way appears to address myriad problems facing trends in rural electrification strategies that involve the dissemination of microgrids and/or individual household energy systems. This paper explores the trends and difficulties facing microgrid strategies for electrification, and then describes, analyzes and tests this bottom-up microgrid approach, which might be coined as "swarm electrification" for a developing country setting.

Objective and Methodology

The present paper builds on an extensive research on microgrid deployments around the globe based on literature review as well as wide-ranging field experience. Based on this approach, it identifies key challenges when it comes to the development, design and implementation of microgrids, which largely account for the lack of success of rural electrification microgrids to date from a range of literature sources including practitioner reports. The overarching objective of the study is to provide tools for alleviating rural and urban energy poverty at the grassroots level and to consequently support the process of reaching the Millennium Development Goals (MDGs) (Ki-moon 2011). To reach this goal, it analyses a model of a sharing-based energy infrastructure, coined as swarm electrification. The objectives are to:

1. analyze the status quo of microgrid deployment in developing countries;
2. develop and analyze the layout of an alternative model where key challenges are addressed;
3. evaluate the designed system based on the factors identified in 1.

The swarm model is discussed in detail in terms of technological design and service delivery scheme through design thinking methodologies, and then tested for financial viability, and finally further analysis necessary for a proof of concept are suggested. The study relies on a diverse set of methods, including literature research, design thinking for model conceptualization and cost-benefit analysis. Baseline energy expenditures and business model characteristics for the case study are collected based on data set analysis, on-site field research and interviews of key stakeholder active in the Bangladeshi rural electrification market. A financial sensitivity analysis model is developed for the swarm concept and tested for viability in comparison with the collected data, including data for un-electrified baselines as well as status quo one-off electrification solutions. Financial viability analysis is carried out through a comparative cash flow analysis in different scenarios for the case of Bangladesh. Moreover, the net present value (NPV) method is applied where each cash flow (incoming and/outgoing) is discounted to its present value and summed up, as shown hereafter:

$$NPV(r,N) = \sum_{t=0}^{N} \frac{C_t}{(1+r)^t},$$

where
t is the time of the cash flow,
r is the discount rate, and
C is the cash flow.

In order to come up with an initial pricing model, the levelized cost of electricity (LCOE) is calculated based on the general formula given below:

$$LCOE = \frac{\sum_{t=0}^{n} \frac{I_t+M_t+F_t}{(1+r)^t}}{\sum_{t=0}^{n} \frac{E_t}{(1+r)^2}},$$

where
I_t is investment expenditure in the year t,
M_t is operation and maintenance expenditure in the year t,
F_t is fuel expenditure in the year t,
E_t is electricity expenditure in the year t,
r is the discount rate, and
n is the life of the system.

Literature Review

The UN General Assembly has declared the years 2014–2024 to be the "Decade of Sustainable Energy for All" (United Nations Foundation 2012), underlining the importance of supporting the roughly 1.3 billion people living without access to electricity. Groh (2014) introduces the concept of an energy poverty penalty arguing that poor energy services rooted in infrastructural handicaps inhibit or at the very least delay people's economic development. He states that "poor energy service quality can refer to insufficiencies, unreliability, dangers in usage, low durability, unfitness, lack of after-sales service and even non-affordability, in the sense of poor financial services" (Groh 2014, p. 83). As a consequence, better energy service quality could serve as an essential tool to fight the energy poverty penalty and ultimately help achieve the Millennium Development Goals (MDGs) (UNDP 2005). The "Energy for All Case" expects that only 30 % of rural areas can be electrified via connection to centralized grids, whereas 70 % of rural areas can be connected either with microgrids or with small stand-alone off-grid solutions (OECD/IEA 2011). Yadoo and Cruickshank (2012) estimate that about half of the 1.3 billion people living off-grid today could be best supplied with decentralized microgrids.

Marnay et al. (2011) note that in more innovative schemes where new sources are added to existing sources, integration of microgrids may be difficult from a centralized planner perspective, which is why the "thinking has moved to a structure wherein there are independent control nodes" (Marnay et al. 2011, p. 4). Although microgrids have been employed for village electrification already for over 30 years, there are only very few examples that can claim to be based on a long term viable model based on financial, managerial and technical criteria (Frearson and Tuckwell 2013). They describe the main barriers as securing a standardized and streamlined procurement system, establishment and governance processes, ready access to suitable finance, appropriate consumer consultation, hardware selection and integration, and developing effective operations and maintenance structures. The large initial capital investment and the related question of refinancing and ownership put a brake on many efforts to implement larger microgrids (Ulsrud et al. 2011). National utilities, that might have the capacity to maintain them, usually lack incentives to do so—being aware that it is often less profitable than the centralized grid considering the disproportionate amount of challenges of maintenance and operation (Goldemberg and Lucon 2010). As an alternative, community managed microgrids have emerged. Literature on these models remains scarce as it is considered a new field (Peterschmidt and Neumann 2013; Rolland and Glania 2011). Still, grey literature, in terms of project reports, indicates that these schemes often do not last very long and fail much earlier than expected. Often microgrids are designed with the goal of an equitable socio-economic development. As a by-product, there is evidence found for theft, non-payment and overuse leading to overall system failure. A paper on mini-grids in China states an overuse case with reduction of service provision from twelve to three hours per day (Shyu 2013).

These processes can be described as a form of the "tragedy of the commons" (Hardin 1968). In contrast, if the ownership is left to small and medium energy enterprises, a severe financing gap limits their capacity to scale to a degree required to run such a scheme (Kebir et al. 2013). Nonetheless, the International Energy Agency (IEA) argues that "smart grids could enable a transition from simple, one-off approaches to electrification (e.g. battery- or solar PV-based household electrification) to community grids that can then connect to national and regional grids" (IEA 2011). However, it must be ensured that the technologies allow for just grids, promoting equality and enable access even to low income households (Welsch et al. 2013). Therefore, starting with a low entry requirement is crucial. Sarker et al. (2012) suggest a DC-based microgrid with distributed generation where Solar Home Systems (SHS) can become connected to the local grid. A first investigation of a concept that starts small and develops step by step has been undertaken by Unger and Kazerani (2012), who have advocated for organically grown microgrids that start with the purchase of small lamps and eventually lead to village cluster-sized grid topologies. This study builds up on the ideas of a sharing-based energy infrastructure and draws conclusions to develop it further, constructing a service model through a case specific analysis for Bangladesh and testing it for financial viability. Good infrastructure can be evaluated based on the four As ("4A") criteria: affordability, accessibility, acceptability and availability (Weijnen and Ten Heuvelhof 2014). The following observations, deducted from the analysis above as well as extensive field experience, represent significant challenges facing centralized planning of microgrids, which so far have not been adequately addressed in implementation models. The authors realize that a successful bottom-up microgrid solution based on distributed renewable energy sources will have to address these challenges in order to comply with the 4A framework:

- Demand tends to grow once electricity is available;
- Pace of growth is hard to determine;
- Oversized systems are not economically viable;
- Undersized systems might fail to adequately perform and therefore hinder social acceptance and economic development;
- Productive use is enhanced with larger electrical loads.

The next chapter aims to take these issues under consideration when discussing the swarm electrification concept based on a case study approach for Bangladesh.

Analyzing the Model: Case Study Bangladesh

According to the World Bank, 40 % of Bangladesh's population, representing 65 million people, has no access to the national grid (World Bank Indicators 2013). Direct Current (DC) Solar Home Systems (SHS), currently consisting of a 20–85 Wp solar panel, battery, and charge controller, have begun to successfully electrify Bangladeshi rural communities through the Infrastructure Development

Company Limited's (IDCOL) national SHS program (IDCOL 2013). Close to three million SHS are already installed through microcredit schemes implemented by Partner Organizations (POs), who are expanding their customer base at a rate of 45,000–70,000 systems per month, making Bangladesh the fastest growing SHS market in the world. However, many households with an SHS do not fully utilize the electricity stored in their battery, resulting in a full battery by midday, and thereby limiting the generation potential of their systems by up to 30 % (Kirchhoff 2014). At the same time, some households may require electricity beyond what their systems can supply, especially during the rainy season, while at the same time others cannot afford a complete SHS at all and remain trapped in energy poverty. Mondal and Klein (2011) further point to the limits of SHS in terms of its potential to directly affect an individual household's ability to improve its income generation. There is a need for more cost effective, reliable and flexible electricity supply. In rural areas of Bangladesh, settlements tend to consist of various clustered areas where households are built closely together in a dense pattern. Hence, it is common to see clusters of households and small businesses with SHS.

Applying the concept of swarm electrification and interlinking these clustered SHS to form a microgrid, end-users could act as "prosumers", forming the core nodes of the microgrid and allowing the end-users to both consume electricity from the microgrid as well as feed electricity into the microgrid. Such an approach enables synergies with network effects generated through the inclusion of localized economies with strong producer-consumer linkages, allowing for local trade. Unlike traditional microgrid approaches, it crucially aims to make the most of the existing infrastructure assets as suggested by Dobbs et al. (2013), which are herein referred to as previously underutilized or unrecognized resources. In this way, participatory inclusion of community members based on existing equipment assets would build upon existing social acceptance of the technology and business models. In addition, up front capital costs associated with green-field microgrid development are heavily reduced. From a technical point-of-view, utilizing systems that are already sized for a particular household would allow nodes of the system to share power and thereby balance out seasonal mismatches. The batteries in the example of Bangladesh are already typically sized with three days of autonomy to bridge cloudy days. This capacity is not required during the non-rainy season and remains under-utilized in the status quo. In the given context, the swarm electrification approach could also be used to interlink multiple households with SHS to households without SHS. By forming a village-scale microgrid through the network connection of electricity-sharing homes, end-users could make use of their differentiated energy generation capacities and consumption patterns to allow for a more efficient and consistent source of energy supply for end-user households compared with the solely stand-alone systems.

Crucially, such a scheme would allow for a microgrid business model in which end-users have the ability to be remunerated for energy that is produced by their system and consumed by other end-users in the microgrid. In the following example, communications and payment management between households is administered by a smart charge controller, referred to as a swarm controller, which meters energy in- and outflows on a real-time display, serves as a data logger, and allows the end-user a basic modicum of control to toggle their system between island mode and microgrid-connected mode.

Regarding the service delivery model, depicted in figure below 1, the concept builds further on existing resources. The great success of the mobile phone industry has brought about an extensive network of local operators for topping–up mobile phones even in remote areas (Nique and Smertnik 2014). Energy delivery mechanisms and innovations have already made successful use of this proven model, by allowing users to top-up their electricity consumption allowance through equipment algorithm keys purchased at mobile phone retail points (Nique and Smertnik 2014). The success of this approach is therefore incorporated into the swarm grid example, allowing users to top-up their electricity consumption allowance by purchasing a numerical code and entering the number into their swarm controller at home. On-site sales, promotion, and after-sales technical services can be performed through identified *local champions* referred to as swarm area managers (SAMs). The SAMs are small local entrepreneurs who have access to a distribution chain and can become a primary provider of electricity for several households (e.g. shop-owners on the central market). They receive a microcredit in order to be able to build up a stock, receive a quota per unit sold and a percentage of the trade volume when handing out the scratch cards to the users to top-up their electricity balance. This implies that only the consumption of electricity is 'taxed' whereas the generation which is consumed directly or fed into the microgrid, is tax free. As the SAMs' revenue is highly based on the network effects, there is an intrinsic incentive for the SAMs to service the evolving grid and generate more sales. The SAMs can visualize, manage, and analyze this data through a back-end software solution provided to them (Fig. 1.1).

In order to allow for a real income generation source, which is crucial to address productive use aspects of successful microgrid designs, end-users should also be able to cash out positive electricity balances that their systems have fed into the grid. At any point of the month, when they need more electricity, they can go to the local SAM and top-up or, in the reverse case of consistent net production, cash out. The latter element expands the pay-as-you-go (PAYG) model (Bladin 2007) by a cash-in-as-you-go (CAYG) element. This ability to cash out also provides direct incentives for efficient electricity use, as their balance on their meter increases as their consumption decreases, once they start feeding into the grid more from their own SHS. In the theoretical application of the swarm model, data loggers built into the swarm controllers will allow for close monitoring of supply and demand within the swarm grid. Depending on supply or demand surplus, additional households without generation capacity can be connected to the grid, or additional generation capacity can be installed by incentivizing entrepreneurial households to buy bigger panels, given that the ability to sell surplus electricity can be considered likely based on past consumption patterns and prices. Service supply areas need to be defined in order to avoid conflict between different SAMs.

Research shows that further implications for the people are the possibility of a more flexible usage of their electricity both in terms of amount of energy in

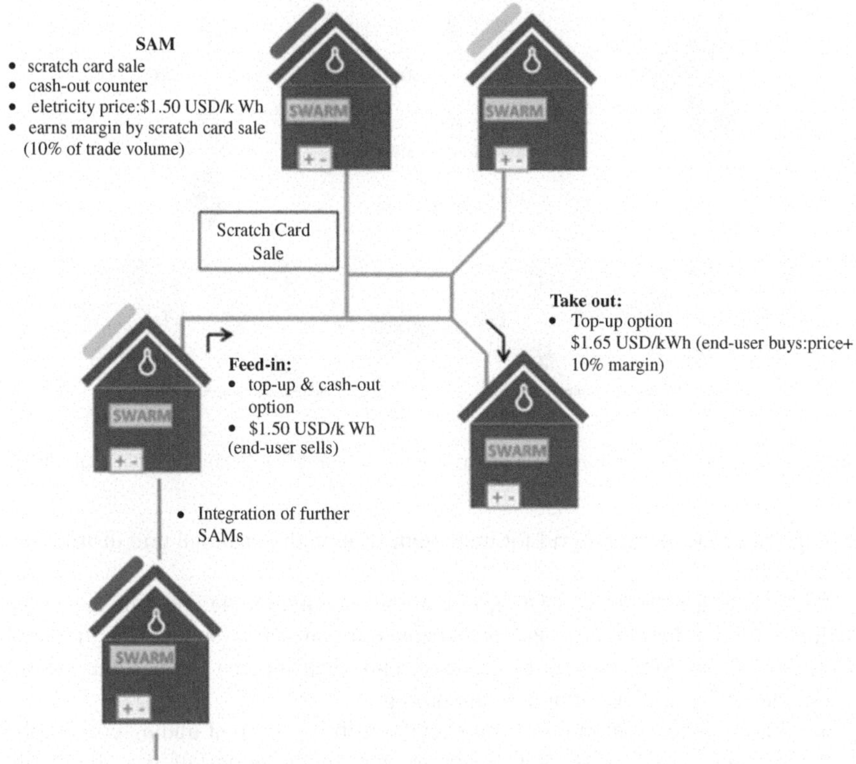

Fig. 1.1 Service delivery model through SAM scheme

Watt-hours (Wh) as well as in terms of time of usage, in addition to a likely improvement in battery state-of-health and prolonging of battery lifecycle based on fewer deep discharge incidents (Kirchhoff 2014). This style of rooftop decentralized generation further implies no centralized solar panel installations occupying large areas of useful land (as commonly seen in top-down microgrid designs), which is a major issue in such a densely populated country (Khan 2012).

The approach represents a democratization of electricity generation provided that the pricing scheme per unit of electricity is designed in a pro-poor approach. In Fig. 1.2, the step-wise approach of swarm electrification is shown. Step one shows individual households equipped with DC SHS as well as houses with neither solar nor grid electricity supply. Step 2 shows the interconnection of households with SHS, whereas in Step 3 the remaining houses are included in the growing DC microgrid. As a final step, the microgrid can be connected to a national or regional grid with minimal points of AC/DC conversion interfaces. With recent advances in smart grid technologies, such a bottom-up interconnected electrification approach becomes feasible (Unger and Kazerani 2012), however a financial and technical analysis must still be performed to fully understand the challenge and implications

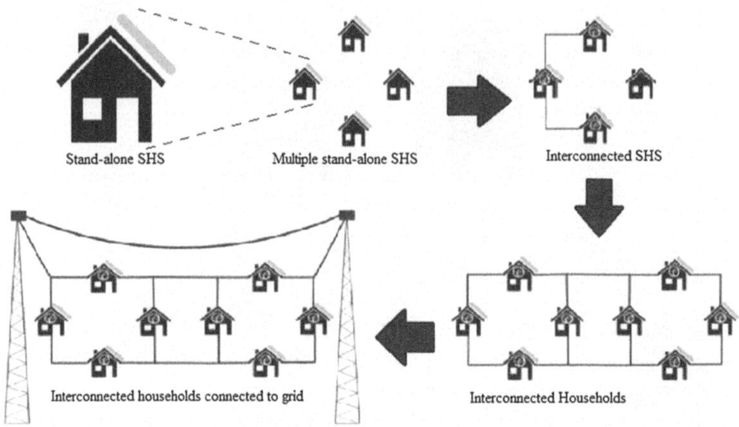

Fig. 1.2 Stepwise approach for swarm electrification. *Source* MicroEnergy International (2013)

of preparing a swarm microgrid for interconnection with a national grid distribution infrastructure.

The resulting network is a DC grid that can facilitate trade and increase usage flexibility and reliability beyond the status quo one-off systems. The trade of electricity allows SHS owners to generate additional income through the sale of excess electricity and consumption smoothing.

In a further investigation into commercial feasibility, current trading costs in the field have been used as baseline scenario. Key assumptions of the underlying models are summarized in Table 1.1 in the appendix. A SHS with a panel size of 50 Wp is modeled. The assumed photovoltaic system derate is set to 0.6.[1] In the presented simulation scenario the household is assumed to sell 29 % of its generated electricity, which is in line with Kirchhoff's previously calculated findings of approximately 30 % excess energy going unused given the limited storage capacity of the stand-alone system (Kirchhoff 2014). It does not, however, include any excess based on the rapidly growing application of more efficient DC-based appliances in the market.

Figures 1.3 and 1.4 show the viability of the approach in a simulation from an economic perspective. Figure 1.3 shows the time dimension measured in years on the y-axis whereas the x-axis indicates cumulative total electricity cost for the average off-grid household. The swarm concept requires an advanced charge controller to enable interconnection of SHS and sale and purchase of electricity between the systems. This scenario with such a controller is indicated with the name "Swarm Controller" in the figures. Assuming the electricity seller is purchasing a new SHS, three scenarios are compared: (1) the costs of continuing to meet electricity

[1] 40 % loss consisting of 20 % due to battery conversion losses, 13 % due to temperature and maximum power point mismatch, 5 % due to maintenance interruptions and 2 % due to cabling losses (Kirchhoff 2014).

requirements through kerosene and car battery, (2) purchasing a "Standard Controller" status quo SHS, and (3) purchasing an SHS with the "Swarm Controller".

Extensive field data[2] is used to estimate the green line indicating annual cost for kerosene and car batteries. The blue line represents the cost based on the SHS sales statistics of the past years. It is worth pointing to the fact that the current SHS microcredit scheme under which more than three million systems were sold does not compete with the present cost of kerosene and car batteries throughout the credit period of 36 months, but rather first breaks even only in year four (c.f. Fig. 1.3).

The red line dotted with squares in Fig. 1.3 indicates the electricity cost for a prosumer with a swarm controller over a lifetime of 10 years, where 30 % of the generated electricity of the 50 Wp SHS is traded/fed-into the microgrid. The green baseline is based on present expenditure for people relying on kerosene and car batteries whereas the blue baseline represents the monthly expenditure under the current microcredit scheme for SHS. The red line mimics the blue status quo SHS path despite its higher initial investment and outperforms the comparative scenarios after the credit has been paid off. Refer to Tables 1.1 and 1.2 in the appendix for details on cost and system sizing used for the simulation. Other advantages such as better system performance due to better battery recharging cycles, more flexible usage of electricity, better system integration and opportunities for increased income generation through acquisition of bigger panel sizes are not taken into consideration. On the other hand, it is assumed that all excess electricity generation can be and is sold within the microgrid. Figure 1.4 takes the perspective of a net consumer (without generation capacity), who pays for electricity consumed from the swarm microgrid.

The calculated NPVs indicate that additional electrification effort can be realized, especially for households who could not afford a full system before. By sharing the power generated from one household located at a particularly sun-exposed location, households that are located in a disadvantaged position for a solar-based system (e.g. in a shaded area) could also gain access to the microgrid electrification.

In that case these people can buy electricity at a lower cost compared to the business-as-usual case (represented by the blue line in Fig. 1.3) while renting a swarm controller and smaller-sized battery through a leasing scheme (represented by the red line). The latter case is designed to bring down monthly cost and initial down payment further down based on a linear depreciation assumption with ten years of expected system lifetime considering depreciation, deterioration and random failures, including a full replacement of the battery after five years. It is further worth noting in the comparison of Tables 1.1 and 1.2 in the appendix that the calculated NPV for a pure consumer (assumed to be a low-income household that could not afford an SHS) is considerably higher compared to a prosumer (slightly more than twice), given that the pure consumer experiences an assumed much stronger prevailing energy poverty penalty over a 10 year timeframe.

---

[2] Underlying data from Rural Electrification and Renewable Energy Development (RERED) II Project Report from the World Bank 2012.

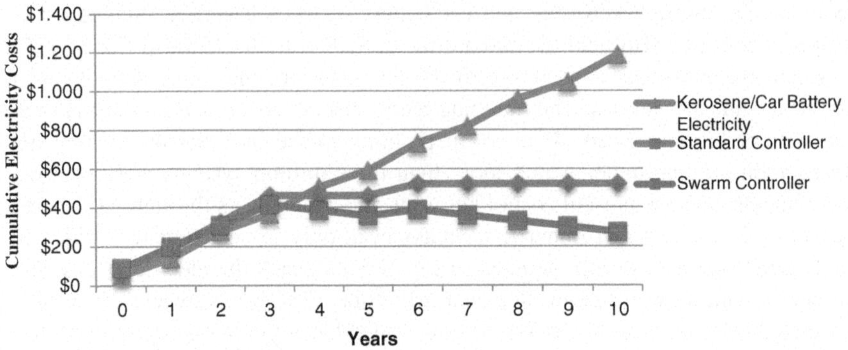

Fig. 1.3 Cost of electricity for an "equipped" electricity prosumer in a swarm scheme (*red line with squares*) versus baseline

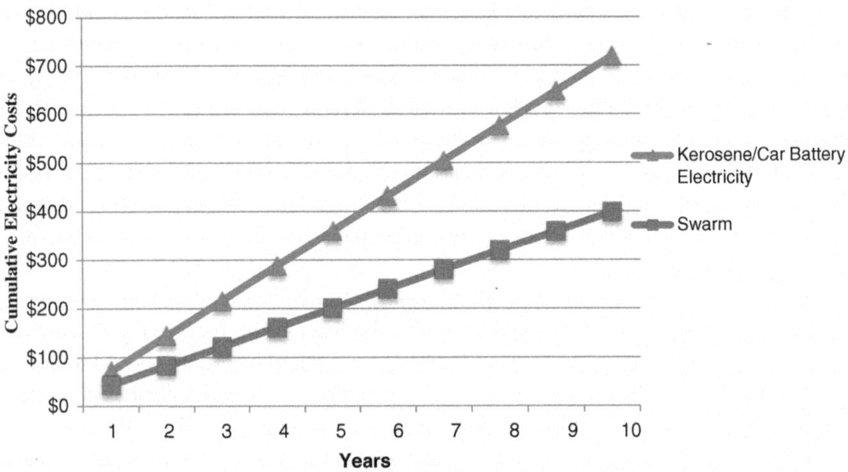

Fig. 1.4 Cost of electricity of an "un-equipped" electricity consumer in a swarm scheme versus baseline

A discount rate of 9 % is assumed, which is reasonable given an inflation rate of 7 and a 2 % risk premium (BBS 2014). Applying the 4A criteria to the concept, observations can be made. The scheme:

1. provides possibilities for flexible provision of electricity based on a mobile retail network already deeply rooted in the rural areas (accessibility),
2. results in favorable economics based on baseline data under a microcredit/-leasing scheme (affordability),
3. builds on network effects and existing resources and increases access to daily and seasonal energy through a balancing network (availability), and

4. utilizes a familiar technology and delivery model that builds on existing social acceptance (acceptability).

The concept allows the user a modicum of control to decide when they are operating in grid-connected mode or island mode, and avoids interpersonal conflicts and cash exchange through a system of digital credits and debits managed by the network of swarm controllers and with payments handled only by neutral mobile phone retail points.

Where the model potentially falls short is in direct comparison with microgrids is the ability to power large loads. Microgrids are designed from the start for increased power use beyond SHS (not just increased energy over time but larger instantaneous power draw for larger appliances). The swarm model might remain dependent on the existing SHS cabling and voltage levels, thereby retaining the instantaneous power draw limits of the SHS even if the overall energy availability and system performance increases. With regard to battery health, while Kirchhoff (2014) has shown that proper State-of-Health and State-of-Charge management are possible for the prosumers in such a scheme, it is not clear that this will be the case for the smaller batteries installed for pure consumers of the microgrid energy. The present simulation price at USD 1.50 per kWh (without any subsidies) is calculated based on the LCOE method considering that there are no running fuel cost, an initial investment based on current numbers in the Bangladeshi SHS program with the additional cost of the swarm controller and battery replacement after five years, 5 % of total investment (approx. USD 500) as yearly operation and maintenance cost,[3] electricity generation as shown in Table 1.1 in the appendix, the same discount rate as applied in the NPV calculations.

With regard to Operation and Maintenance (O&M), requirements largely synergize with the O&M requirements for status quo SHS (cleaning panels, refilling batteries with distilled water, replacing fuses), which are well-understood and regularly practiced by the target Bangladeshi communities. However, O&M of the grid infrastructure itself (running new cabling, preventing theft, repairing cabling, installing and checking safety devices, etc.) would need to be carefully considered for a sustainable business model, and could potentially also be conducted by the same POs instead of the SAM. It should be noted that a safe extra low voltage (SELV) can be chosen as the network voltage to mitigate the need for extensive safety training and equipment, as these voltages fall into the touch safe range. It is further important to note that the model is sensitive to changes in the electricity trading price per kWh, as well as the available and tradable amount of Wh (see Figs. 1.5, 1.6 and 1.7 in the appendix). The baseline scenario for the sensitivity analysis stands at a 50 Wp system, where 35 % of excess energy is generated[4] and

---

[3] Values based on ten year historical data of the Bangladeshi SHS program.

[4] This is higher than in the economic analysis in Fig. 1.3 (30 %), as well as that the purchaser consumes now 40 Wh (instead of 20 Wh per day). This is in line with a trend where excess energy will tend to increase due to appliances with higher efficiency built into existing systems as well as consumption will.

fully sold at a price of USD 1.50 per kWh for a system that is running for five years. The purchaser is assumed to buy 40Wh of electricity per day in order to cover her electricity needs. Figure 1.5 illustrates the trade-off between a very pro-poor approach with a trading price range of approx. USD 0.50 and USD 1.75 as the border conditions, displaying both prosumer and purchaser as well as the status-quo system owner. A price of USD 0.50 (considerably lower than the calculated LCOE of USD 1.45) puts the advantage on net consumers and people not able to afford a system of their own. A price of USD 1.75 sets the incentives completely on the net producer side to buy more generation capacity and sell off the excess while keeping purchaser still at par with the traditional monthly SHS system cost (note: consuming, however, only 40 Wh per day). Figure 1.6 shows that a SHS only needs to produce a little over 10 % of excess capacity in order to for this model to become feasible. Under the baseline of 35 % excess capacity, accordingly, at least 30 % of the excess needs to find a buyer in order to break-even.

If either supply or demand falls short, mitigation mechanisms exist through incentives for prosumers to become more of a producer when choosing panel size given the existence of a business opportunity. For the latter case, additional electrification of households too poor to afford a system can close the demand gap. The respective SAM might decide to utilize price differentiation that depends on factors such as distance from the nearest connecting household, the rate of power coming directly from the solar module or the battery, time of usage, or other variables. Further investigation is needed here.

Discussion and Conclusions

Despite the current trend toward traditional microgrids and one-off SHS solutions for rural electrification, the authors show that under presented conditions and assumptions the concept of swarm electrification may present a better fit to meet the combined goals of universal energy access for all and fostering rural economic development. The approach requires neither a large initial capital investment nor top-down system sizing. The key barriers addressed in section Literature Review appear to be adequately addressed in theory, as it builds on an existing and proven technology, end-user financing, delivery mechanisms and social acceptance trends, thereby meeting the criteria set out in the 4A evaluation scheme for good infrastructure (affordability, accessibility, acceptability, availability). Moreover, a tragedy of the commons problem is unlikely to occur in this case as the majority of individuals have their own system or supply, with the ability to choose to utilize their energy generation and storage equipment as income generating assets, monitored on an individual metering system without a centralized capped storage capacity, or to decide to run their system in an independent island mode.

The theoretical case study for Bangladesh indicates that the swarm concept is able to create win-win situations. Some simple cost-benefit calculations suggest that the process can be designed in a financially mutually beneficial way for end-users who are able to afford a complete SHS, as well as for end-users who are unable to afford a complete SHS and currently pay high prices for baseline energy sources such as kerosene and car batteries. Comparable calculations and simulations could be run for different settings on the one hand to test the degree of scalability and impact for the Bangladeshi off-grid sector but also for a feasibility assessment of potential replications in different country settings in terms of their existing resources and expenditure patterns, respectively. The financial model is based on network effects, and thus dependent on initial sales of swarm controllers gaining momentum, as well as being vulnerable to the potential occurrence of unexpected critical social acceptance issues. Furthermore, the concept requires a smart pricing mechanism in order to simultaneously incentivize consumers and producers as well as a local operator. The topic of potential solutions for ownership schemes of such a microgrid remains to be proven in real world implementation examples. From a technical perspective, solutions to enable increased power and thereby larger loads should be addressed if the infrastructure is limited by fixed SHS cabling and voltage levels. Safety and switching of DC distribution voltages need to be further investigated. An adequate agent-based control scheme, as outlined by Kirchhoff (2014), needs to be developed and field tested. These questions need further in-depth research. Although, the concept has a built-in opportunity for scalability, the issue of replication potential for other perhaps less densely populated countries remains to be seen.

The authors conclude that changing the mindset of prohibitive last mile cost (centralized perspective) to an end-user perspective and the peoples' own development capabilities may lead to increased success in rural electrification and pro-poor economic development schemes. As such, a paradigm change from top-down planned centralized microgrids toward a bottom-up microgrid approach where the decision and managing power is up to the people and their existing resources themselves without creating a common pool resource could have a positive impact on the development of economically and technically viable localized electricity distribution infrastructures. In this scenario, people are no longer obliged to wait for a utility grid extension, but start building a local grid themselves, beginning with individual household-level systems afforded through inclusive end-user financing and delivery mechanisms. The authors expect that in the future, microgrids based on these concepts will play an important role for decentralized energy supply in order to foster rural development. A paradigm shift in both research and practice could break down the traditional dualistic conception of rural electrification, where utility grid extensions or one-off stand-alone energy systems are pitted in competition, and allow for a productive exploration of innovative bottom-up energy access models.

Acknowledgments A previous version of this study was presented at the MES-BREG 2014 Conference Innovating Energy Access for Remote Areas: Discovering untapped resources in Berkeley, California, in April 2014. Thank you to all participants who provided valuable feedback. You all contributed to the improvement of this study. We thank Hans-Böckler Foundation for their continuous financial support of the Postgraduate School Microenergy Systems. We thank Raluca Dumitrescu and Daniel Ciganovic from MicroEnergy International for their support in the financial analysis.

Appendix

See Tables 1.1 and 1.2; Figs. 1.5, 1.6 and 1.7.

Table 1.1 Cost scheme "prosumer"

SHS owner	
System size (W)	50
PV system derate	0.60
Daily system PV generation (Wh/day)	135
System voltage (V)	12
Battery size (Ah)	80
Max. battery % depth of discharge	50 %
% of generated electricity sold	30 %
Average Wh available to sell (Wh/day)	40
Average Wh available to sell/month (Wh/month)	1200
Levelized cost of electricity (USD/kWh)	$1.5
Monthly revenue from electricity sales (USD/month)	$1.80
Existing charge controller cost (USD)	$0
Swarm controller cost (USD)	$40
System sharing wiring cost (USD/prosumer)	$3
Swarm controller simple payback (years)	2.59
NPV (USD)	**$98.62**

Table 1.2 Cost scheme "consumer"

Electricity purchaser	Credit	Leasing
Swarm controller	$40	
Battery (10 Ah)	$25	
System sharing wiring (USD/consumer)	$3	
Hardware cost (USD/month)	$1.93[a]	$1.75[b]
Levelized cost of electricity (USD/kWh)	$1.5	
Daily usage (Wh/day)	20	
Monthly cost of electricity used (USD/month)	$0.90	$0.90
Total cost (USD/month)	$2.78	$2.65
NPV (USD)	**$321.38**	

[a] Calculated based on a 12 % yearly service charge (flat)
[b] Based on 5-year lifetime with linear depreciation model

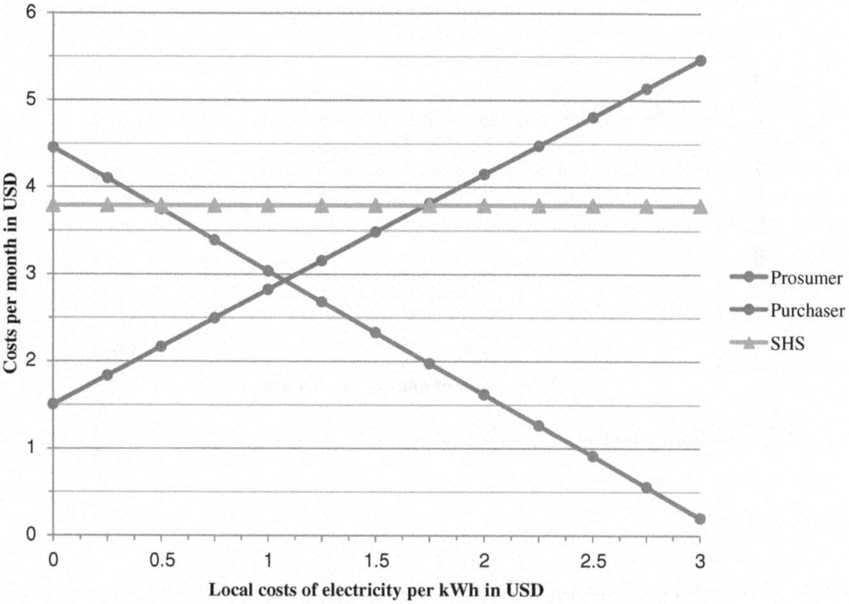

Fig. 1.5 Electricity trading price scenarios

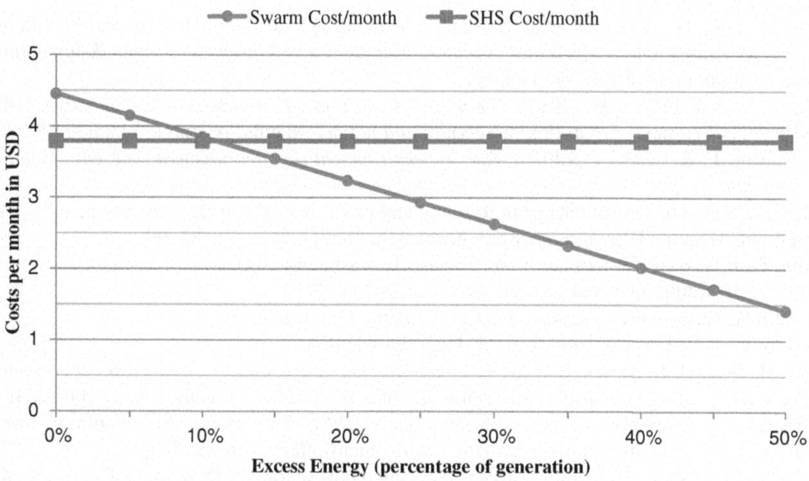

Fig. 1.6 Excess energy generation scenarios

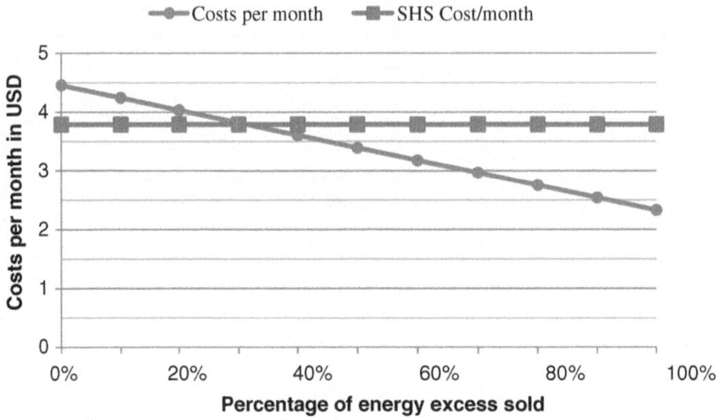

Fig. 1.7 Excess energy trading amount scenarios

References

AGECC. (2010). *Energy for a sustainable future*. New York: Secretary-General's Advisory Group on Energy and Climate. (Summary Report and Recommendations).

Bangladesh Bureau of Statistics (BBS). (2014). *Latest indicator*. Available http://www.bbs.gov.bd/home.aspx. Accessed 20 Mar 2014.

Bladin, P. (2007). Improving microfinance through telecommunications. *Journal of Microfinance/ESR Review, 9*(1), 14–19.

Dobbs, R., Pohl, H., & Lin, D., et al. (2013). Infrastructure productivity: How to save $1 trillion a year. McKinsey Global Institute. Available http://www.mckinsey.com/insights/engineering_construction/infrastructure_productivity

Frearson, L., & Tuckwell, M. (2013). *The future of minigrids: From low cost to high value. Using demand driven design to maximise revenue and impact*. Manila: Asian Development Bank.

Goldemberg, J., & Lucon, O. (2010). *Energy, environment and development* (2nd ed.). London: Earthscan.

Groh, S. (2014). The role of energy in development processes—the energy poverty penalty: case study of Arequipa (Peru). *Energy for Sustainable Development, 18*, 83–99.

Hardin, G. J. (1968). *The tragedy of the commons*. Washington, D.C.

IDCOL. (2013). http://www.idcol.org/. Accessed 21 Mar 2014.

IEA. (2011). *Technology roadmap smart grids*. Paris: OECD/IEA.

IEA. (2012). *World energy outlook 2012*. Paris: OECD/IEA.

Jakob, M., Steckel, J., Klasen, S., Lay, J., Edenhofer, O., Grunewald, N., Renner, S., & Martínez-Zarzoso, I. (2013). Feasible mitigation actions in developing countries, in preparation. Available https://www.kfw-entwicklungsbank.de/PDF/Entwicklungsfinanzierung/Umwelt-und-Klima/Konferenzen-und-Veranstaltungen/Kohlenstoffarme-Entwicklung-Expertenworkshop-2013/Entdeken-Feasible-Mitigation-Actions-in-Developing-Countries.pdf

Kampwirth, R. (2009). *SchwarmStrom- Die Energie Der Zukunft*. Hamburg: LichtBlick.

Kaygusuz, K. (2011). Energy services and energy poverty for sustainable rural development. *Renewable and Sustainable Energy Reviews, 15*(2), 936–947.

Kebir, N., Spiegel, N., Schrecker, T., Groh, S., Scott, C., & Aliaga, G. (2013). *Exploring energy SME financing in emerging and developing countries*. Oestrich-Winkel: Sustainable Business Institute.

Khan, M. R. (2012). A stand alone DC microgrid for electricity supply in rural Bangladesh. In *Proceedings of the 2nd International Conference on Developments in Renewable Energy Technologies (ICDRET) 2012*, Dhaka (pp. 1–4).

Ki-moon, B. (2011). Sustainable energy for all: A vision statement. United Nations Organization. New York. Available www.sustainableenergyforall.org. Accessed 31 July 2014.

Kirchhoff, H. (2014). Hidden resources in solar home systems as the base for bottom-up grids. In *Proceedings of the International Conference on Innovating Energy Access for Remote Areas: Discovering untapped resources*. Berkeley: Microenergy Systems and Berkeley Rural Energy Group. ISBN 978-3-7983-2694-1.

Kothari, A., & Shrivastava, A. (2013). *Economists on the wrong foot* (Vol. XLVIII, No. 33). Mumbai: Economic and Political Weekly.

Mandelli, S., & Mereu, R. (2013). Distributed generation for access to electricity: "Off-main-grid" systems from home-based to microgrid. In *Renewable energy for unleashing sustainable development* (pp. 75–97). Berlin: Springer.

Marnay, C., Nordman, B., & Lai, J. (2011). Future roles of milli-, micro-, and nano-grids. In *CIGRÉ International Symposium the Electric Power System of the Future—Integrating Supergrids and Microgrids*, Bologna, Italy, 13–15 Sept 2011.

MicroEnergy International. (2013). *Swarm electrification*. Available http://www.microenergy-international.com/index.php?id=884. Accessed 21 Mar 2014.

Mondal, A., & Klein, D. (2011). Impacts of solar home systems on social development in rural Bangladesh. *Energy for Sustainable Development, 15*(1), 17–20.

Nique, M., & Smertnik, H. (2014). The Synergies between mobile phone access and off grid energy solutions. In *Proceedings of the International Symposium Innovating Energy Access for Remote Areas: Discovering untapped resources*. Berkeley, USA: Microenergy Systems and Berkeley Rural Energy Group. ISBN 978-3-7983-2694-1.

Nussbaumer, P., Bazilian, M., & Modi, V. (2012). Measuring energy poverty: Focusing on what matters. *Renewable and Sustainable Energy Reviews, 16*(1), 231–243.

OECD/IEA. (2011). *Energy for all: Financing access for the poor*. Special early excerpt of the World Energy Outlook 2011, Paris.

Pachauri, S. (2011). Reaching an international consensus on defining modern energy access. *Current Opinion in Environmental Sustainability, 3*(4), 235–240.

Peterschmidt, N., & Neumann, C. (2013). *Scaling up successful micro-utilities for rural electrification*. Oestrich-Winkel: Sustainable Business Institute.

Rolland, S., & Glania, G. (2011). *Hybrid minigrids for rural electrification: Lessons learned*. Brussels: Alliance for Rural Electrification.

Sarker, M. J., Asare-Bediako, B., Slootweg, J. G., et al. (2012). DC microgrid with distributed generation for rural electrification. In *47th International Conference on Universities Power Engineering (UPEC)*, London.

Shyu, C. W. (2013). End-users' experiences with electricity supply from stand-alone mini-grid solar PV power stations in rural areas of western China. *Energy for Sustainable Development, 17*(4), 391–400.

Sovacool, B. K. (2012). The political economy of energy poverty: a review of key challenges. *Energy for Sustainable Development, 16*, 272–282.

Tenenbaum, B., Greacen, C., Siyambalapitiya, T., et al. (2014). *From the bottom up—how small power producers and microgrids can deliver electrification and renewable energy in Africa*. Washington, D.C: The World Bank.

Ulsrud, K., Winther, T., Palit, D., Rohracher, H., & Sandgren, H. (2011). The solar transitions research on solar minigrids in India. *Energy for Sustainable Development, 15*(3), 293–303. http://www.sciencedirect.com/science/article/pii/S0973082611000433.

Unger, K., & Kazerani, M. (2012). Organically grown microgrids: Development of a solar neighborhood microgrid concept for off-grid communities. In *IECON 2012—38th Annual Conference on IEEE Industrial Electronics Society*, Montreal (pp. 5663–5668).

UNDP. (2005). *Energizing the millennium development goals*. New York: United Nations Development Programme.

United Nations Foundation. (2012). *Energy access practitioner network-towards achieving universal energy access by 2030*. New York: United Nations Foundation.

Welsch, M., Bazilian, M., Howells, M., Divan, D., Elzinga, D., Strbac, G., et al. (2013). Smart and just grids for Sub-Saharan Africa: Exploring options. *Renewable and Sustainable Energy Reviews, 20*, 336–352.

Weijnen, M. P. C., & Ten Heuvelhof, E. F. (2014). *Next generation infrastructures*. Edx. Online course. Week 1: Introduction. TU Delft.

World Bank Indicators. (2013). http://data.worldbank.org/country/bangladesh. Accessed on 21 Mar 2014.

Yadoo, A., & Cruickshank, H. (2012). The role for low carbon electrification technologies in poverty reduction and climate change strategies: A focus on renewable energy minigrids with case studies in Nepal, Peru and Kenya. *Energy Policy, 42*, 591–602.

Chapter 2
Identifying Hidden Resources in Solar Home Systems as the Basis for Bottom-Up Grids

Hannes Kirchhoff

Abstract In the context of the modern energy access challenge, one new pathway towards electrification is to make use of hidden resources already in the field through small microgrids. In particular, this paper analyses the amount of excess energy of medium sized (65 Wp) solar home system (SHS) located in Bangladesh. The SHS is modeled using synthetic load curves and a sophisticated battery model that accounts for battery ageing. The simulation shows that more than 30 % of the electricity potentially generated by the SHS remains unused.

Keywords SHS · Hidden resources · Microgrid · Bottom-up

Introduction

The energy access challenge presents itself on several fronts at the same time. While there are still more than 2.3 billion people without reliable access to electricity (United Nations Foundation 2012), a carbon lock-in is on the horizon if the power sector does not turn green quickly (IEA 2011, p. 3). For individuals who currently do not have access to modern energy services that are "affordable, clean, reliable, and safe" (Legros et al. 2009, p. 3) this translates into high prices and uncertain timelines for access.

As pointed out by Tenenbaum et al. (2014), it is time for the "second", decentralized track in electrification. A track that does not rely on central entities but can grow from the bottom-up. The "swarm electrification" approach (Groh et al. 2013), shows a pathway for how this can be done, building on the existing infrastructure, such as the 2.6 million solar home systems (SHS) already installed in Bangladesh (IDCOL 2013). SHS are linked to form small microgrid clusters that can gradually be extended by going through different stages, c.f. Fig. 2.1. The process is a step by

H. Kirchhoff (✉)
Microenergy Research Group, TU Berlin, Hardenbergstraße 16-18, 10623 Berlin, Germany
e-mail: hannes.kirchhoff@microenergy-systems.de

S. Groh et al. (eds.), *Decentralized Solutions for Developing Economies*,
Springer Proceedings in Energy, DOI 10.1007/978-3-319-15964-5_2

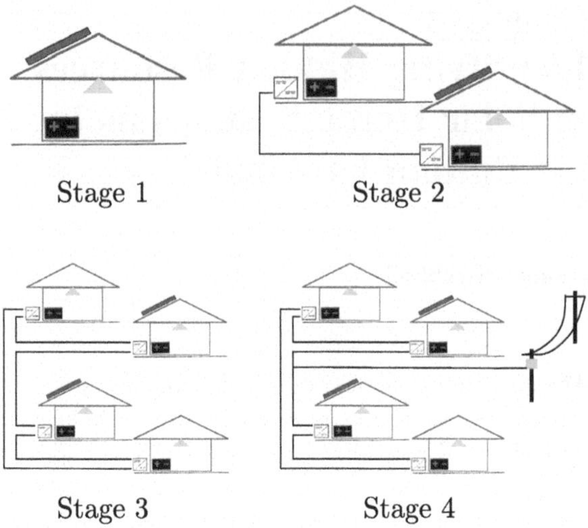

Fig. 2.1 Step by step bottom-up electrification

step bottom-up approach that eventually leads to coverage of a whole village. The formed microgrid can, if desired, even be connected to the main grid. There is limited literature available on interconnected SHS, however, one simulation showed a 15 % saving in generation capacity required when eight large SHS located in Tanzania were connected (Unger and Kazerani 2012).

Research Objectives

This paper addresses the perspective for a single SHS in the initial process of swam electrification, showing the potential for energy sharing. Hence, the hidden resources in terms of excess energy that remains unutilized in normal SHS operation is investigated on.

Methods

The 65 W_p SHS package under the IDCOL solar program is simulated for the course of one year using Matlab Simulink with 5 min intervals. The SHS package consists of a 65 W_p solar panel and a 100 Ah battery. Details of the respective component modeling are as follows.

Table 2.1 Parameters used in the PV module model

Parameter at standard test conditions, from Kyocera (2007)	Value
V_{MPP}	17.4 V
I_{MPP}	3.75 A
V_{OC}	21.7 V
I_{SC}	3.99 A
K_v	−0.0821 V/K
K_i	0.00159 A/K
Parameters obtained using Villalva et al. (2009)	Value
a	1.3 [–]
N_s	36 [–]
R_p	266,565 Ω
R_s	0.2454 Ω

Solar Module Simulation

For the model in this paper, the SHS is equipped with a 65 W_p multicrystal PV module, fitted to data sheets from Kyocera KC65T panel (Kyocera 2007), which is modeled using a mathematical model developed by Villalva et al. (2009). This model is an improvement of the single diode equivalent circuit and can be easily adapted to different panels. The work presented in the paper includes an algorithm to determine the best fitting parameters for the shunt resistance R_S and the parallel resistance R_P requiring only datasets at open and short circuit as well as at the maximum power point (MPP). These parameters are then used to model the characteristic current behavior as in (2.1) where V_t is the thermal voltage and a the ideality factor of the diode. The parameters used are listed in Table 2.1.

$$I_{PV} = I_0 \cdot \left[\exp\left(\frac{V + R_S \cdot I}{V_t \cdot a}\right) - 1\right] \tag{2.1}$$

The temperature influence on the panel are reflected on through the temperature coefficient for voltage K_V and the temperature coefficient for current K_I.

The model is verified with the electrical performance at nominal operating cell temperature which is given in the datasheet (Kyocera 2007), producing an error of less than 0.4 %.[1]

The influence of ambient temperature is most critical for the simulation of the PV panel where energy output is decreased for higher temperatures. Hence, a conservative assumption with a constant ambient temperature of 30 °C is taken. The temperature of the cells is modeled as being directly proportional to the irradiation G, as in (2.2). This approach is proposed by Hansen et al. (2000, p. 14).

[1] Input values are G = 800 W/m², T_{cell} = 47 °C and V_{PV} = 15.3 V. The datasheet states 46 W output and the model calculates 46.15 W.

$$T_C = T_A + 0.03 \frac{\mathrm{C} \cdot \mathrm{m}^2}{\mathrm{W}} \cdot \mathrm{G} \tag{2.2}$$

Real solar data for Dhaka of the year 2001 is used, specifically global horizontal solar insulation data from NREL (2005). The panel's orientation is assumed as perfect south and 24° tilt to allow for maximum generation. Correction factors for this tilt are based on data from Boxwell (2013).

Battery Model

The battery modeling has to cover a wide range of parameters and constants, but there are essentially only two variable inputs into the model during the simulation: temperature and current. Corresponding to the ambient temperature, the battery temperature is assumed constant at 30 °C. This assumption is based on the fact that the current flows to and from the battery are relatively small and combined with the large surface to volume ratio of the battery should not lead to mayor warming effects (Boldt 2012).

Thus, the only remaining input variable is the current. The output parameters are the terminal voltage of the battery the state of charge (SOC) and, as a lifetime monitoring variable, the remaining capacity of the battery. Degradation and corrosion lead to reduction in capacity, which is reflected in the state of health (SOH). For batteries in electric vehicles the SOH is often defined so that it reaches zero when the remaining capacity is 80 % (Lam 2011). This does not necessarily hold true for batteries in SHS as the reduced capacity can still be used. Hence, in this paper the definition as in (2.3) is utilized, details on the calculation of the remaining discharge capacity $C_d(t)$ are given in Bindner et al. (2005) and Boldt (2012).

$$SOH = \frac{C_d(t)}{C_d(t=0)} \tag{2.3}$$

The terminal voltage U_{bat} is obtained by multiplying the cell voltage U_{cell} by the number of cells (six). The cell voltage is derived by using a set of two Shepherd Equations, one for charging (2.4) and one for discharging (2.5), where the SOC and DOD are the state of charge and depth of discharge respectively.
For I > 0:

$$U_{cell}(t, I) = U_0 - g \cdot DOD + \rho_c(t) \cdot \frac{I(t)}{C_N} + \rho_c(t) \cdot M_c \cdot \frac{I(t)}{C_N} \cdot \frac{SOC(t)}{C_c - SOC(t)} \tag{2.4}$$

For I ≤ 0:

$$U_{cell}(t, I) = U_0 - g \cdot DOD + \rho_d(t) \cdot \frac{I(t)}{C_N} + \rho_d(t) \cdot M_d \cdot \frac{I(t)}{C_N} \cdot \frac{DOD(t)}{C_d(t) - DOD(t)} \tag{2.5}$$

The Eqs. (2.4) and (2.5) can be both be rearranged so that each constitutes of one portion independent of the battery current and the other part dependent on the battery current. These are the values used for a simple equivalent circuit of the battery in the SHS model with a controlled voltage source (U_0) and an internal resistance (R_i) in series. Ageing effects are reflected in the model that are related to the following stress factors:

- *Time between full charges*, in particular modeling the build- up of irreversible sulphation,
- *Partial cycling* by taking into account the end of charge voltage when crossing over a SOC value of 90 % but not reaching 100 % which is critical for corrosion,
- *Ah-throughput* by evaluating the total discharge current in a weighted number of cycles which is an indication for the loss of active material as well as increased stratification,
- *Low discharge rates* which enhance stratification,
- *High temperature* is reflected in all relevant factors, in particular the build-up of the corrosion layer and the effect of gassing.

Charge Controller

The charge controller was modeled with an overcharge and a deep discharge protection. The threshold values for theses operations were taken from the IDCOL SHS program specifications and translated from voltage cut-off values to SOC-values. As shown in Fig. 2.2, the deep-discharge protection is activated after crossing 30 % SOC and the load is reconnected at 50 % SOC. Once the battery is

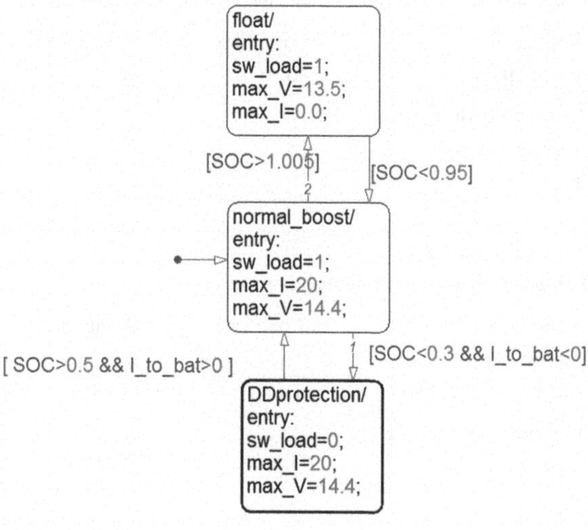

Fig. 2.2 Model for the charge controller

fully charged, the charge controller changes the charging mode from boost to float, limiting the charging voltage to 13.5 V.

Load Profile

As no real data on the usage pattern was available, literature values are used and combined to form a synthetic load curve. According to UNFCCC (2012, p. 6), 65 W_p SHS in Bangladesh are typically equipped with loads given in Table 2.2. A aggregate load of 43 W is used between 3 and 4 h per day, resulting in a daily energy consumption of 162 Wh.

However, other authors have argued that demand can be higher due to daily operation of up to 5 h per day (Khadem 2006, p. 4). To account for this discussion, an extensive use case of 240 Wh/d with longer usage hours and additional devices is also considered (c.f. Table 2.3). It is assumed that this extensive use case is applicable once in a week. This is simulated by giving a 6/7 probability to a 162 Wh load day and a 1/7 probability to a 240 Wh load day. User behavior, such as decreased demand when charge controller indicates a low SOC is not taken into account. The load distribution over the day is taken from Khadem (2006, p. 4).

In addition to the above, a daily noise level of 15 % and an hourly noise level of 20 % is added as suggested by Hafez and Bhattacharya (2012, p. 8), as shown in the exemplary profile in Fig. 2.3.

In the simulation, loads are modeled as a variable resistor. The value of the resistor is calculated by taking an average battery voltage of 13 V into account.

Table 2.2 Loads for normal use day (UNFCCC 2012, p. 6)

Appliance	Load (W)	Usage (h/d)	Daily demand (wh/d)
5 CFL lights	30	4	120
Black and white TV	10	3	30
Mobile charger	3	4	12
Total			162

Table 2.3 Loads for an extensive use day

Appliance	Load (w)	Usage (h/d)	Daily demand (wh/d)
5 CFL lights	30	5	150
Black and white TV	10	5	60
Radio/cassette player	5	2	10
Mobile charger	2 × 3	5	30
Total			240

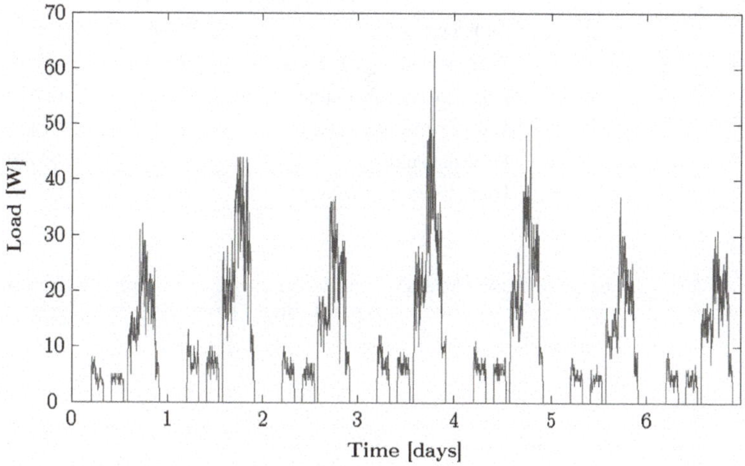

Fig. 2.3 Exemplary weekly load profile for one SHS

Results

For the simulated SHS located in Bangladesh, generation significantly exceeds the demand. Figure 2.4 shows how the daily generated energy is much larger than the daily demand for large parts of the year. Only in the beginning of the second half of the year does the daily demand occasionally exceed the daily generation.

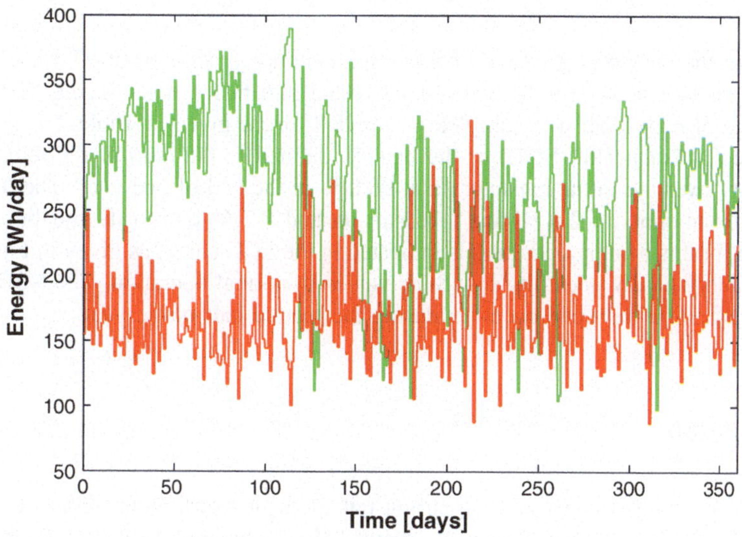

Fig. 2.4 Daily generation (*green*) and demand (*red*)

Table 2.4 Simulation results

Parameter	Value	Unit
Total potential generation	97.5	(kWh)
Total excess generation	31.1	(kWh)
Ratio excess generation	31.9	(%)
Total demand	62.0	(kWh)
Total disconnected load	0.0	(kWh)

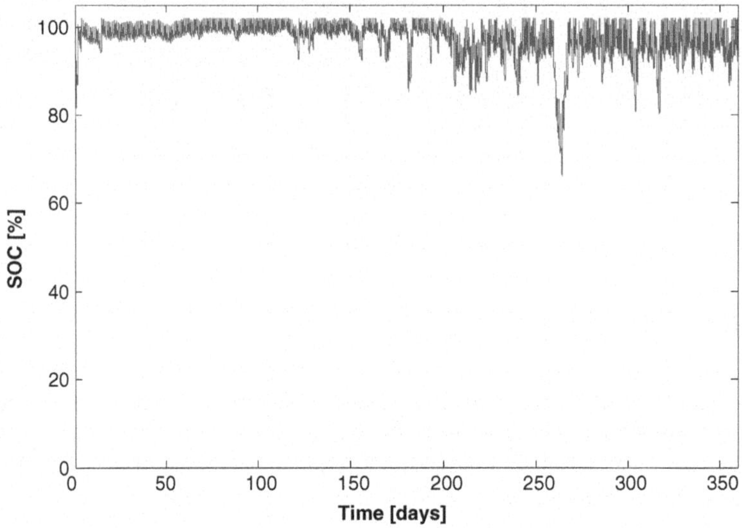

Fig. 2.5 SOC over the course of a year

Over the course of a year a total potential energy generation of 97.5 kWh was obtained, of which 31.9 % was not harvested and hence not utilized. The total demand of 62.0 kWh was completely supplied for, as given in Table 2.4.

The state of health (SOH) of the battery, defined as the ratio of the remaining capacity over the initial capacity, is at 96.8 % at the end of one year. The loss of capacity amounts to 2.5 % due to corrosion and 0.7 % due to degradation. The reason for this healthy operation is indicated in the SOC curve, as given in Fig. 2.5. The battery hardly goes into low SOC values in normal operation. The SOC falls under 80 % on only one occasion during the second half of the year.

Discussion

The results of this paper show a strong potential for interlinking SHS with other units. More than 30 % of the energy generated was dumped. However, there are a number of losses that have not been accounted for including reflection, dirt, shading or losses in the domestic wiring which could influence the outcome of this value. The assumptions made for the load profile play a critical role. In particular, seasonal load variation is not being accounted for which could have a large impact.

Nonetheless, there is a significant amount of energy hidden and available to be sold from the SHS owner to a small microgrid cluster. The initial connection to another household with a small battery that can store and use this energy and thus does not require any separate generation capacity could be the starting point for such a bottom-up electrification scheme.

Acknowledgments The author would like to thank the *Gesellschaft von Freunden der TU Berlin e.V.* for their support and the opportunity to present this work at the *MES-BREG 2014 Symposium: Innovating Access for Remote Areas | Discovering Untapped Resources* in Berkeley.

References

Bindner, H., Cronin, T., & Lundsager, P. (2005). *Lifetime modelling of lead acid batteries.* Roskilde: Risø National Lab, Wind Energy Department.

Boldt, F. (2012, March 8). *Simulation of micro electricity systems for on and off grid applications-focusing on the analysis of battery ageing and design of a system controller* (M.S. thesis, Dept. Eng.). Berlin: Hochschule für Technik und Wirtschaft Berlin.

Boxwell, M. (2013). Dhaka average solar insolation. Retrieved August 27, 2013, from http://solarelectricityhandbook.com/solar-irradiance.html.

Groh, S., Philipp, D., Lasch, B. E., & Kirchhoff, H. (2013). Swarm electrification—suggesting a paradigm change through building microgrids bottom-up. In *Energy systems in transition: Inter- and transdisciplinary contributions* (pp. 212–217). Karlsruhe: Helmholtz Association.

Hafez, O., & Bhattacharya, K. (2012). Optimal planning and design of a renewable energy based supply system. *Renewable Energy, 45*, 7–15. Retrieved from http://www.scribd.com/doc/121420013/Optimal-planning-and-design-of-a-renewable-energy-based-supply-system-for-microgrids.

Hansen, A., Sørensen, P., Hansen, L., & Bindner, H. (2000). *Models for a stand-alone PV system.* Risø: Risø National Laboratory.

IDCOL. (2013, November). Renewable energy projects—IDCOL solar energy program. Retrieved January 30, 2014, from http://www.idcol.org/energyProject.php.

IEA. (2011). *World energy outlook 2011—executive summary.* Paris: International Energy Agency.

Khadem, S. K. (2006). *Feasibility study of wind home system in coastal region of Bangladesh.* Colorado: Homer Energy.

Kyocera. (2007). *KC65T—high efficiency multicrystal photovoltaic module* (Datasheet). Retrieved from http://www.kyocerasolar.com/assets/001/5170.pdf.

Lam, L. (2011, August 23). *A practical circuit-based model for the state of health estimation of Li-ion battery cells in electric vehicles* (M.S. thesis). Delft: University of Technology Delft, Department of Electrical Sustainable Energy.

Legros, G., Haver, I., Bruce, N., & Bonjour, S. (2009). *The energy access situation in developing countries: A review focusing on the least developed countries and Sub-Saharan Africa.* New York: United Nations Development Programme and World Health Organisation.

NREL. (2005, March 5). Solar: hourly solar (direct normal (DNI), global horizontal (GHI), and diffuse) data for select Bangladesh stations from NREL.

Tenenbaum, B. W., Greacen, C., Siyambalapitiya, T., & Candelaria, J. (2014). *From the bottom up: How small power producers and mini-grids can deliver electrification and renewable energy in Africa.* Washington, DC: The World Bank.

UNFCCC. (2012, May 29). Installation of solar home systems in Bangladesh. UNFCCC.

Unger, K., & Kazerani, M. (2012). Organically grown microgrids: Development of a solar neighborhood microgrid concept for off-grid communities. In *IECON 2012—38th annual conference on IEEE industrial electronics society* (pp. 5663–5668). Montreal, doi:10.1109/IECON.2012.6389060.

United Nations Foundation. (2012, June). Energy access practitioner network—towards achieving universal energy access by 2030.

Villalva, M. G., Gazoli, J. R., & Filho, E. R. (2009). Comprehensive approach to modeling and simulation of photovoltaic arrays. *IEEE Transactions on Power Electronics, 24*(5), 1198–1208.

Chapter 3
A Concept of DC Nano-Grid for Low Cost Energy Access in Rural Bangladesh

M. Rezwan Khan and Edward D. Brown

Abstract Energy access is one of the major impediments in popularizing renewable energy based energy solutions for rural areas of the developing countries like Bangladesh. Despite the success of Solar Home System, energy availability for economic activities are still limited. Very small scale solar PV based grid systems, less than 10 kW, named as nano-grid, seems to be an attractive solution. This paper presents the idea of using DC for the nano-grids to reduce cost, retaining most of the advantages of the regular AC grid systems for the household applications. Small cottage industries and/or low power economic activities can also be supported from the nano-grids that can bring in qualitative changes in the livelihood of the rural community.

Keywords Nano-grid · DC-grid · Economic activity · Irrigation · Household load in Bangladesh

Introduction

Cost of Renewable Energy (RE) based energy access solutions face an uphill task in the developing world as affordability is a major issue. In the recent years, price of solar PV has fallen significantly and the trend remains unabated. It is expected that in near future solar PV will become one of the most popular energy source in the offgrid/rural areas for the countries where availability of sunshine is not too low. Although, the energy cost of a PV based system is still high compared to that of grid power, there is ample scope for PV based energy generation for offgrid regions

M.R. Khan (✉)
United International University, Dhaka, Bangladesh
e-mail: rezwanm@uiu.ac.bd

E.D. Brown
Loughborough University, Loughborough, UK
e-mail: E.D.Brown@lboro.ac.uk

S. Groh et al. (eds.), *Decentralized Solutions for Developing Economies*, Springer Proceedings in Energy, DOI 10.1007/978-3-319-15964-5_3

where the grid extension seems unlikely due to economic or geographical considerations. Studies of off grid rural areas in Bangladesh (Khan et al. 2012) indicate that the main fuel for lighting in the off-grid areas are kerosene based lamps (Kupi or Harricane) and a family on average spend Tk. 200 per month on the cost of the kerosene. Besides the high cost of this imported oil to families on low incomes, these lamps produce significant fumes posing health threats and are prone to fire hazards (Lam et al. 2012; Mills 2012). In the last few years, Bangladesh has experienced an unprecedented growth of the Solar Home System (SHS) (IDCOL 2015) in offgrid rural areas. Initially, the energy cost of the SHS was very high in comparison to grid electricity due to high PV price, but the actual cost for lighting a house with efficient fluorescent lamps via the SHS were comparable to the expenditure that families had been making on kerosene purchases, with the advantage that the health hazards from the fumes and the fire hazard from the open flames had effectively been eliminated. Additionally, people had the opportunity to charge mobile phones and watch TV in the case of the installation of systems with panel sizes of higher than 50 Wp (IDCOL 2015) which constituted another very important incentive for families to opt for a SHS. As the price of the PV started to fall, the rate of penetration of SHS in the market soared accordingly. In the recent years higher efficiency in LED lighting and low energy LED TVs has reduced the average PV size in SHS. Five years back, 50 W system was the most popular one with provision for 3 fluorescent lights and a black and white TV. Today, 40 W systems can accommodate larger number of LED lights with a low power colour TV. Lower PV price and reduced PV size have compounded to make the systems more affordable to the rural people. Today, more than 50,000 SHS are being installed every month in Bangladesh (IDCOL 2015) and it is considered as one of the fastest growing and most successful solar programs in the world.

Despite the successes of the Bangladesh SHS programme, there still remains a sizeable proportion of the rural population for whom the costs of such systems remain beyond their reach. In this context, solar lighting products similar to what have been promoted via the Lighting Africa initiative (Lighting Africa 2015) have also been proposed as a very low cost solution for these types of households. Typically, these types of solutions use very small PV panels, sometimes termed Pico-PV. The typical size of the PV panels vary from 5–10 W, with a small storage battery and 2/3 LED lights. As these systems are very small, the typical cost is low and the system is well within the affordable range of the population remaining at the bottom of the pyramid (BOP) in Bangladesh. Recently, there are several propositions from a variety of international organizations and fund giving agencies to implement similar systems in rural Bangladesh to promote these types of solutions to the lighting needs of the poorest Bangladeshi communities However, before taking a hasty decision to roll out similar programmes in Bangladesh based on the success story of Lighting Africa, it is important to analyze socio-economic conditions in rural Bangladesh which is quite different from Africa. One of the most striking dissimilarities between communities in rural Bangladesh and most parts of rural Africa is the much higher population density in the former which leads to people living closely in houses clustered in a small area. This may offer possibilities for pursuing an alternative solution in Bangladesh. It is a common observation that the electricity generated by smaller PV panels is more expensive than that generated by larger panels (as, for example, the price ratio of a 100 W panel to a 10 W panel is more than 10) as there are some fixed overhead costs like framing and the connectors at the back of the panel. In other words, the larger the system the lower is the cost per watt. Very small systems also require small sized batteries and small sized lead-acid batteries are usually not very good quality meaning that the practical options left are Nickel or Lithium based batteries. These Nickel or Lithium based batteries are 4–5 times more expensive than an equivalent lead-acid battery which also adds to the relative cost of small systems. It is estimated (Khan et al. 2012) that the energy cost of a small system with this type of more expensive battery is almost two times than that for a conventional lead-acid battery based energy storage system. Bangladesh economy has been growing at a rate of around 6 % for last 12 years and demand for electricity also grows accordingly. Pico-PV systems does not provide any opportunity for growth other than installing an additional system.

Despite all the positive features of the SHS, it has a very serious limitation in terms of its potential to directly affect an individual household's ability to improve its income generation and overall quality of life (Blunck 2007; Mondal and Klein 2011, p. 17). The energy output of a SHS is so small that it can only support the meeting of basic lighting necessities and perhaps the charging of mobile phones and the running of a TV as a basic entertainment for rural people. While people are debating about the limitations of the SHS in boosting economic activities due to their low energy output, Pico systems with even lower energy output will face further severe criticism. Studies (Khan et al. 2012) made on energy needs in rural Bangladesh, however, indicate that there is demand for further very important facilities. One of these is a fan for comfort during the hot summer months and the others include income generating activities like irrigation, saw mill, flourmill for rice or wheat crushing, rice husking machine etc. Solar PV based irrigation in particular seems to have a very good prospect as 2/3rd of the total number of irrigation pumps driven by diesel in the country are located in off-grid areas. Demand of refrigerator is still very limited due to economic limitations of the households.

In this paper we propose a community based small sized centralized solar PV system, termed 'nano-grid' for its small size (in contrast to the larger networks envisaged under terms such as mini-grids), that can provide energy to 15–20 households and can support small scale developmental activities like irrigation. At the same time, we raise an important debate of DC versus AC for these small scale grid systems keeping in mind the additional cost of conversion from DC to AC.

The Debate of DC Versus AC

Appropriateness of DC versus AC for the grid lines is a century old debate and apparently resolved in favour of AC as the distribution system is accepted as AC worldwide. With the development of Power Electronics, the debate has again revived and there are many instances of High Voltage DC (HVDC) transmission lines where advantages of DC transmission are exploited. Unlike AC, DC systems do not suffer from voltage drops due to inherent line inductance and charging currents due to line capacitance. At the same time ‘skin effect’, a predominantly AC effect that increases line loss, is absent in DC. Moreover, high efficiency DC machines like Brushless DC Motors are getting more popular in recent years despite of their higher cost. With increasing grid energy price throughout the world, more efficient DC machines off set their higher price within a time span of 5 years of usage by dint of their lower power consumption. There are already strong voices being raised regarding the advantages of DC mini/micro grids (Khan 2012).

So far the safety and protection is concerned, DC is considered to be less safe than AC as it is more prone to sustained arcing resulting in more frequent fire hazards. In case of AC there is a natural tendency for arc quenching at the time of zero crossing, which is not the case with DC. So, the conventional DC circuit breakers are designed more rugged and are more expensive. However, recent researches indicate that the arcing problem can be overcome quite easily by putting capacitors across the circuit breakers (Khan and Khan 2014) and at convenient locations in the transmission/distribution lines.

Solar PV Based Irrigation in Bangladesh

Irrigation is one of the major economic activities carried out in rural Bangladesh as Bangladesh is predominantly an agricultural country. Bangladesh has a monsoon climate where there is plenty of rainfall during the months from June to September. Although water is one of the essential components for agriculture, the rainy season in Bangladesh is not the main cropping season as flooding is a common phenomenon and most of the low lying lands are inundated. The main cropping seasons are ‘Aman’ harvested during the month of November and ‘Robi’ harvested during the month of May. The ‘Robi’ season is predictably dry with plenty of sunshine and with proper irrigation this is the most important crop producing season for Bangladesh. However more than 60 % of the rural area is not connected to the national electricity grid and irrigation mainly depends on diesel based engines in these areas. The cost of diesel in the city areas is about US$0.88 per litre and it is at least 20 % higher in the rural areas due to the additional of transportation and storage costs. Bangladesh imports US$1 billion worth of diesel every year for use in the agricultural sector alone (http://www.nationmaster.com/). The cost of diesel based irrigation at the field level is US$0.27 per kW-h equivalent of electricity. The irrigation pump owners charge 20–25 % of the crop produced for irrigating the land for a season depending upon the location and nature of the agricultural land.

Under the prevailing condition, our studies presented here, indicate that with the present price of PV panels and available solar radiance, solar PV based irrigation can be competitive to diesel based irrigation in Bangladesh. However, it is important to understand that irrigation requirements in any particular season vary

Table 3.1 House hold consumption (summer months)

	Light	Fan	TV
% House hold (HH)	100	100	25
No./HH	3	2	1
Watt/unit	5	20	30
Hrs of usage, h	4.5	8	8
Diversity	0.8	0.8	0.8
Energy/day, W-h	54	256	48
Total/day, W-h			358
Avg. load/HH, W			62.5

with the rain fall, the type of crops grown etc. Any dedicated irrigation scheme, where PV does not supply energy to any other load, face energy wastage during the non irrigating months (from June to September). So, for any PV-based alternative that can replace diesel based irrigation, it is very important to integrate the irrigation scheme along with a rural grid so that PV energy will have alternative usage. In a rural household in Bangladesh light, TV, fan and mobile charger are the main loads with the fans accounting for more than 70 % of the energy demand (please see Table 3.1). So it is easily foreseen that the load demand is higher during the summer months due to wide spread demand for fans. Hence, seasonal variation of household load and availability of sunshine plays a very important role in designing a cost effective and efficient solar PV based system. During the main irrigating months (February to May) sunshine is higher than the annual average by around 20 %, the temperature is relatively low and household demand for electricity is also low. So, we would expect that the higher sunshine during the irrigating months can very effectively be utilized for irrigation without causing shortages for household energy consumption. During the months of June to September, the weather is cloudy and there is plenty of rainfall. Lower sunshine during this season results in less production of PV energy but needs to cater only the household demand as irrigation is not required. Hence, if we consider the distribution of sunshine and the seasonal variation of energy demand that includes irrigational demand, electricity requirement would appear to match the availability of PV output. This ensures maximum possible utilization of the PV power leaving very little option for wastage.

The Concept of DC Nanogrid: Its Appropriateness in the Context of Bangladesh

Drawing together the discussion in the previous sections, it can be seen that the nano-grid idea is based on the fundamental concept of the Solar Home System, where the basic necessities of households are met, but at the same time some small scale applications like irrigation can also be incorporated. This concept takes advantage of the fact that houses in Bangladesh are usually clustered together in

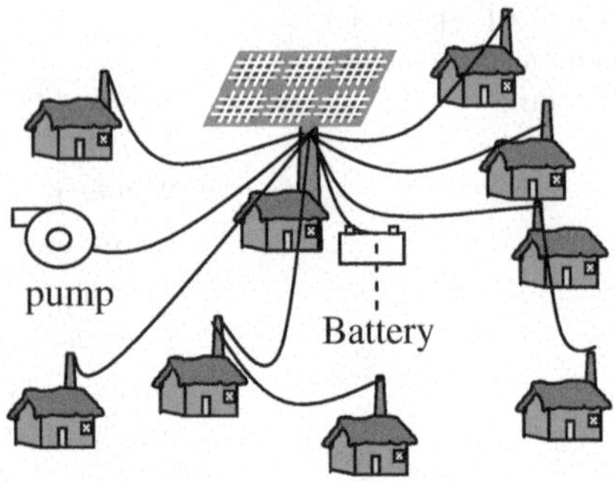

Fig. 3.1 Schematic diagram for the nano-grid

rural areas in a group of 15–20 houses (within a diameter of less than 150 m). A schematic diagram depicting the concept is shown in Fig. 3.1. In the proposed nano-grid system, something like a 1.5–2 kWp PV system is installed in a small cluster of households within a radius of 60–70 m and power is distributed to 10–15 households from this system. The PV panels and the battery used in the proposed nano-grid will be connected in series in such a way that the grid voltage is 220 V DC (nominal) and the households are supplied with this voltage. Roof top locations of one or two houses will be chosen for PV installation and the storage battery will be placed in a convenient location close to it.

As described in the later part of this section, the main household loads for this system are likely to be light, TV, fan and mobile phone. The LED or CFL lights and TV have their own built in controller circuits that make them insensitive to DC or AC supply. These days, brushless DC fans are widely available in the market. Although the brushless DC fans are more expensive, they are much more efficient ($\sim$80 %) than the usual induction motor based AC fans (efficiency $\sim$60 %). The higher cost of the brushless DC fans is likely to be compensated within three years considering the lower power consumption of the fans. However, in case of irrigation pumps or some other income generating activities a separate inverter is likely to be needed. The advantage of the DC grid is its low cost, as no inverter is used and at the same time this avoids inverter energy losses. Our preliminary estimates indicate that there can be around 25 % cost saving by avoiding inverters in the grid (assuming an inverter cost of US$0.7/Watt for small inverters and an average conversion loss of 5 %).

The average typical household (HH) load in rural Bangladesh in the summer and winter seasons are given below in Tables 3.1 and 3.2. While 100 % of the households ask for light and fan, only 25 % can afford to have TV. The demand for refrigerator is not considered at this stage as it goes beyond the affordability of the average households.

Considering the typical load in a household, the expected summer time (May to September) load is around 2.5 times the expected winter (November to March) household loads. This, as explained above, is due to the fact that there is expected to be high usage of fans in summer due to hot weather condition. From Tables 3.1 and 3.2 it can be clearly seen that fans constitute a significant part of the household load

Table 3.2 House hold consumption (winter months)

	Light	Fan	TV
% House hold (HH)	100	100	25
No./HH	3	2	1
Watt/unit	5	20	30
Hrs of usage	6.0	0	10
Diversity	0.8	0.8	0.8
Energy/day, W-h	72	0	60
Total/day, W-h			132
Avg. load/HH, W			22.5

in summer whilst the zero fan load in the winter months reduces the total load to less than half. Although the mobile charger is an important component, its actual energy consumption is very small and has not been included in the tables. The surplus generated from the significant fall in the household demand is proposed to be used for developmental activities like irrigation during the dry months.

Figure 3.2 shows the average daily solar radiation on a horizontal flat surface in Bangladesh. The main irrigation season is from February to 1st week of May and the weather remains reasonably dry and cool until the end of April. It is interesting to note that sunshine is higher during these months, which together with relatively low demand for fan-use in the house holds during this period generates sufficient surplus energy to divert it for irrigation. Our preliminary calculations show that a small pump of 1.1 kW (1.5 HP) can be easily run to irrigate the fields. During the months after June, the rainy season starts and demand for irrigation is reduced to around 5–10 %.

At this stage a question may be raised regarding the motor driver of the pump—should it be AC or DC. It is a well accepted fact that DC pumps are more expensive than AC pumps with simpler driver circuit. On the other hand an AC pump will require an inverter as the nano-grid is proposed to be DC. If the option is for an AC pump, an inverter is needed any way and it may appear that the merit of the

Fig. 3.2 Month wise average daily insolation in Bangladesh

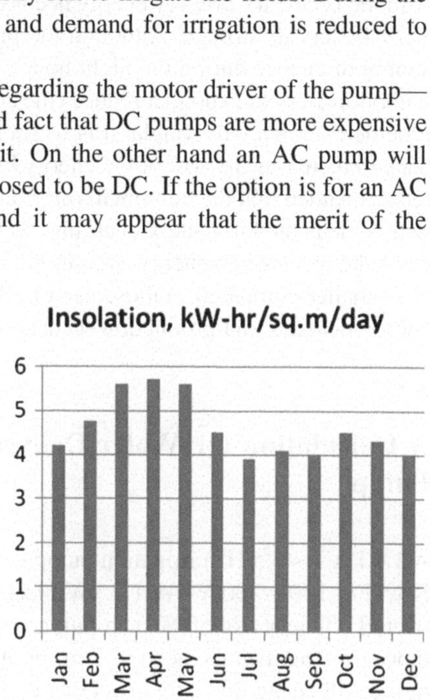

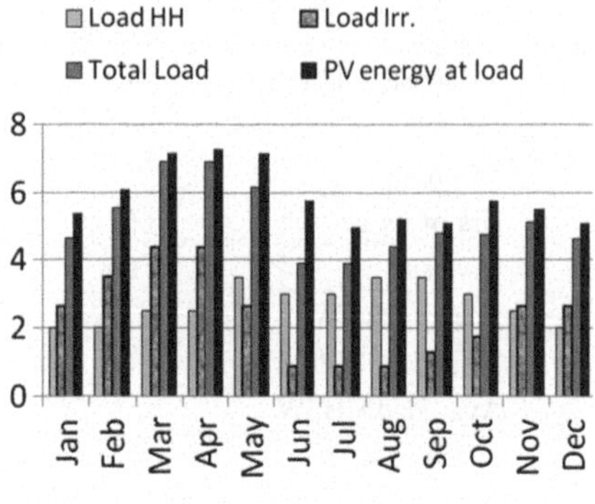

Fig. 3.3 Month wise average daily energy (kW-h) budget for a 1.9 kWp PV nano-grid supplying power to 10 households and an irrigation pump of 1.1 kW

proposition for a DC nano-grid would look less attractive. It is important to understand that the actual size of the inverter for the pump is much smaller than the size of the inverter required in the grid and at the same time isolated motor driving inverters will cost less per watt than the grid type inverters.

Although the sunshine is lower during the rainy season (June-September), absence of irrigation makes the overall energy demand lower and a well-designed system should be able to cope quite satisfactorily with household demand. For a very small system, we estimate that a 1.1 kW pump and 10 households can be supplied with the necessary primary energy needs using a 1.9 kWp PV system with a 480 AH, 12 V battery as storage. Although the proposed nano-grid uses battery backup for supply of energy during the night hours, irrigation can be done during the day time without battery backup that reduces the cost of electricity for irrigation can make the scheme commercially viable. It is worth mentioning that the requirement of battery backup increases the cost of PV energy by more than 60 %. During the day time it is recommended that the pump will run from 10.00 am to 2.30 pm. The rest of the time will be dedicated to battery charging. At the same time, the panel size is such that it can generate enough energy even in rainy, cloudy or foggy days to keep the battery size smaller compared to those used in SHS. Figure 3.3 below gives a month wise energy demand and production scenario for such a system.

A Calculation for Water Delivery for a Small Irrigation Pump

A 1.1 kW ($\sim$1.5 HP) irrigation pump can run quite satisfactorily if the input power is within 1000–800 W. A 1.1 kW electric pump driven by an inverter, having an overall efficiency of 50 %, can pump around 90,000 l of water per day (4 h of run time) from an average head of 7 m. In the case of a submersible deep tube well the figure will be significantly lower due to the higher water head. Most of the irrigation pumps in Bangladesh use shallow tube wells not exceeding a depth of 6–7 m. Rice is the most important crop in Bangladesh that needs irrigation during the dry months, 90,000 l per day can irrigate about 6–7 acres of rice field for a whole season (3 months). In the case of other crops like vegetables, wheat or maize the area of irrigated land will be much larger as they require less water than rice cultivation.

The Cost Calculation for the Proposed Nano-Grid

Cost of the system and electricity

The description of a model system is given below

Size of the PV—1.9 kWp
Size of the battery bank—480 AH at 12 V
One irrigation pump (AC)—1.1 kW (1.5 HP)
No. of households—10
Summer load per house hold—360 W-h per day
Winter load per house hold—135 W-h per day

Estimated total cost of the system including installation and accessories is US $4652. Based on IDCOL model of financing (50 % grant, 30 % loan at 6 % interest rate for 8 years and 20 % equity) energy cost for house hold electricity is US$0.35/kW-h and energy cost for irrigation is US$0.20/kW-h.

Considering a monthly connection charge of US$1.25 electricity bill for an average household for summer months is US$5.0 and that for winter months is US $2.7. Corresponding irrigation energy cost per season is 10.7 % of the crop produced for rice fields. As mentioned earlier, the usual charge for diesel based irrigation is 20–25 % of the crop produced in the irrigated fields. It is worth mentioning here that the average energy cost in a SHS under a similar financing model is close to US $0.50. As SHS has only fixed load of light, mobile charger and/or TV, it cannot accommodate the seasonal variation of sunshine resulting in underutilization of the available PV energy. The nano-grid provides energy at a lower cost with more options for household gadgets. Apart from household use, it can incorporate a small sized irrigation pump that can have significant impact on agriculture.

Conclusions

The paper points out the fact that many of the household gadgets are insensitive to DC or AC and does not require any additional modification when used with DC. DC machines, usually more efficient than the AC machines, are getting less expensive and becoming more affordable. So, in an off grid rural area DC becomes the natural choice in the case of a newly built PV system for the supply of electricity. The paper proposes the concept of DC nano-grid highlighting its technical advantages and some of the economic and social considerations involved in their development. These are not capital intensive solutions as their size is quite small and can be easily implemented in a small community overcoming the financing difficulties; unlike, for example, the case of large scale mini-grids (Ulsrud et al. 2011). At the same time, supply of DC voltage instead of AC at the household level avoids the relatively high cost of inverters eliminating inverter losses, particularly when the load demand is less than 10 % of the peak during late night hours. Incorporation of irrigation ensures limited scale of economic activities involved with nano-grids that will bring qualitative change in the livelihood of the rural population.

References

Blunck, M. (2007). Electricity and sustainable development: Impacts of solar home systems in rural Bangladesh. Diploma thesis, Department of Geography of Johannes Gutenberg. Germany: University of Mainz.

Khan, M. R. (2012). A stand alone DC micro-grid for electricity supply in rural Bangladesh. In *Proceedings of International Conference on the Developments in Renewable Energy Technology*. Dhaka, Bangladesh, Jan 2012.

Lam, N. L., Smith, K. R., Gauthier, A., & Bates, M. N. (2012). Kerosene: A review of household uses and their hazards in low- and middle-income countries. *Journal of Toxicology and Environmental Health, Part B: Critical Reviews*, *15*(6), 396–432.

Mills, E. (2012). Health impacts of fuel—based lighting. In *3rd International Off-Grid Lighting Conference and Trade Fair*. Dakar, Senegal, 13–15 Nov 2012.

Khan, M. R., Khan, M. F., & Chowdhury, S. A. (2012). *24 kW Solar-diesel hybrid power system for off grid Shonatala Island of Nasir Nagar upazilla of Brahminbaria districts*, GIZ–UIU study report.

Khan, M. R., & Khan, Md. F. (2014). Arc Suppression in moderate voltage DC grid for solar PV applications. In *3rd International Conference on the Developments in Renewable Energy Technology* (*ICDRET 2014*). Dhaka, Bangladesh, 29–31 May 2014.

Mondal, A., & Klein, D. (2011). Impacts of solar home systems on social development in rural Bangladesh. *Energy for Sustainable Development, 15*(1), 17–20.

Ulsrud, K., Winther, T., Palit, D., Rohracher, H., & Sandgren, H. (2011). The solar transitions research on solar mini-grids in India. *Energy for Sustainable Development*, *15*(3), 293–303 (minigrid).

IDCOL (2015) Infrastrcutre Development Company Limited. http://www.idcol.org/old/bd-map/bangladesh_map/. Accessed 15 Feb 2015.

Lighting Africa (2015) An Innovation of World Bank Group. IFC/ The World Bank. http://www.lightingafrica.org. Accessed 15 Feb 2015.

Gas diesel oil consumption in agriculture statistics. http://www.nationmaster.com/graph/ene_gas_oil_con_in_agr-gas-diesel-oils-consumption-agriculture.

Chapter 4
Experience from First Solar Mini Grid Service in Bangladesh

Hasna J. Khan, Asma J. Huque, Kaysar Ahmed Sagor, Tanvir Ahmed, Rashedul Alam and M. Riazul Hamid

Abstract A 100 kW Solar mini grid service in the remote off-grid rural market of Sandwip island of Bangladesh has shown technical and commercial viability for the service provider. Financial planning of solar mini grid requires an optimum blend of consumer categories to ensure maximum socio-economic return on investment. Demand and supply side energy management is an important component of stable power supply from the solar minigrid. Policy-wise, subsidized financing with attractive incentives for the private service providers is essential at this early stage of solar mini-grid deployment.

Keywords Solar minigrid · Sandwip island · Bangladesh

Introduction

Despite its large population, Bangladesh has achieved substantial economic growth over the last 15 years. Deficient infrastructure prevents the country from achieving its full growth potential. This situation is particularly evident in the economically disadvantaged remote and rural areas. Whereas access to energy is a priority in the development framework, much work needs to be done on basic infrastructure for rural electrification. Remote and dispersed areas are currently adopting off-grid electrification as a viable alternative to the national grid service. In a country where

H.J. Khan (✉) · A.J. Huque
Prokaushali Sangsad Limited, House 4, Road 6, Block C, Banani, Dhaka, Bangladesh
e-mail: hasnajkhan@gmail.com

A.J. Huque
e-mail: psldhaka2@gmail.com

K.A. Sagor · T. Ahmed · R. Alam · M.R. Hamid
Dept. of Electrical and Electronics Engineering, Ahsanullah University of Science and Technology, Dhaka, Bangladesh

S. Groh et al. (eds.), *Decentralized Solutions for Developing Economies*,
Springer Proceedings in Energy, DOI 10.1007/978-3-319-15964-5_4

Solar Home Systems (SHS) was established as a complementary solution to grid electrification, national interest has proven it to be the most viable off-grid electrification today. Prokaushali Sangsad Limited (PSL) was engaged in providing the original concept leading up to the final SHS program design (World Bank 2001). Currently Bangladesh deploys over 70,000 SHS per month, with over 2.6 million (IDCOL 2015) in total, under the national solar home program executed by Infrastructure Development Company Limited (IDCOL). This is one of the most successful off-grid SHS programs, with the highest installation rate in the world today.

Similar to SHS, solar minigrid can also offer reliable service for rural off-grid areas in Bangladesh (Khan and Huque 2012). With the initiative and technical support of PSL, private utility company Purobi Green Energy Limited (PGEL) has invested in the nation's first commercial solar minigrid in a southern coastal island of Bangladesh with financing from IDCOL. Germany based Asantys Systems, with local partner Energy Systems, installed the solar minigrid technology in un-electrified Enam Nahar market in Sandwip island. Various productive end-use revealed the potential economic benefits from clean solar power compared to the previous micro diesel generators. Commercial enterprises switched from diesel service to solar. Similar application of solar mini grids for the vast number rural markets in the off-grid locations can offer significant improvement of socio-economic condition of this underserved population. Although rural markets show the maximum potential for solar mini grid application, clustered households adjacent to the markets are also additional beneficiaries. The experience of PGEL shows that a suitable mix of consumers is important for the success of a sustainable business, while also ensuring social and environmental benefits. Additionally, policy-wise, subsidized financing with attractive incentives for the private service providers is essential at this early stage of mini-grid deployment.

This paper outlines the technical and business viability for operating solar mini grid in the remote, off grid locations of Bangladesh. Having the first installation of its kind, currently PGEL is the only utility company in the country which operates as a solar minigrid service provider, receiving technical support from PSL. Hence, this pioneering experience has become a learning platform for other investors, financing and developing agencies, in addition to the policy makers. Some of the national clean energy policy decisions for off-grid rural electrification business models are being made using the experience gained from Enam Nahar Market.

System Design and Planning

Consumer Profile in Enam Nahar Market

A survey was performed by PSL in 2008 to collect information on power demand of about 200 small and large enterprises in the project area. Initial survey showed that Sandwip Island had a dynamic population with several diesel electrification services providers supporting the general public. Several educational institutions, health service centers, hospitals, major markets, computer service centers were present in Enam Nahar market. Although major businesses operated with their captive power, general shops in the markets received electricity from private diesel micro-grid service providers. Such services operated on an adhoc basis for about 5 h a day. Households were found to be un-electrified in general, unless using individual solar home systems. Power lines used by the private service providers were of poor quality since they were not implemented considering efficiency and line loss. The service providers also did not account for their own time spent, therefore the cost of power production was arbitrary for the private business men. A diesel generator owned and operated by the government occasionally operates in Sandwip island for 3–4 h a day serving local government offices in the administrative center, which is situated about 5 km from Enam Nahar Market.

Business Plan

The planning of the project envisioned that a solar mini grid could serve the basic needs of commercial enterprises within Enam Nahar Market and some adjacent households, in addition to local schools and health centers. PGEL was hence formed with an objective of serving this remote location with state of the art mini grid technology. The level of services to be offered were to be based upon the needs of its clients, and quality of service would follow the standards of the national distribution grid.

Total estimated cost of 100 kW PV plant with 40 kW Diesel backup would be $730,000 (2009 price), including hardware, site development, civil works for an office building, distribution line, and 5.4 % soft cost for technical assistance etc. The financial plan consisted of (a) equity, (b) soft loan, and (c) grant. The business forecast with financial analysis showed the service to be financially viable under the loan scheme of 30 % at an interest rate of 6 %, with a tenor of 11 years with 2 years grace period offered by IDCOL. In addition to the loan, a grant of 50 % (KfW grant for Pilot Project) and PGEL's equity of 20 % of cost constituted total investment in the mini grid. Keeping an affordable tariff with service hours of 9 am to 11 pm year-round, while retaining adequate revenue, was key to the financial viability.

Tariff for Electricity

People residing in remote locations of Bangladesh pay one of the highest rates for electricity from diesel powered micro-grids. Surveys showed that most of the commercial enterprises of Enam Nahar Market were using diesel micro-grid services. Depending upon the type of appliance being used, daily rates being converted to tariff ranged from $0.56 to 0.96/kWh as shown in Table 4.1. Such high rates from

Table 4.1 Tariff prior to solar in Enam Nahar market

Appliance	Watts	kWh/day	$/kWh
CFL lamp	24	0.132	0.96
Ceiling fan	60	0.33	0.77
Florescent tube	40	0.22	0.58
Television	40	0.22	0.58

diesel generated power was endured due to the high demand of electricity for commercial activities in the market.

Preliminary surveys suggested that the potential consumers of a solar mini grid in Enam Nahar Market were eager to pay an equivalent amount to cover their cost of service, considering its reliability and scope of modernization with alternative energy. Based upon tariff prior to solar electrification, financial analysis and project financing criteria of IDCOL, PGEL set the tariff at 32 Taka/kWh ($0.40/kWh) with a connection charge of about $67 for all new consumers.

Hardware and System Configuration

In September 2010, a 220 V mini-grid of 100 kW solar PV coupled to a 40 kW Diesel backup system was commissioned to meet partial demand of electricity for the people of Enam Nahar Market of Sandwip island. The layout of the grid connected and bidirectional inverters, with solar modules and battery bank is shown in Fig. 4.1.

For nominal power of 64.8 kW, the PV modules are connected to 6 grid tied 11 kW inverters (Sunny Mini Central of SMA with MPPT), supplying directly to the 220 V mini-grid distribution line. Three phase AC distribution line is configured through the multi-cluster box, which is the interface for all connectors and controls. The unused portion of the generated power is stored in the batteries through 12 bidirectional inverters called Sunny Islands, distributed in 4 clusters. Additional

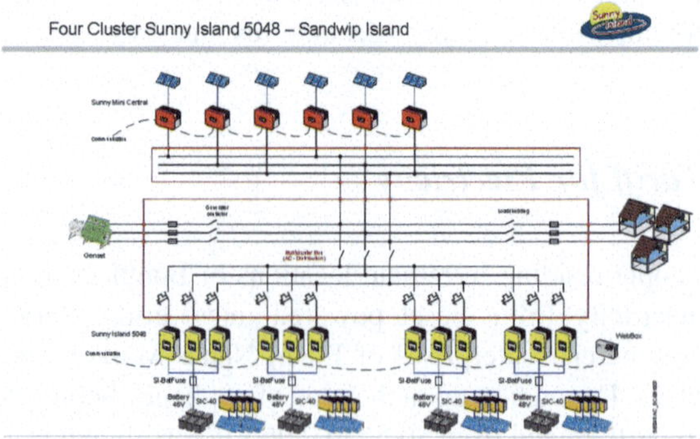

Fig. 4.1 Configuration of solar power plant (Courtesy SMA, Germany)

40 kW of PV are generated and stored directly into the same battery bank through DC battery chargers (SIC40 with MPPT). The power plant has a total 96 batteries in 4 battery banks with total 12,000 Ah (48 V). The battery bank is sufficient to cover the evening load of the market with average insolation. During the worst season of solar irradiation and low state of charge of the batteries, backup power is provided by a 40 kW Diesel generator, which also provides the equalization charge to the battery bank.

Total land used by PGEL for the solar power plant is 0.61 acres. Area requirement for 100 kW solar PV (Kyocera brand) was about 1500 m^2, which includes the ground mounted solar park along with the roof of the office building. Beyond solar plant design, site development, various licensing, civil and fieldwork supervision were done by PSL. Attempt was made to use the pre-existing power grid lines belonging to Bangladesh Power Development Board (BPDB). The BPDB lines were never energized in 20 years due un-economic production cost from diesel fuel. Unfortunately the beaurocracy became complex, and cable condition was not guaranteed. Moreover, since adoption of state owned property would not be practical within a non-government project, PGEL procured its own 5 km distribution lines for the solar plant.

A three phase distribution line with 220 V AC bus line is connected to the bi-directional inverter-battery assembly through a multi-cluster interface for mini-grid. The power station being located in the central part of Enam Nahar Market, three separate phases extend in different directions, with a balanced load from the consumer end.

As shown in Fig. 4.2, estimated load profile of the plant had a peak power exceeding 40 kW from 7 to 10 pm in the evening, where lighting contributes mostly to the load.

Much attention was then given to Demand Side Management (DSM) of the service. In order to minimize evening load, the consumers were encouraged to use only Energy Saving compressed florescent lamps (CFL), and appliances with standard specifications. A major campaign followed to introduce efficient LED lamps to replace all other types of lighting devices, in addition to energy efficient fans and motors. Distribution line, circuit breakers, service drop lines, and household meters follow the specification standards of the rural electrification cooperatives of the Rural Electrification Board (REB) of Bangladesh. Each consumer is billed on a monthly basis based upon individual electric meter readings.

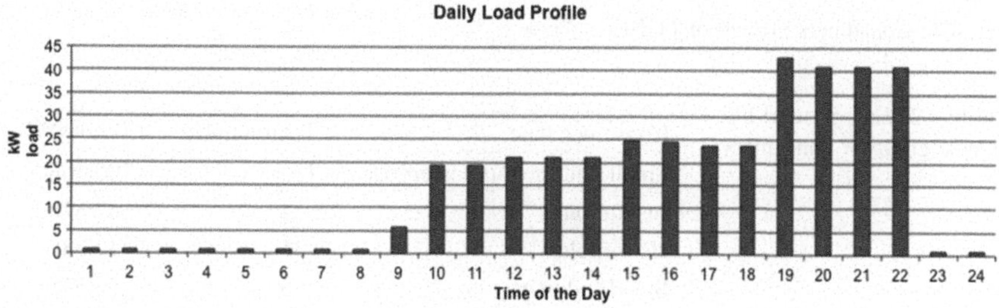

Fig. 4.2 Planned load distribution for service

Results and Discussion

Since October 2010, the solar mini-grid of PGEL serves areas surrounding the Enam Nahar Market to provide electricity for basic lighting, fans, computers, printers and various appliances used on a daily basis. To date the plant has generated more than 320 MWh from solar PV, with a specific yield of 104 kWh/kWp (monthly average for 2013), displacing 222,230 kg of CO_2. Generation from backup Diesel is below 10 % of total supply of PGEL during last six months of 2013.

Up to December 2013 the service was being offered to a total of 230 consumers who had taken connection gradually over three years, as shown in Fig. 4.3.

The major focus was to serve enterprises that can benefit from profitable income generation through the use of reliable electricity. It took about one year for general consumers to build confidence in reliability of the services of PGEL before they collectively decided to disconnect from the earlier diesel services. Uninterrupted electricity for 13 h from PGEL proved to be very attractive for the consumers, which is not even available in major cities of the country.

Based upon the type of activities performed, all the consumers fall into diverse groups of four categories as shown in Table 4.2, where SMEs form the major 58 %. Based upon average consumption pattern, the consumers are classified into three classes as shown in Table 4.3. It is noteworthy that each class nearly equally share the total energy produced and sold by PGEL.

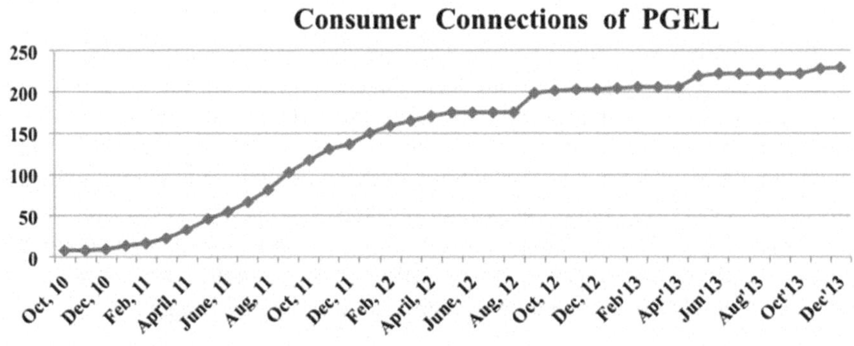

Fig. 4.3 Consumers growth of PGEL

Table 4.2 Energy used by diverse group of consumers

Consumer type	Total number	Total kWh
Small and medium enterprise	136	105,308
Institutions	54	82,876
Households	35	12,538
Small industry	5	4515

Table 4.3 Classification of consumers of PGEL

Consumer class	No.	kWh/mo./cons.	% Total sales
Large cons. (>100 kWh)	16	173	36.9
Medium cons. (41–100 kWh)	36	59	28.3
Small cons. (1–40 kWh)	178	15	34.8

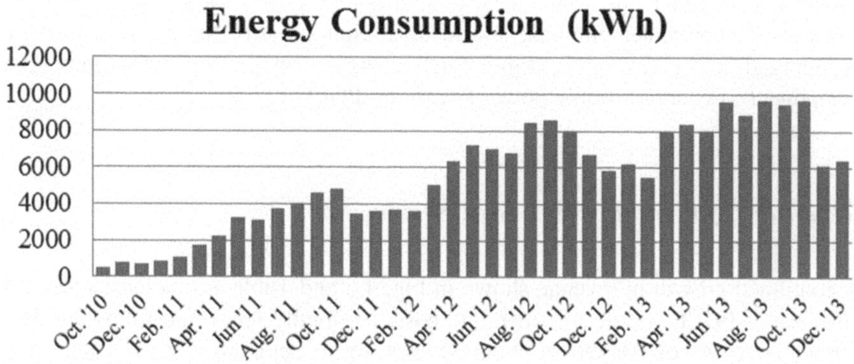

Fig. 4.4 Monthly load pattern and energy consumption

Monthly load pattern: Seasonal variation of load can be seen through the annual energy consumption shown in Fig. 4.4. Overall trend in energy consumption is related to the increase in consumer base of PGEL (Fig. 4.3). Between November and February fans are not used, therefore energy consumption is low during this period. With the onset of hot days in summer, electricity usage increases from April, followed by cooler temperature and less use of fan during the monsoon in July, and a peak demand for electricity appears in August of each year.

Daily load pattern: Although the initial survey indicated a peak load in the evening (Fig. 4.2), according to the actual daily load pattern of an average day shown in Fig. 4.5, there are two peaks in total energy usage. One peak appears

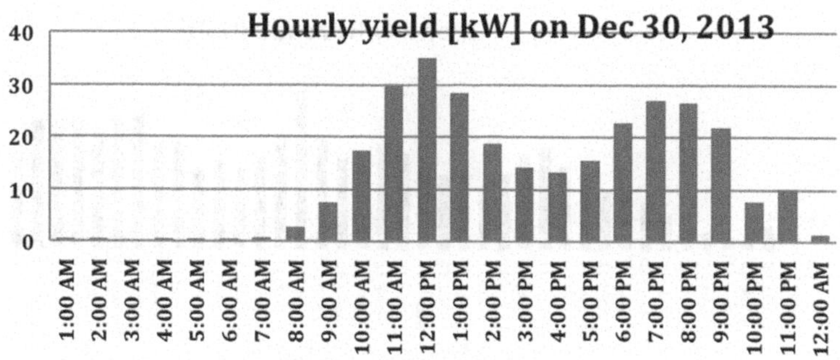

Fig. 4.5 Daily load pattern of PGEL

around 12 PM when the offices, health centers and institutions are in full operation. Another peak appears between 7 PM and 10 PM, mainly for evening lighting and other commercial activities of the SMEs in the market. Since the nighttime use of electricity is mainly from battery storage, the load distribution restricts the use of power-intensive equipment for commercial activities during night. Specifically, the effort is to ensure complete charging and protection of the batteries from over discharge. According to the daily charging pattern, the batteries are mostly charged between 6 am and 1 pm, which leaves the afternoon power for various productive end uses. Hence water pumping and medical equipment of the health center are encouraged to operate at daytime. Such energy management is an important component to ensure a stable power supply by PGEL.

Revenue from Solar Mini Grid Service

A continued growth in revenue shown in Fig. 4.6 and Table 4.3 indicate sustained business of PGEL, where the operating costs are totally recovered following debt service coverage of IDCOL of about \$74,000 up to September 2013. Total operating cost (not including debt service) of PGEL is about 25 % of monthly revenue. With the growth of consumer size, the services of a local bank is being used for monthly bill collection (Table 4.4).

Challenges Faced by Mini Grid Service

Policy Uncertainties: Many potential risks remain with the first time investors and project implementers, and PGEL has experienced several such ordeals. Overcoming regulatory inertia, including undefined policies, has been a major drawback. Given that solar mini grid is a new technology, it is experiencing special scrutiny from development organizations, financing institutions, and BPDB, to gauge its true

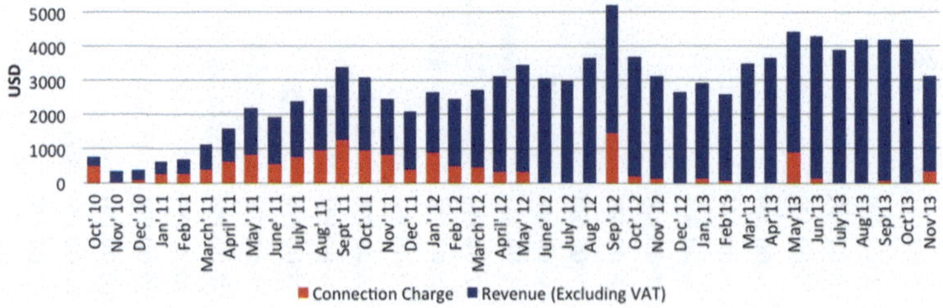

Fig. 4.6 Monthly revenue of PGEL

Table 4.4 Revenue trend of minigrid service

	2011	2012	2013
Total revenue ($)	24,446	39,138	44,319
Connection charge (%)	33	11	4
Electricity sale (%)	67	89	96
Total operating cost ($)	4891	9248	11,527
Diesel (%)	1	20	25
Others site expenses (%)	43	31	15
Salary	56	49	60

merit as remote area power provider. Environmental clearance for renewable energy based power plant needs to be simplified and streamlined. Policy for allocating concessionary market for private sector needs to be in place to avoid future threat from highly subsidized public services to enter in the same area. And finally, financing agreements between project implementer and investors need to be standardized. A recent document "Guideline for the Implementation of Solar Power Development Program" (MEMR 2013) has looked into some of the issues.

Project Cost: Estimated budget for the project was $730,000, where 89 % of the capital cost was for hardware. The high level of investment on hardware is typical for renewable energy projects, especially for solar, where the first cost is high and operating cost is minimal.

Due to the nature of load distribution, the peak evening load of PGEL is supported with batteries. The energy storage system through flooded lead acid batteries was 26 % of total hardware cost. These are industrial type batteries with a service life of 10 years, whereas the plant life is estimated to be 20 years. Total replacement of the battery bank at the end of its useful life is a challenge to PGEL, which will also increase the life-cycle cost of the system. Recurring backup diesel fuel cost must also be managed over time by spreading the peak power load, especially the energy intensive loads in the market. Hence, the cost incurred for energy storage with batteries, and use of Diesel fuel for backup power are two significant costs in financial viability of the plant.

Subsidies and Incentives for RET: IDCOL plans to support 50 solar mini grids by 2016, with similar financing scheme as PGEL. It should be noted that 50 % grant fund reducing the capital cost is an essential element for financial viability of such services. Moreover, there must be reasonable certainty that the public grid will not extend its services to the area in the near to medium term. Otherwise, sponsors would demand compensation for their assets in favorable terms, since their customers might opt for the drastically reduced subsidized tariff, even if reliability is low. PGEL's minigrid being first of a kind power project, it is used as a demonstration that feeds into subsidy policymaking, which, however, in this case, is a risk to the private investors.

Conclusion

The data presented in this study undertaken by PSL provides a reference for future solar mini grid configurations for remote rural markets in off-grid areas of Bangladesh. It provides an overview of commercial viability of solar mini grid service with special financing, keeping reliability of service as the priority. In the long run it may be worthwhile for the REB/PBSs to consider solar mini grid service in the remote un-electrified locations. Economic viability of such services is open for discussion on public-private partnership.

References

World Bank Report. (2001). Bangladesh Rural Electrification-Solar Home Program Preparation. Prepared by Prokaushali Sangsad Limited (PSL) for The World Bank.

Khan, H. J., & Huque, A. J. (2012). Reliable power supply with solar mini grid in sandwip island of Bangladesh. In *Proceedings of 2nd International Conference on the Development in Renewable Energy Technology* (*ICDRET'12*), Dhaka, Bangladesh, 5–7 Jan 2012.

Ministry of Power, Energy & Mineral Resources (MEMR). (2013). Govt. of Bangladesh. Guideline for the Implementation of Solar Power Development Program. Power Division.

IDCOL (2015) Infrastrcutre Development Company Limited. http://www.idcol.org/old/bd-map/bangladesh_map/. Accessed 15 Feb 2015.

Chapter 5
Prospects for Electricity Access in Rural India Using Solar Photovoltaic Based Mini-Grid Systems

Leena Chandran-Wadia, Shruti Mahajan Deorah, Sameer Nair and Anshuman Lath

Abstract Solar photovoltaic (PV) based mini-grid systems have the potential to be an environmentally friendly and sustainable long term solution for electricity access in India. However, the high upfront costs of these mini-grids present policy makers, entrepreneurs and consumers alike with difficulties in financing them. Other challenges to their implementation stem from socio-economic issues and from the lack of adequate support from government agencies. We assess the potential for deployment of solar PV based mini-grids in rural India, for the provision of on demand electricity access, 24 × 7, beyond just lighting. We describe one very high-quality off-grid installation in detail, in operation for over 2 years now, that exemplifies the challenges involved in providing sustainable and scalable solutions. We highlight some of the best practices that have been brought out in this installation. We review the policy measures of the Indian government in the context of scaling out such innovative solutions, not just in remote off-grid villages and hamlets but also in grid-connected villages, and argue that the elusive goal of 100 % electrification can in fact be achieved by the year 2022, when India turns 75, given a conductive policy environment.

Keywords Energy access · Solar photovoltaics · Mini-grids · Rural india

---

L. Chandran-Wadia (✉)
Observer Research Foundation (ORF), NKM International House, 178 Backbay Reclamation, Babubhai Chinai Marg, Mumbai 400 020, India
e-mail: leena.wadia@orfonline.org

S.M. Deorah
Goldman School of Public Policy, University of California, Berkeley, USA
e-mail: shruti.deorah@berkeley.edu

S. Nair · A. Lath
Gram Oorja Solutions Private Limited, NCL Innovation Park, Pashan, Pune 411 008, India
e-mail: sameer@gramoorja.in

A. Lath
e-mail: anshuman@gramoorja.in

S. Groh et al. (eds.), *Decentralized Solutions for Developing Economies*,
Springer Proceedings in Energy, DOI 10.1007/978-3-319-15964-5_5

Introduction

India is blessed with tremendous solar potential, given its daily average solar reception[1] of 4–7 kWh/m. Solar PV based mini-grid systems are relatively easy to install, can be scaled with increasing demand as needed and require minimal day-to-day intervention, which make them particularly suitable for use in rural India. Solar energy based lanterns, small home systems, and water pumps are already being widely used around the country. A compelling reason to deploy mini-grids[2] however, is the opportunity to provide '*on-demand electricity*, 24 × 7', beyond just lighting, to rural homes and businesses so that livelihoods can be enhanced.

Of the 300 million or more Indians who do not have access to electricity today (Census of India 2011), it is estimated that approximately 10 million[3] live in villages and hamlets that are too remote for the grid to reach. Estimates of the count of such villages or hamlets vary from anywhere between 11,000[4] to about 50,000. There is consensus that in these standalone villages and hamlets, renewable energy based electricity access can be a long term solution rather than a stop-gap one—'till the grid arrives'. The goal here would be to provide electricity that is available 24 × 7, for both household and commercial consumption, using appropriately sized solar PV mini-grids, either AC or DC, optionally hybridized with other renewable sources such as wind, biomass or micro-hydro.

The surprise is that the bulk of Indians living without electricity, some 290 million or more, reside in villages that are considered electrified! The Ministry of Power (MoP) of the Government of India (GoI), which holds the mandate for electrification, requires just 10 % of the homes in a village to be connected to the grid in order to declare the village as electrified. This is why, despite having achieved over 96 % electrification, we have on average 44 % of rural households that are still not connected to the grid. As many as 84 % households in Bihar, 63 % in Uttar Pradesh and Assam and 54 % in Jharkhand are not electrified.

Although the bulk of the villages are electrified, they are still not 'energized'—there is insufficient supply of electricity. Metering, billing and collection is another traditional weakness in the system. Many households choose not to get connected to the grid because they are wary of unreasonable billing and the high upfront fees. Although several households resort to tapping electricity illegally through unofficial connections, there is no real incentive for them to get connected because the electricity supply from the grid is so erratic. Other reasons for many households not being electrified include issues related to accessibility and the large distances

---

1 http://www.mnre.gov.in/schemes/grid-connected/solar/.

2 The Government of India distinguishes between micro-grids ($<$10 kWp) and mini-grids (between 10 kWp and 250 kWp). For the purposes of this paper we refer to both as mini-grids.

3 Assuming there are 50,000 small, remote, villages each consisting of 40–50 households and accommodating 200 residents on average.

4 Private conversation with officials from the Ministry of New and Renewable Energy, Government of India.

between clusters of homes within a village, and 'right-of-way' issues that prevent cables being laid to more homes. Grid-connected mini-grids can be set up in such locations provided the policy environment is made conducive.

Research Objectives

We attempt to understand the difficulties associated with implementing solar PV mini-grids at scale in rural India, and to highlight best practices in mitigating these challenges. Although there are some technological and sociological barriers to ensuring the sustainability and longevity of installations, the main challenge is financial—the high upfront costs for the panels, the balance-of-system, and the costs for installing, servicing and maintaining these in remote locations. The Ministry of New and Renewable Energy (MNRE) of the Government of India (GoI) subsidizes 30 % of the costs of such systems under the Jawaharlal Nehru National Solar Mission (JNNSM). Another program of the GoI for remote, off-grid, locations called the Decentralized Distributed Generation (DDG), from the MoP, provides 90 % subsidy which is disbursed through the State governments.[5] Despite this seemingly generous subsidy very few private entrepreneurs have set up solar PV mini-grids so far. This is primarily due to two reasons: (1) the complex mechanisms used by the government ministries for disbursing subsidies that do not inspire the confidence of entrepreneurs. This is borne out by estimates that suggest that subsidies to the tune of Rs. 10 billion (Rs. 1,000 crore) are yet to be paid out to vendors by MNRE, despite installations having been completed some years earlier; and (2) unclear policy with respect to connectivity to the grid should it were to come. The objective of this paper, which is based on a detailed research report (Deorah and Chandran-Wadia 2013), is to identify design considerations and characteristics of successful PV mini-grids installations that adequately address the issues facing attempts to scale.

Methods

We have visited and gathered data from many mini-grid installations around the country—the technology and financial models used and the challenges faced by them. These include some of the installations by Husk Power Systems, Mera Gao Power, DESI Power, SunEdison and Chhattisgarh Renewable Energy Development Agency (CREDA) among others.

[5] http://powermin.nic.in/whats_new/pdf/Guidelines_for_Village_Electrification_DDG_under_RGGVY.pdf.

Gram Oorja has also conducted detailed surveys in about 100 villages, on behalf of Shakti Foundation, to assess the potential for deploying mini-grids. They interviewed villagers in a group of 10–12 residents, on average, at each location to try and gauge their capacity and willingness to pay for electricity, based on their baseline expenditure on lighting and mobile charging. The results clearly indicate a strong desire for electricity beyond lighting and little or no resistance to the idea of paying for electricity services. An average monthly billing of Rs. 100–150[6] is completely feasible and the numbers are even higher in some of the more prosperous villages. Indeed the fact that villagers are already spending such amounts on kerosene and mobile charging is being used by most mini-grid operators around the country as a baseline for creating their business models.

We present a detailed case study of a 9.36 kWp solar PV AC mini-grid system installed by Gram Oorja at Darewadi, a small breakaway hamlet nestled in the Western Ghats in Maharashtra, not recorded as a separate hamlet/village in the Census of India. It is inhabited by 39 families of Mahadev Koli tribals, totaling approximately 200 residents, who practice rain-fed farming (growing just one crop a year) and gathering wild herbs for the Ayurvedic industry for their livelihood. These villagers have been enjoying access to on-demand electricity, 24 × 7, for about 2 years now. Each home has been given basic lighting (2–3 LED bulbs totaling 10 W) and one plug point for use in mobile charging, for installing a TV, computer, or other appliances. Street lights and lighting for common areas have also been provided. The installation, which cost Rs. 3 million, was funded by Bosch Solar as part of a techno-commercial pilot and the MNRE subsidy was not availed.

System Design, Maintenance and Management

The system cost at Darewadi is relatively high, partly due to its remote location but partly also due to special local considerations. One of these is that Darewadi receives three months of very heavy rainfall each year, during the monsoons, along with strong winds and continuous rain spells lasting for over a week sometimes. Gram Oorja therefore set the solar panels in concrete to withstand winds of up to 200 km/hr.

Keeping in mind the fact that electricity generation can be relatively low for long intervals during the monsoons, Gram Oorja also created three separate feeder lines —one each for households, street lights and commercial loads. A local caretaker, trained by them, manually prioritizes electricity to households during periods of low generation switching off even the street lights if necessary. The strategy has worked well and the villagers have had to go without electricity for as little as 5 days in

[6] The exchange rate is approximately Rs. 62 to 1 US Dollar.

2013. Commercial[7] loads are only allowed during the day when the panels are generating electricity. This allows the system to be provisioned with minimally-sized battery banks.

Sizing the system with a larger battery could have avoided the days without electricity but it would have driven the costs up substantially, both initial capital expenditure as well as replacement costs. Gram Oorja sized the battery system to provide backup for just one night, rather than the three nights that most designs provide for. The battery bank comprises a 600 Ah, 48 volt, system that can store 28.8 kWh of electricity at 100 % capacity. At a prescribed depth of discharge of 50 % this provides usable electricity of 14.4 kWh which is sufficient to meet the overnight needs of households as well as street lights. With the street lights switched off it can last much longer, a few days. A battery of this size costs approximately Rs. 250,000 and needs to be replaced every 4–5 years.

Each home has a metered connection and is fitted with a circuit breaker that cuts off unusually high loads. The distribution wiring at Darewadi has been done completely professionally, as per the regulations of Maharashtra's State Electricity Regulatory Board. Notwithstanding the ease of installation that comes from simply stringing wires over trees (Ferris 2014), it is best for the safety of the village and for long-term sustainability of the assets, that wiring to individual homes is done as per the standards prescribed for the grid. Doing this has increased the initial capital expenditure substantially, but it has also ensured that Darewadi has a completely grid-like installation that balances out the higher initial cost with longevity and reduced long term maintenance costs since frequent interventions for repairs are avoided. We believe that such a safe, grid-like installation must be adopted as a best practice by the government and incorporated into its policy framework.

A village trust, consisting of 7 members, has been created to take charge of the assets and to oversee their care. Everyday maintenance for the panels and battery, and management of the feeder lines when necessary, is done by a young man from the village who has been trained for the purpose and is paid by the trust. The land for the panels and for the construction of the safe room for balance-of-system has been taken on a 25-year lease from a family, who are given free electricity in lieu of rent. It is critical that villagers are made stakeholders in this way through initial ground-work aimed at building trust based relationships. Ownership by the community not only ensures safety of the installation, but also minimizes payment default as is evident in Darewadi.

The Darewadi system was designed to accommodate substantial growth in usage over time. The initial usage, during the 1st year, was just 30 % of capacity. With the recent addition of a couple of water pumps, utilization has touched 50 %. Since commercial loads are relegated to daytime use only, there is still plenty of room for growth for several years. The exact pattern of growth remains to be seen. The actual

[7] The original thinking with regard to commercial loads was that local entrepreneurs could use electricity to provide value-added services to the community. However, only shared community resources have come up so far. No differential pricing for this 'commercial' load has been envisaged.

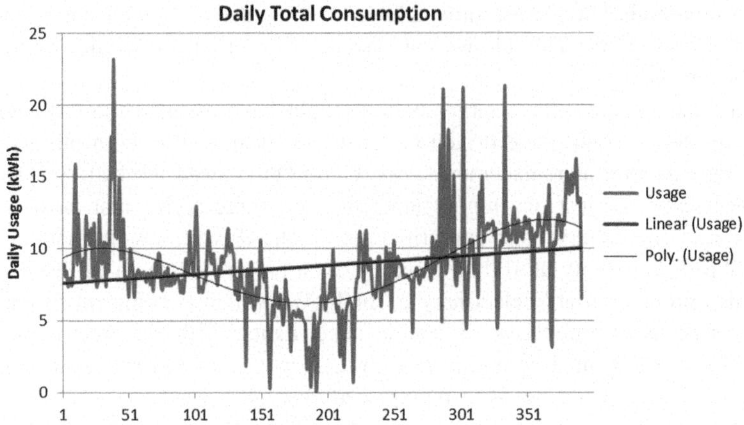

Fig. 5.1 Daily usage of electricity for a period of 400 days. Two trends are visible—the lower generation and usage during the monsoons and the overall increase in consumption over a period of one year

pattern of consumption recorded at Darewadi for a period of 400 days is shown in Fig. 5.1. Two trends are clearly visible—the lower generation and consumption of electricity during the 3 months or so of the monsoon period, and the overall increase in consumption over a period of 1 year. The fortnightly average for consumption grew by 47 % during this period.

Costs and Tariff

The tariff at Darewadi has been designed to create a corpus that will cover the cost of maintenance and battery replacement every 4–5 years.[8] While the effective tariff being charged is Rs. 20/kWh,[9] it must be noted that the amount collected is retained within the village, in a trust account, so that villagers do not have to rely on further external/government funding for battery replacements. The monthly bill for a typical household in Darewadi is approximately Rs. 120–150, including a fixed charge of Rs. 90 to cover street lights and common usage. For homes with televisions the cost goes up to Rs. 150–200. The monthly revenue being generated towards the corpus is presently approximately Rs. 5,500.

Darewadi is fortunate in that it is a hamlet with a high degree of coherence among its population. There have been no payment defaults so far even though the system has been in operation for more than 2 years. Poor payment collection has resulted in

[8] A 10 kWp system is expected to generate approximately 12,000–14,000 units (kWh) of billable electricity annually.

[9] Four times the cost paid by urban residential consumers.

high transaction costs for DESI power and several other for-profit players. Creating a village level committee for tariff collection ensures self-discipline and significantly reduces collection costs for entrepreneurs. The Darewadi committee has been empowered to punish the defaulters with disconnection if necessary.

Passing on capital expenditure—equipment and installation costs—to the users, even amortized over 25 years which is the lifetime of the panels, would have meant charging a minimum of Rs. 36 per unit of electricity, and much higher (Rs. 50 or more) if the capital expenditure is sought to be recovered within a shorter period, say 8–10 years. Such a high tariff is unconscionable, given that urban users pay just Rs. 5 per unit. Many rural consumers are in fact being asked to pay these kinds of prices and are therefore being forced to keep their consumption low. The monthly consumption of a family of five at Darewadi is typically just 3–4 units of electricity whereas the electricity Act, 2003, of the Government of India seeks to provide 1 unit (kWh) per day per family. Villagers cannot however afford to increase their consumption unless the price is kept in check.[10]

The State government of Chhattisgarh has the largest number of mini-grids deployed in the country today, approximately 700 of them, mainly DC mini-grids of sizes ranging from 1–5 kWp designed to replace the usage of kerosene 'till such time as the grid arrives'. CREDA subsidizes the capital expenditure entirely, and levies only a nominal charge of Rs. 5 per month per household to cover operational expenses. There is no metering of usage and every user pays the same fee. Over time this model has resulted in indiscriminate use by a section of the villagers making it difficult for the government to sustain the level of service originally envisaged. In several of the mini-grids electricity is provided for only a few hours each day because the present loads are much higher than the planned loads.

In our view, this issue is best addressed by introducing a tariff and billing system based on metering the usage of each individual connection as is being done at Darewadi. Metered usage based billing serves multiple purposes. It

a. creates a sense of ownership of the asset among the villagers;
b. creates a demand for quality service from the mini-grid operator; and
c. ensures a degree of self-regulation with respect to usage. Indiscipline with respect to usage in these off-grid systems will inevitably result in project failure.

Livelihoods and Long-Term Sustainability

The fact that collections from villagers in Darewadi for electricity usage each month is going towards a corpus, that will be utilized to pay for battery replacements, will help ensure the long term sustainability and uptime of the installation. Existing

---

[10] Many State governments set an upper limit on the tariff that is chargeable to consumers, depending on whether they are above or below the poverty line.

battery technology (lead-acid batteries in this case) needs replacement 5–6 times over the lifetime of the panels. If the onus of paying for battery replacement and also repairs and maintenance were left to external financial support then the system would inevitably fall into disuse for long periods.

On-demand electricity is critical for creating livelihood opportunities, as well as for the provision of healthcare, education, clean water supply and entertainment. While DC micro-grids (typically of size less than 5 kWp) are providing lighting and mobile charging to households in several villages in power starved states, the flat-rate price/unit they charge cannot cope with increase in consumption e.g., with the installation of a television or other appliances. It has been observed time and again in many mini-grid installations around the country that once electricity is supplied, villagers tend to discover latent needs and new applications driving up the average consumption per household quickly. This is true also of Darewadi where a flour mill was installed soon after electrification. Several households now have a television and two computers have been donated by an NGO although these are not yet being optimally used. Two water-pumps added recently by the villagers have brought much joy to them and contributed to increasing consumption greatly. These pumps will enable several farmers to graduate from growing just a single-crop to two or even three crops a year. Another example is the village of Meerwada in Madhya Pradesh, where SunEdison set up a 14 kWp solar PV mini-grid. This plant provides electricity to approximately 400 people in 70 households. Funded by SunEdison as a demonstration project, utilization of this plant climbed quickly from the initial 15 % to about 70 %, largely due to the installation of a plant for chilling milk, within just a few months.[11]

Utilization of the mini-grid inevitably increases, at first for entertainment purposes and later for livelihood options. Many of the latter uses are knowledge intensive activities in addition to being infrastructure intensive. Computer education, milk chilling plants, value added services to do with food and agriculture such as preparation and sale of baked products, cold storage for fruits and vegetables, are just some examples of livelihood enhancing options that have followed in the past. In fact it is these very opportunities that are potentially more likely to generate revenues for the energy entrepreneurs, should they choose to stay engaged with the local community and get involved in provisioning some of these services.

Results

Darewadi was designed to be a self-sufficient and self-sustaining installation. Gram Oorja has succeeded in this goal because they have been able to move out, handing over control of the mini-grid to the community, which was in fact their original intention. They continue to remain available on call as needed. Their work

[11] Rahul Sankhe, MD SunEdison, speaking at ORF Mumbai, July 2012.

illustrates that if the up-front costs of a high-quality installation can be covered through a one-time investment, it is possible to make these off-grid installations self-sufficient, even covering battery replacement costs. End users are happy to receive reliable electricity supply and are therefore very willing to pay for it. This is an important input to policy makers and government agencies, such as CREDA of Chhattisgarh that has put in considerable energy into scaling out mini-grids. In another DDG tender, Madhya Pradesh Urja Vikas Nigam (MPUVN) has also imposed unrealistically low flat tariffs—Rs. 15 a month for consumers who are below the poverty line and Rs. 30 for those above—without providing any clarity on how future battery replacements are to be funded.

In this context some of the key innovations made by Gram Oorja to ensure quality and long term viability of the standalone mini-grid in Darewadi ought to be adopted as best practices and scaled out. These include:

- Charging based on metering individual connections—to ensure disciplined and careful usage;
- Feeder line separation—for prioritizing electricity for lighting and to help create a differential tariff for commercial loads should it become necessary;
- Minimal battery support—for reducing installation as well as maintenance costs, and for minimizing environmental impact;
- Ensuring community buy-in—and also their involvement in ownership and care of assets;
- Grid-like installation—for the safety of the community and for the longevity of assets, through reducing the need for frequent interventions, thus also contributing to keeping lifetime costs lower.

Employment of local youth can make a tremendous difference to the last mile implementation and servicing of mini-grids. However, given the lack of trained manpower, entrepreneurs installing small standalone mini-grids cannot leverage economies of scale while training a few youth at a time. A national program, perhaps as part of the National Skill Development Mission,[12] will be required to train tens of thousands of technicians, and also thousands of local youth who can work alongside the private entrepreneurs.

Barring Darewadi and Meerwada, which are both non-commercial installations, neither the 'for-profit' mini-grid operators nor the state agencies are in fact seeking to provide 24 × 7 electricity access to rural consumers. The private entrepreneurs are constrained by the need to control their costs, and the state agencies see these mini-grids as a stop-gap solution to be used largely only for lighting till the grid comes. This is very unfortunate because it is clear that when state agencies get involved it is possible to achieve scale quickly, as illustrated by the 700 or so mini-grids installed by CREDA in Chhattisgarh. Not just in remote locations, but even in grid-connected villages, these mini-grids have a tremendous potential to provide electricity on demand, 24 × 7, to help enhance livelihoods, a potential that must be leveraged.

[12] See http://www.nsdcindia.org/index.aspx.

A recent report titled 'Microgrids[13] for rural electrification' (Schnitzer et al. 2014) which presents a critical review of best practices based on seven case studies, five of them in India, corroborates and extends several of the observations made in this paper. This comprehensive report covers the installations of 5 agencies in India including the Renewable Energy Development Agencies of the States of West Bengal (WBREDA) and Orissa (OREDA), besides Chhattisgarh, and two private entrepreneurs DESI Power and HPS, who deal in biomass based installations. The report also throws up several additional points for discussion. It cautions that State agencies need to find a way to avoid pitfalls, such as the overuse due to the flat fee charged by CREDA and the cutting short of the life of mini-grids set up by WBREDA due to grid extension. The report cites lack of 'community cooperation' and 'village cohesiveness' as one of the challenges facing Indian micro-grids, such as those operated by HPS & OREDA. As acknowledged earlier, Darewadi does not suffer from conflicts within the community that could compromise the operation of their mini-grid. Nevertheless, this emphasizes the importance of initial ground-work and fine tuning of the model of community involvement in mini-grid projects.

Discussion

The policy issues that come in the way of setting up mini-grids in grid-connected villages are much more complex. There is lack of clarity with respect to two critical issues: (1) aspects of integration with the grid, and (2) the difference in tariff between electricity supplied from the grid vs. that generated by renewable energy systems. Currently, there is no policy framework that provides a favorable exit option to the entrepreneur should the grid become available (energized with adequate and reliable supply of electricity) before their investment in setting up the mini-grid is recovered. This is the single largest deterrent for entrepreneurs and investors wishing to enter the field. A policy that safeguards existing investments and creates the appropriate framework for integrating mini-grid systems with the main grid could greatly accelerate roll-out of more mini-grids.

The gap in tariff also needs to be tackled before mini-grids can proliferate. Ideally mini-grid installations in grid-connected villages ought not to be accompanied by battery backup. This would reduce installation costs considerably and consequently also the gap in tariffs. The tariff for electricity from solar PV mini-grids can be Rs. 20/kWh as in Darewadi, or even higher, or come down to Rs. 6.5–7 in installations without battery storage. However, given that electricity supply from the grid is highly erratic, it is unlikely that the use of storage capacity can be eliminated if consumers are to have access to 24 × 7, reliable, electricity. A policy environment that provides tariff subsidy to these consumers, particularly to those below the poverty line, would certainly be money well spent. Entrepreneurs who set

[13] The authors refer to grid sizes below 100 kWp.

up mini-grids must in turn be compensated for their investments through a policy regime that provides them with adequate subsidies in a transparent manner. The primary method of awarding subsidy—to the lowest cost bidder—compromises the quality of installations. Making funds available to entrepreneurs directly, on the basis of representative benchmark costs, will substantially lower barriers to entry.

In terms of the cost of subsidies to the government, installing 10 kWp systems with battery backup at the benchmark cost of Rs. 3 million per remote, off-grid, village would imply that the upfront capital outlay on say 50,000 mini-grids of this size would be a total of Rs. 150 billion (i.e. Rs. 15,000 crores). Staggered over a period of 5 years, the allocation would be Rs. 30 billion or Rs. 3,000 crores per annum. The Indian government spends several times this figure on annual kerosene and diesel subsidies (Teri 2012). However, if solar PV mini-grids with battery backup have to be used to provide electricity to the 290 million Indians living in grid-connected villages the costs could become prohibitive.

A Roadmap to 100 % Electricity Access for India@75

The year 2022 will mark the 75th anniversary of India's independence. We urge the governments at the Centre as well as the States to take aggressive steps in order to ensure that the hitherto elusive target of '100 % Electrification' is achieved by then. The newly elected Government of India has in fact gone a step further and announced an ambitious target of bringing electricity to every home before the end of their current term in office i.e., by 2019.[14] Estimates vary but it is believed that up to 100,000 villages and hamlets many need to be touched in order to reach this goal. Implementation of tens of thousands of solar PV mini-grids, preferably without battery backup, could be a definitive path to achieving 100 % electrification. The time period available, of 5–8 years, makes it difficult but not impossible to achieve the goal. However, given all the challenges highlighted earlier, governments will have to learn quickly from the experience gathered so far to scale out with determination and speed.

We believe that this will require an energy revolution to be orchestrated by the government, working in tandem with all stakeholders. As in the case of the Green Revolution that was brought about in India in the 1970s to improve agriculture and eliminate famine, this will require the creation of an entire ecosystem around potential implementers of these mini-grids—assisting them with the requisite policy environment, access to finance, technology and tools, and adequately trained human resources. We believe that the ecosystem is not impossible to put together once there is the political will to get the job done.

---

[14] http://www.bloomberg.com/news/2014-05-19/modi-to-use-solar-to-bring-power-to-every-home-by-2019.html.

We argue that two levels of 'assisted entrepreneurship models' will be essential in this context. The first is the assistance that must be given to entrepreneurs such as Husk Power Systems, Gram Oorja, DESI Power, and others so that they can take on the scale challenge of reaching electricity to the millions of Indians who are going without electricity today. The financial assistance required by them, chiefly in the form of capital subsidies, should rightly be considered as an infrastructure investment rather than as a pure industry subsidy. Constitutionally, the responsibility for provision of electricity to citizens rests with the state governments. Assistance from the state run distribution companies (discoms) and the State Electricity Boards (SEBs) would be extremely useful in securing right-of-way for wiring to homes in the grid-connected villages and for potential grid integration. Unfortunately, these state actors are not involved in the renewable energy effort which is being driven largely by the Central government, through MNRE and its state nodal agencies such as CREDA, OREDA etc.

At another level a large number of local entrepreneurs, adequately trained local youth, can be assisted by these larger players in different kinds of franchisee models (such as the ones being used by Husk Power Systems and Gram Power for example) so that they can provide myriad services—billing and payment collection, maintenance, and provision of value-added services. This would also create some much-needed jobs and economic activity at these locations.

The successful implementation of tens of thousands of solar PV mini-grids in India is an attractive option for enhancing livelihoods and catalyzing development in a sustainable manner. We believe that the innovations called out in this paper can go a long way towards overcoming some of the challenges and helping to make the installations successful, thus bringing development and prosperity to rural India.

References

Census of India. (2011). *Household amenities and assets: Source of lighting*. Office of the Registrar General and Census Commissioner.

Deorah, S. M., & Chandran-Wadia, L. (2013). *Solar mini-grids for rural electrification*. Mumbai: Observer Research Foundation. Retrieved from http://www.arthaplatform.com/resources/162/solar-mini-grids.

Ferris, D. (2014, January 16). Indian microgrids aim to bring millions out of darkness. Retrieved Jan 25, 2014, from Yale Environment 360. http://e360.yale.edu/feature/indian_microgrids_aim_to_bring_millions_out_of_darkness/2729/.

Schnitzer, D., Lounsbury, D. S., Carvallo, J. P., Deshmukh, R., Apt, J., & Kammen, M. D. (2014). *Micro-grids for rural electrification: A critical review of best practices based on seven case studies*. United Nations Foundation.

Teri. (2012). *A Citizens' guide to energy subsidies in India*. The International Institute for Sustainable Development.

Chapter 6
The Case for Solar-Diesel Hybrid Minigrids in Bangladesh: Design Considerations

Shahriar Ahmed Chowdhury, Shakila Aziz and Sebastian Groh

Abstract Bangladesh is experiencing the most successful solar home system based rural electrification program in the world. Yet the present scheme has revealed several shortcomings including the lack of a clear link to productive commercial activities as well as of a deeper energy inclusion. With sparsely distributed loads and limited energy demand, grid extension has not been found to be economically feasible as an electrification approach for the remote areas in Bangladesh. As it is the common tendency of the rural people to live in household clusters of high density, minigrid approaches have recently been identified as an attractive alternative to the prevalent stand-alone scheme. There is a perceived need for a wider understanding of the roles and applications of minigrids as tools towards achieving greater energy and resource efficiency. At the same time, there is a paucity of publications which explore the potential contributions of minigrids and their potential contribution with respect to resource efficiency. In order to address this perceived need, this paper describes approaches and methods used in implementing diesel-based minigrids on the one hand, and the contribution of renewables on the other. This paper highlights design considerations for developing solar-diesel hybrid minigrids in off grid rural areas of Bangladesh.

Keywords Solar PV · Hybrid minigrid · Load factor · Storage system

S.A. Chowdhury (✉)
Centre for Energy Research, United International University (UIU), Dhaka, Bangladesh
e-mail: shahriar.ac@gmail.com

S. Aziz
School of Business and Economics, United International University (UIU),
Dhaka, Bangladesh

S. Groh
Research Focus Microenergy Systems, Technische Universität Berlin, Berlin, Germany

S. Groh et al. (eds.), *Decentralized Solutions for Developing Economies*,
Springer Proceedings in Energy, DOI 10.1007/978-3-319-15964-5_6

Introduction

Approximately 1.3 billion people around the world lack access to an electricity grid and yet another billion suffer from a severely intermittent supply (IEA 2013). In Asia alone, there are some 800 million people with inadequate access to electricity (Intellecap 2012). In Bangladesh 40 % of the population, representing 65 million people, have no access to the national grid (World Bank 2013). Most of these people live in remote rural areas, partly on islands, where there is no electricity utility grid infrastructure. For a country like Bangladesh, the extension of the grid network to remote places is expensive due to high transmission costs in comparison to the small load demand.

Solar Home Systems (SHS), currently consisting of a 20–85 Wp solar panel, battery, and charge controller, are being installed at a rate of up to 75,000 per month in rural communities through the Infrastructure Development Company Limited's (IDCOL) national SHS program, amounting to 3.2 million systems in total as of today (IDCOL 2014). Despite this success, taking the perspective of the end-user, the current SHS scheme has revealed the following shortcomings:

- Not inclusive enough to reach the poorest segments;
- Lack of flexibility in terms of usage patterns and payment methods;
- Productive use remains very limited;
- Excess capacity: generated energy is lost while battery is full (Kirchhoff 2014).

Commerce and industry are not yet developed in the rural areas and the share of electricity consumption from these sectors is not expected to be high. With increased awareness of the adverse effect of global warming due to greenhouse gas emissions, efforts are being made all over the world to curb climate change by increasing energy generation from renewable energy (RE) resources. In line with the global effort, Bangladesh has established a plan to generate 5 % of the total power generation ($\sim$500 MW) from renewable sources by 2015 and 10 % ($\sim$2,000 MW) by 2020 (GoB 2008). As renewable energy conversion technologies are still expensive, it is not clear the extent to which RE sources can be tapped within the next two years to meet the set target. Until now solar has proven to be the main RE resource in Bangladesh that can be harnessed effectively and has wide spread application at this stage. However, even with the current downward trending PV price in the world market, PV based grid power generation cost is significantly higher than the gas/coal-fired grid energy price prevalent in the country. Thus, the grid expansion, based on RE resources, is expected to be slow for the next few years, unless there is major governmental support. However, with the gradual increase in the price of fossil fuels and declining price of solar PV, the cost of generation of electricity from solar PV will become competitive with the cost of generation of electricity from fossil fuels. In rural areas there is demand for power for meeting the basic needs (such as lighting and cooling) of the rural household. Surveys from the potential areas for minigrids show that people in off-grid areas are prepared to pay a much higher price than the price of national grid electricity. This opens up the opportunity to install small sized minigrids for localized populous areas in off-grid areas or villages. Such small-scale projects will provide electricity to the rural people and can contribute significantly to improve their quality of life (Khan and Huque 2012). At the same time, these projects, when designed properly, can be sustainable and economically viable. Even if the grid is extended in the future to the area of a stand-alone minigrid, the solar PV generators can be connected to grid via grid-tied inverters. Still, as of today, there is only one solar minigrid of 100+ kW in operation in the rural market of Sandwip island (Khan and Huque 2012). Moreover, there is a paucity of publications that explore the potential contributions of minigrids and their potential contribution with respect to resource efficiency. This paper aims to open up the discourse on minigrid design considerations in Bangladesh in order to support the developments of a wider uptake in the country, particularly just before the roll out of multiple pilot projects, which are trying to follow-up on the success of the prior scheme, which has focused on stand-alone systems.

Local Functional Project Structure

The solar-diesel hybrid minigrid system for rural electrification program in Bangladesh is supported by the Infrastructure Development Company Limited (IDCOL) of Bangladesh, which is a non-banking financial institution, established by the Government of Bangladesh (IDCOL 2014). The program is financed by the international donor community represented by loans from WB and JICA and grants from GPOBA, USAID, KfW, DFID, ADB. The funds are then channeled via soft loans and grants to the developers of solar minigrid projects. Project developers are encouraged to submit proposals for the financing of solar hybrid minigrid projects to IDCOL. These developers choose a remote off-grid area, which has the potential for the installation of a minigrid. They also appoint consultants to develop the project profile for IDCOL financial support. Based on the market survey conducted by the project developers, IDCOL, in consultation with the sponsor and the consultant, conducts load assessments. The technical consultant then designs the hybrid system and prepares a bill of materials for the project and also supports developers to select suppliers based on the bill of material for the project. After arriving at the project costs based on submitted price quotations, IDCOL approves the project and starts the documentation process for financing the project. Please see Fig. 6.1 for an illustration of the scheme.

The financial scheme for the solar-diesel hybrid minigrid project developed by IDCOL is 50 % grant, 30 % soft loan (interest rate 6 %) and 20 % equity from the project developer. The soft loan is given for tenure of 10 years with 2 years grace period. Under this financial scheme one project has successfully been installed in Sandwip island in the estuary of the Bay of Bengal in 2010. Initially, the project aimed to provide electricity to a rural market and its adjacent households from 9am in the morning to 11 pm at night. IDCOL further approved three more solar minigrid projects, which are under currently construction.

Seven other projects have already been submitted to IDCOL for financing considering uninterrupted power supply for 24 h. IDCOL has a target of installing 50 solar-diesel hybrid minigrids all over Bangladesh by 2016 (EMRD 2012).

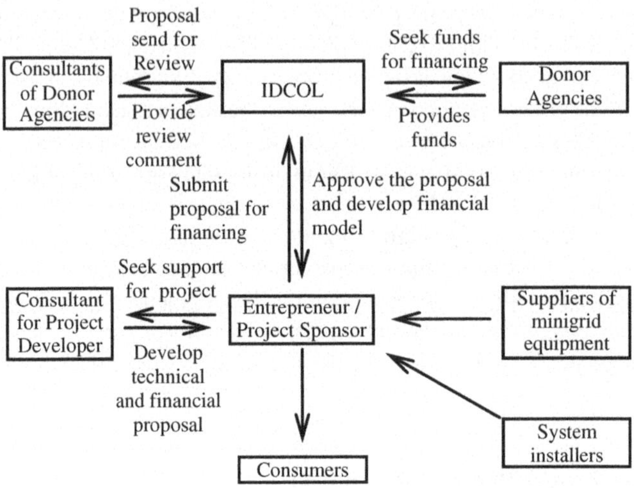

Fig. 6.1 Functional structure of the solar-diesel hybrid minigrid projects in Bangladesh

Key Design Considerations

The key parameters for designing a minigrid in a specific location are referred to as the minigrids design considerations. Some considerations can be made according to the actual scenario of the project location in order to minimize the cost of energy and increase sustainability of the project. Design steps of solar-diesel hybrid minigrids are given below:

1. Load estimation and development of load duration curve.
2. Estimation of day load and night load.
3. Calculation of battery size considering supply of night energy demand, battery depth of discharge and considering no autonomy days.
4. Sizing of diesel generator considering peak load, maximum allowable runtime per day.
5. Length of distribution feeders considering 5 % voltage drop at the end of distribution line.
6. PV panel capacity of maximum 250 kWp (as per government regulation a minigrid of less than 250 kWp of capacity does not need government approval (Solar Guide Book 2013).

Following criteria are given special attention while designing the minigrids:

Selection of technology Different technologies for minigrid power systems are available worldwide. Biomass is the most widely used renewable energy source in Bangladesh, but it is still used in a primitive way. In on-shore areas, wind energy provides sufficient potential for small-scale power generation in Bangladesh, on the other hand during winter seasons there is almost no wind for power generation.

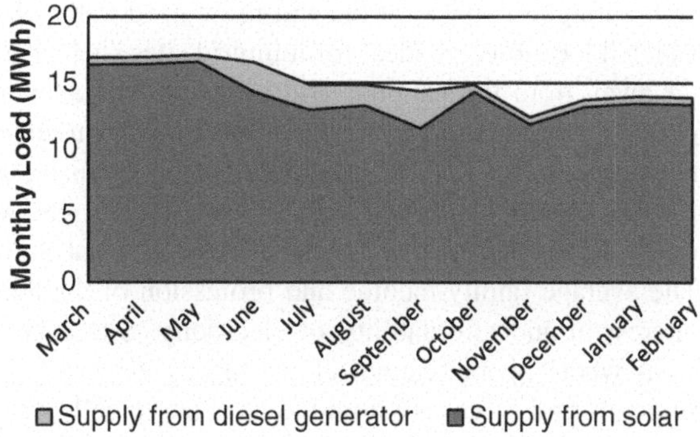

Fig. 6.2 Typical energy mix for a solar diesel hybrid minigrid in rural Bangladesh

Solar PV systems have proven to be the most successful RE technology to be disseminated in Bangladesh. Considering the availability of resources, solar-diesel hybrid minigrids are considered to be the most suitable solution.

Renewable energy fraction A higher renewable energy fraction in the annual energy mix minimizes the running cost for the energy supply system. Transportation of diesel oil to remote places is also challenging for the rural community. For hybrid projects in Bangladesh, therefore, the RE fraction should be considered to be more than 90 %. A higher RE fraction also maintains options for future demand growth (additional loads can be served by running the diesel generator for longer durations). Estimated energy mix of a typical solar-diesel hybrid minigrid is shown in Fig. 6.2.

Productive use of solar power During the daytime, income generating activities such as solar irrigation, cottage industries, husking mills, sawmills, grinding mills (for spice), welding machines, lathe machines, and ice factories in rural market places can be supported through the solar minigrids. These day loads do not require energy storage facilities as they run during the day time. The cost of energy for day loads is cheaper than the cost of energy for night loads. Therefore, day loads should be encouraged when designing minigrids.

The addition of a generator reduces battery requirement A diesel generator helps to reduce the size of the storage system for a minigrid and no autonomy days need to be considered.

Cost reduction potential The increase of day loads reduces the levelized cost of energy.

Site Survey and Survey Items

The site survey is the initial step in the planning process in order to find an adequate location. Survey findings show the feasibility of a project in the present and future context. Parameters for the survey vary from one location to another. At first, ability and willingness to pay of the end-users need to be addressed. Secondly, remote and isolated locations are ideal for minigrid sites such as islands and habitations that are far away from the national grid network. The sites are selected in such a way to prefer sites where the government has no plans to extend a grid line within the next ten years. As for IDCOL requirements the levelized tariff of electricity considering IDCOL model stands at US$0.37 for a typical 150 kWp hybrid system, which is much higher than the energy tariff (US$0.11) of grid electricity (Intellecap 2012). The average family income and profession of the adults of the location reflect the interest in and affordability of electricity. Site surveys show that there are some areas where people cannot afford to buy electricity. On the other hand, however, some off-grid places are found to be economically developed and people are willing to pay a high price for electricity year round. People use kerosene to meet their lighting demand in off grid areas while in some off grid areas diesel generators are used to provide electricity for some hours per day. In those areas, mini-utilities based on diesel generators often supply neighbouring houses as well as small businesses. An exemplary case study is included in the appendix with the kWh price based on simple calculations and resulting in US$3.10 and US$3.60 per kWh. Finally, load assessment is the most indispensable tool to determine the size of the power plant. In the context of rural Bangladesh, lighting, cooling fans, and mobile phone chargers are the main loads. Televisions, refrigerators and DVD players are also found to be potential loads. Rural sites for minigrids are selected to cover one or two rural villages with a rural market place.

Demand Assessment

Demand assessments in rural areas need to be performed with great care as rural people are not always familiar with electrical loads other than lighting and cooling. Their expectations are influenced by family or acquaintances who live in or frequently visit urban or city areas. They often express desire to use luxury loads such as TVs, DVD players and refrigerators irrespective of their energy consumption. Careful assessment is needed to assess the real and rational load demand. It is wise to count the number of rooms in the households and the monthly income of the households if possible. Present expenditure for energy and expressed willingness to pay for better lighting should be taken into account (Groh 2014).

Packages To overcome actual load assessment barriers, different packages should be defined for different types of users according to their income level. The packages should be designed using energy efficient loads, for example LED lamps, higher efficiency cooling fans, etc. Furthermore, to complete the load as a package, a connection should be drawn between monthly energy price and the present monthly expenditures for energy, so that the people can better understand the relative price. Based on survey findings in seven rural off grid areas, consumer packages were developed and are shown in Table 6.1. The table also shows the breakdown of each

Table 6.1 Typical loads for different packages

Package	Gadget type	Watt	Quantity	Total watt
Package 1 (Small household)	LED light	7	3	46
	Cooling fan	25	1	
Package 2 (Medium income household)	LED light	7	4	128
	Cooling fan	25	2	
	TV	50	1	
Package 3 (Well off household)	LED light	7	6	142
	Cooling fan	25	2	
	TV	50	1	
Package 4 (Rich household)	LED light	7	6	262
	Cooling fan	25	3	
	TV	50	1	
	Refrigerator	100	1	
Package 5 (Shops)	LED light	7	2	39
	Cooling fan	25	1	
Package 6 (Big shops)	LED light	7	4	228
	Ceiling fans	25	2	
	TV	50	1	
	Refrigerator	100	1	
Package-7 (Rural school and college)	LED light	7	20	640
	Cooling fan	25	20	
Package-8 Industrial (Saw mills, Lathe m/c, others)	LED light	7	1	5,032
	Cooling fan	25	1	
	Motor	5,000	1	
Package 9 (Irrigation pumps)	Motor	2,000	1	2,000

package according to the demand of rural households. The ratio of the packages varies with the variation of socio-economic condition of that locality.

Load factor The rural economy of Bangladesh is mostly dependent on agriculture. People of the rural area work on their land all day long and go to sleep early. However, the scenario may not remain the same post electrification; therefore, the planned load factor should consider the predicted post electrification scenario.

Day Load The cost of energy in a solar-diesel hybrid minigrid system is high. The main reasons behind this are the high price of the storage system and the cost of diesel. It is possible to reduce the energy cost to a tolerable limit by increasing the day load. Energy can be supplied directly to the day loads from the solar PV system except during rainy or foggy weather. Thus, we can avoid the requirement of higher storage and thereby reduce the overall cost.

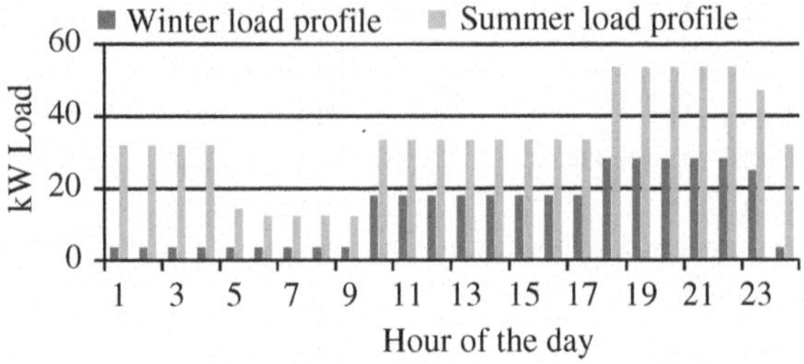

Fig. 6.3 Typical load profile of a 150 kWp hybrid minigrid

Night Load A primary challenge in running an RE-based minigrid is the decline in energy production during the early evening. To maintain the supply of power after sunset, it is essential to continue the power supply either from a storage system or by running a diesel generator. On the other hand, the main load demand actually rises in the evening because lamps for lighting and other electric gadgets are switched on simultaneously. Therefore, it is imperative to calculate the night load accurately to determine the optimum size of the storage system and the run time of the generator.

Peak Load In general, peak loads are experienced in the evening. The size of off-grid inverters and generators should therefore be sized depending on the night load.

Seasonal variation During summer, the demand for electricity is at its highest as cooling load adds to the lighting load. Moreover, the run time of the cooling load is much higher than that of the lighting load. On the contrary, during winter, the cooling load goes off. Therefore the overall demand for electricity is lower in winter than in summer. Furthermore, in the context of Bangladesh, cool temperatures last longer in rural than in urban areas. Considering these seasonal variations, the plant size will vary, and a source of power should be designed that can provide reliable power throughout the year. Figure 6.4 below shows the estimated load profile in summer and winter of Bangladesh for a 200 kWp solar-diesel hybrid power plant according to survey data. During the summer, irradiance in Bangladesh is higher than in winter; on the contrary during summer the irrigation demand is highest. So, additional power generation from solar PV systems can be used for irrigation (Fig. 6.3).

System Design

Survey data shows that demand reaches its peak after sunset and generally lasts till 10 pm. Solar power is the main source, and the generator is kept as a backup or standby for such a minigrid. The daily average generator run time depends upon the shortfall in solar energy and amount of storage in the battery bank. Overall, solar-diesel hybrid minigrids should be predominantly powered by their solar PV systems. Monthly generation from PV can be determined from the hourly averaged irradiation data. Hourly average expected demand data is determined from the site survey. The day load is considered from 8 am in the morning until 5 pm in the evening.

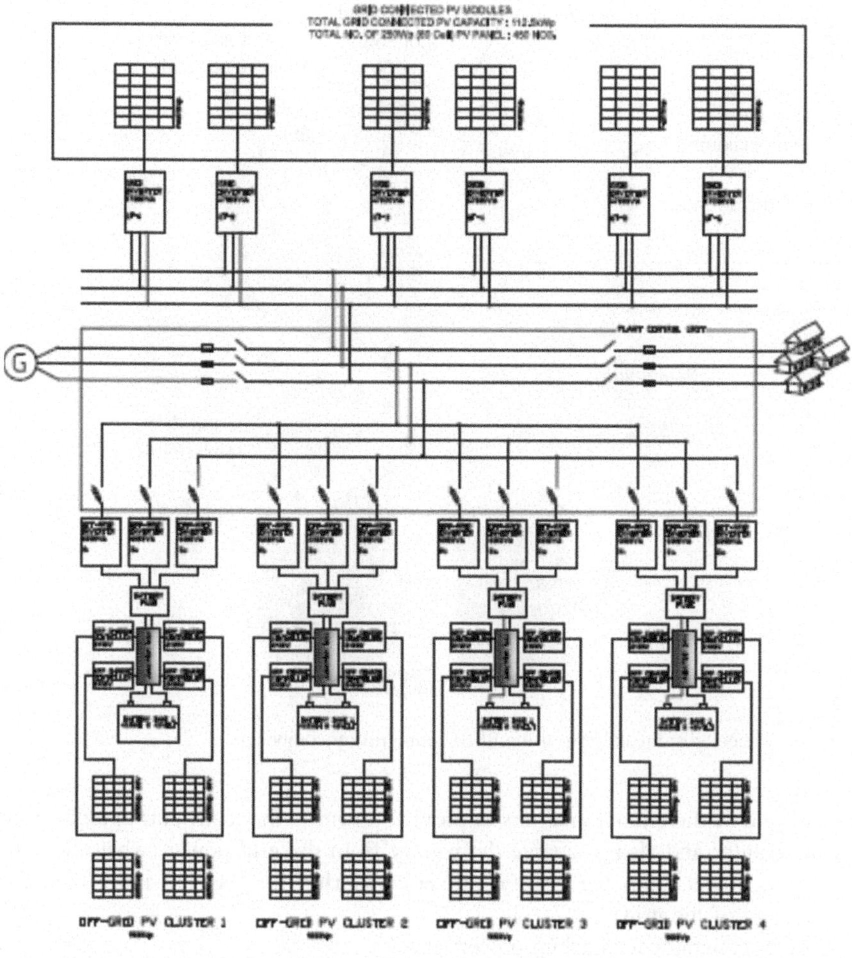

Fig. 6.4 Single line diagram of a typical 150 kWp solar-diesel hybrid minigrid system

Generally the demand of the day load is lower than the generated PV power, so in the battery capacity calculation the day load is excluded. The night load is served by stored energy in the battery through the bi-directional off grid inverters. The generator will run only when the stored energy in the battery is not sufficient to serve the night load. As the system contains a diesel generator, no autonomy days are considered. Figure 6.5 shows the schematic diagram of a typical solar-diesel hybrid minigrid. The primary hardware components of the solar hybrid PV plant consist of:

- Solar PV panels to convert solar energy into direct current (DC) electricity;
- Solar charge controllers to charge the battery directly from the solar PV panels;
- Grid-tied inverters to convert the solar energy to alternating current (AC) main power to serve the day loads;

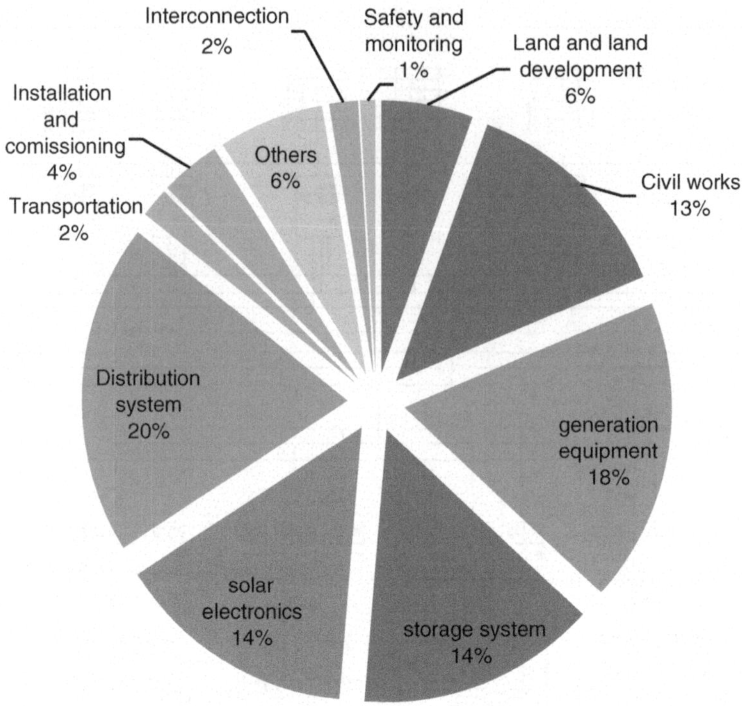

Fig. 6.5 Project cost breakdown of the SEBL minigrid by component

- Off grid bi-directional inverters to provide power to the loads during night from the battery and also to charge the battery from the grid power (when the generated power from the grid-tied inverters during the day are more than the demand of the grid);
- Diesel generator for backup power; and
- Batteries to store energy.

Cooling fans are one of the major loads in rural areas of Bangladesh. It has been found from various technical designs that a significant portion of solar energy remains unused during the winter. To reduce the portion of unused energy, some seasonal industrial loads are sought in order to utilize that excess energy. In some of the cases the minigrids provide energy to rural irrigation pumps only during the winter.

Financial Modelling for Minigrids

The financial models include cost and revenue assumptions for each site for the minigrids. It is the policy of IDCOL to keep the tariff within 30 BDT (US$0.38) per kWh, and projects where the tariff exceeds this are not considered viable. Cost assumptions include estimates for project fixed costs and running costs. For example, the project developed by Solar Electro Bangladesh Limited (SEBL) in Monpura, an island in the estuary of the Bay of Bengal, can be considered a typical solar minigrid project. This project is meant to bring electricity to 661 consumer units, comprising of 500 households, 155 shops and some mills and workshops. The plant capacity is 177 kWp, storage system of 332,640 Ah at a system voltage of 2 V and including a diesel generator of capacity 80 kVA. Fixed costs include the following items:

- Land and land development: Bangladesh is a flood prone country and while designing the minigrid in the majority of cases the land has to be filled so that the level is above the recorded flood level. Generally, this varies from two to three meters above the existing level. In areas where land is cheap, it is purchased outright. However, where the land price is so high as to make the project non viable, the land is leased for 20 years.
- Civil works: These include a boundary, control room, office building and mounting structures for solar panel installation. Some sites require a barbed wire boundary to prevent theft and vandalism.
- Generation equipment: These include solar PV panels, generators and generator protection accessories.
- Storage system: This includes batteries. Minigrids in Bangladesh use 2 V tubular positive plate industrial lead acid batteries. These batteries have warranties for 7 years. During the 20 years of the project life, batteries are considered to be replaced twice (in the 8th and 15th year).
- Solar electronics: These include grid-tied inverters, off grid inverters and charge controllers (refer to Fig. 6.4). All solar electronics have warranties of 10 years. They are replaced once, in the 11th year of the project.
- Distribution systems: These are three phase, four wire distribution systems. In some areas single phase systems are also considered. In this SEBL project, a distribution network of 7 km of three phase and 3 km of single phase had to be developed to meet the demands of the consumers.
- Transportation: Some equipment requires transportation to the project site all the way from the bay port. The minigrid sites are usually in remote locations and often have no proper road access. In these cases the equipment has to be carried manually or on animal driven carts. This accounts for one reason why the transportation costs are significantly higher than a similar project in grid-connected areas.
- Installation and commissioning: Installation of the panels and distribution network, configuration of solar electronics, and testing and commissioning of the system are considered in this category.
- Interconnection: cables and accessories for the interconnection of panels, inverters, batteries, charge controllers and system bus are considered in this category.
- Safety and monitoring: This includes a lightning protection system, surge protection system, solar radiation, temperature and wind velocity measurement equipment. It also includes neutral grounding and potential earth grounding.
- Other costs: Includes management and administration costs such as office set up costs, consultancy, preliminary expense and contingency.

The project assumes that 70 % of the total customer base will be connected to the minigrid in the first year, and the rest by the second year.

Land and land development can vary from 5 to 7 % of the total project cost, depending on the location. The project cost can vary according to location and load pattern. If the night load is more than the day load, the storage cost will likely be greater than the generation equipment cost. In the SEBL project, the day load is higher than the night load, so the storage cost is lower. The aim while designing a minigrid is to increase the day load relative to the night load. This can be done by including the village markets among the customers. These markets often have saw mills, rice husking mills, grinding mills and small workshops that ordinarily run on diesel engine or diesel engine driven generators during the day. Reducing the storage cost will bring down the overall cost of electricity. Densely populated villages with a large village market are attractive sites for minigrids.

A load profile is calculated based on the demands for different types of packages. Each minigrid project has different packages with different capacities based on whether it is to be used for domestic, commercial, organizational or irrigation purposes. Three types of packages are for households of different sizes, three types for shops, two types for the local schools and mosques and one package type for irrigation. The number of customers for each minigrid range from 600 to 1,200. 85 % of loads are households, 12 % are shops and the remaining are organizations and irrigation pumps. It is assumed that households, shops and organizations operate all year, but irrigation pumps operate on a seasonal basis with varying demand for electricity.

The household packages are expected to have loads drawing electricity for the full 24 h. Lights will remain on during the evenings and TVs are assumed to run during the waking hours. In case of the commercial packages, there will be no loads during the night when the business is closed. Highest loads are in the months of March, April and May (summer months due to the demand of cooling fans), followed by the rainy months of June–December. Lowest loads are in the winter months of December–February. Solar water pumps are supposed to run between November to May. Peak irrigation demand is from March to April.

The revenue calculations are based on a starting rate of 30 BDT per kWh. The tariff and line rent is increased 10 % every 2 years, up to the 20th year of operations. Fans and TVs are the highest load groups in each minigrid, accounting for 30–45 % of the revenue for each project. Lighting, the most basic service, accounts for 10 % of revenue (minigrid users are suggested to use LED lights for efficient use of electricity). Some loads vary from minigrid to minigrid. For example, some sites use water irrigation facilities and some do not. Some minigrids will run a rice mill, saw mill or ice mill.

The ranges for some of the financial indicators for the projects developed for rural electrification in Bangladesh are given below:

Internal Rate of Return (IRR): 11–19 %
Return on Equity (ROE): 12.4–14.9 %

Equity IRR: 13.5–19.5 %
Weighted Average Cost of Capital (WACC): 9.6 % (excluding grant)
Average Debt Service Coverage Ratio (DSCR): 2.13–2.87

Conclusion

Solar diesel hybrid minigrid systems, if designed properly, can serve off-grid areas with grid quality electricity. The proportionate increase of day load reduces the requirement for a storage system, which in turn reduces the cost of electricity. The addition of a diesel generator further reduces the size of the storage system as well as helping to avoid the need for autonomy days. Demand assessment is one of the most important criteria for designing a minigrid for rural areas. The demand or consumption pattern of a particular area is largely dependent on the socio-economic condition of that area. One model designed for a particular area cannot be directly used for another area. Each area should be individually surveyed for demand assessment. Installation of solar PV based minigrids not only provides electricity to the rural off-grid areas but also reduces GHG emissions. The cost of electricity in a minigrid can be kept within affordable limits with a careful and proper design. There should be a provision for future expansion to avoid system overload. The cost of electricity from isolated minigrids is more than that from national grids, so the use of higher efficiency electrical gadgets can reduce the energy consumption.

Appendix

Exemplary case study on mini-utilities based on diesel generators *Hatikata bazaar*

S. No.	Queries	Answer
1	Location	23°39′11″N, 89°59′21″E
2	District	Manikgong
3	Sub-district	Upazilla
4	Union	Azim Nagar
5	Police Station	Horirampur
6	Name of village market	Hatikata
7	Name of owner	MilonGazi
8	Mobile number	01711 702532, 01971 702532
9	Year of DG purchased	2007
10	Capacity of diesel engine	20 HP
11	Client	
11.1	Shops	25 (generally one light per household and shops)
11.2	House (family)	50

(continued)

(continued)

S. No.	Queries	Answer
12	Tariff	
12.1	Light	BDT 20/CFL light (15 W)/day
12.2	TV	BDT 70/TV-DVD (50 W)/day [11 connections in total]
13	Running hours (from after sunsets)	
13.1	Summer/rainy season	5 h
13.2	Winter	6 h
14	Fuel consumption	12 litres/day [fuel price at the island is BDT 72/litre]
15	Daily revenue of the entrepreneur	BDT 700–800/day

The entrepreneur also takes care of the supply of electric appliances (e.g. the TVs)
Basic calculation for the kWh price for CFL light
BDT 20/(15 W * 5.5 h/1,000) = 242 BDT/kWh = USD$3.10/kWh
Basic calculation for the kWh price for TV
BDT 70/(50 W * 5 h/1,000) = BDT 280 = USD$3.60/kWh
assuming 5 h of running time for a TV
There are seven diesel generator based mini-utilities on the island with a similar scheme varying in size: 2 * 20 HP + 3 * 12 HP + 6 * 3 HP

References

Bangladesh Energy Regulatory Commission, Ministry of Power Energy and Mineral Resources (EMRD). (2012). Retail electricity tariff for all utilities, 29 Mar 2012.

Government of Bangladesh (GoB). (2008). Renewable Energy Policy of Bangladesh, Power Division, Ministry of Power Energy and Mineral Resources.

Groh, S. (2014). The role of energy in development processes—the energy poverty penalty: Case study of Arequipa (Peru). *Energy for Sustainable Development, 18*, 83–99.

Infrastructure Development Company Limited (IDCOL). http://www.idcol.org. Accessed 10 Apr 2014

Intellecap. (2012). Lighting Asia: Solar off-grid lighting—market analysis of: India, Bangladesh, Nepal, Pakistan, Indonesia, Cambodia and Philippines. International Finance Corporation.

IEA (2013). International Energy Agency. World Energy Outlook 2013. Paris: OECD/IEA.

Khan, H. J., & Huque, A. J. (2012). Reliable power supply with solar mini grid in sandwip Island of Bangladesh. In *Proceedings of 2nd International Conference on the Development in Renewable Energy Technology (ICDRET'12)*, Dhaka, Bangladesh, January 5–7, 2012.

Kirchhoff, H. (2014). Hidden Resources in Solar Home Systems as the Base for Bottom-Up Grids. In *Proceedings of the International Conference of Innovating Energy Access for Remote Areas: Discovering untapped resources*, Microenergy Systems and Berkeley Rural Energy Group. Berkeley, USA. ISBN: 978-3-7983-2694-1.

Solar Guide Book. (2013). Ministry of Power Energy and Mineral Resources, Government of Bangladesh.

World Bank (2013). World Development Indicators. Bangladesh. http://data.worldbank.org/country/bangladesh. Accessed 15 Feb 2015.

Chapter 7
A Simulation Gaming Approach to Micro-grid Design and Planning: Participatory Design and Capacity Building

Maizakiah Ayu Abdullah and Scott Kennedy

Abstract Existing micro-grid design and planning approaches tend to emphasize techno-economic assessments and lack community engagement, necessary for effective planning and implementation. New approaches must be employed to not only include significant social impacts of micro-grids beyond technical components, but prioritize human development objectives, participation and capacity building. A newly proposed simulation gaming approach to micro-grid design provides an innovative, participatory tool and process that incorporates social, organizational, technical and financial factors for improved design and planning. Additionally, the approach represents an experiential learning and capacity building exercise that teaches shared resource management and collaborative decision-making.

Keywords Energy design · Micro-grid planning · Participatory design · Simulation game · Capacity building

Introduction

Recent advances in distributed generation technologies have intensified interest in decentralized electricity delivery models, particularly in its potential to meet rural users' energy and development needs through greater flexibility in technology options and organizational structures. However the design and planning of a micro-grid system for comprehensive development purposes requires careful and further

M.A. Abdullah (✉)
Abu Dhabi, UAE
e-mail: mabdullah@masdar-alumni.ac.ae

S. Kennedy
Boston, MA, USA
e-mail: sknndy@gmail.com

S. Groh et al. (eds.), *Decentralized Solutions for Developing Economies*,
Springer Proceedings in Energy, DOI 10.1007/978-3-319-15964-5_7

consideration. With a wide range of possible technologies and delivery models, micro-grid design and planning is highly dependent on available resources, local social context, markets, institutional structures and users' energy requirements.

Planning a community-level, micro-grid system generally consists of a needs assessment, a resource assessment, technical design, and economic and financial analyses. Energy needs assessments for rural electrification are commonly conducted using participatory methods such as surveys, questionnaires and focused group discussions (FGDs) in order to assess electricity demand and ability to pay (Howells et al. 2002). Data from these methods are then used as a baseline for techno-economic assessments, employing energy design tools such as HOMER, LEAP, and RETScreen (Connolly et al. 2010) to design a sufficiently reliable technical system at minimal cost. Tariff schemes are then designed around cost recovery and end users' financial abilities.

The majority of energy design and planning approaches are largely focused on the technical system, or on techno-economic assessments that do not enable sufficient or effective community participation beyond initial base-lining. Techno-economic assessment tools are typically operated by non-community members, who may not have a thorough understanding of local energy needs and the system's impacts on the community. Additionally, rural community members with limited technical skills and understanding of micro-grid operations (or rural electrification in general) are unable to use these tools to fully participate in the design process, which restricts community engagement and does not leverage nor build their capacities.

Conventional techno-economic assessment tools and methods also tend to exclude critical, social micro-grid success factors such as organizational or institutional considerations, demand side management and community participation, all of which affect uptake and operational sustainability. A narrow system boundary also inadvertently limits the micro-grid's scope for human development impact, as systems are installed to mainly deliver electricity rather than support energy use for facilitating development. Furthermore, systems that do not account for social dynamics and local complexities may cause unanticipated and unintended outcomes (e.g. electricity or equipment theft) that consequently limit and hinder energy access for development. In order for rural electrification to enable development, system design and planning must reprioritize human development objectives over electricity delivery, and encompass an expanded system boundary to account for broader social and financial considerations targeted at development-related outcomes.

Meanwhile, participatory methods used to collect data for energy planning are challenging to conduct and may provide limited useful information. Inaccurate demand forecasts directly affect system size and capacity, which in turn produces higher installation costs for an oversized system, or inadequate electricity supply if undersized.

Surveys and questionnaires on household energy use elicit information through a mainly one-way transfer, and do not induce learning or capacity building. Obtaining useful responses depends on participants' understanding of the questions (which also means the interviewer has to ask participants meaningful questions) (Cross and

Gaunt 2003; Howells et al. 2002) and on the power dynamic between interviewer and respondent, which may further affect accuracy of responses. Additionally, communities with limited exposure to electrification may have difficulty in predicting their own electricity consumption and behavior. To account for this, surveys and questionnaires are used to forecast electricity demand based on the community's existing energy use (including traditional energy use). This creates uncertainty and inaccuracies as past energy use does not necessarily translate directly into electricity use, and moreover cannot predict new uses with electrification.

We propose a novel participatory design approach that merges the technical aspects of a techno-economic assessment tool with the emphasis on the social system from participatory processes. We utilize a simulation game to address the aforementioned shortcomings by building capacities in decision-making and resource management. This process expands the system boundary beyond a techno-economic assessment by including organizational considerations, participation and social interaction. It provides a simulation gaming environment in which community members have equal opportunity for participation (Chua 2005) and induces dynamic learning so they can contribute informed decisions and input into the design process (Brandt 2006).

This new approach can be used to both enhance the design process (i.e. build a sustainable micro-grid that delivers adequate, reliant and affordable electricity supply for enhancing human capabilities) and serve the more fundamental objective of building technical and governance capacity of the community.

Research Objectives

The main objective of this work is to create a useful participatory tool and process that can be used to elicit a rural community's energy needs and system design parameters, in order to facilitate design and planning. Meanwhile, the participatory process is also aimed at empowering and building community members' capacities by enabling a greater understanding of household and system load profiles, technical limitations of a micro-grid, the importance of system cost recovery and managing the micro-grid as a shared resource.

Methods

The Simulation Game as a Tool and Artifact

The participatory game is a simplified representation of an operating micro-grid, with individual players representing households (or other load center forms) and a facilitator acting as a system operator. The game may be used to explore both planning and operational decisions. The facilitator prompts the players to choose their end-use loads (e.g., appliances), as well as to play out their periods of operation. Players are networked so individual and aggregate behavior can be visualized real-time.

In the current version, the simulation has been implemented using the multi-method simulation software, AnyLogic. A game-round is initialized with system capacity, number of households and levelized cost of energy (LCOE), which may be calculated using HOMER or similar tools. System cost may be updated in later rounds if the community is allowed to change system capacity. The example shown here has reference data from a Malaysian case study (Abdullah 2013). The game shows two windows that represent individual household consumption and system-wide consumption. The game has a graphical user interface where players can choose household appliances from a pre-defined set or a customized "other". The current choice of appliances and power ratings are also based on the Malaysian case study. Depending on the focus of given game play, a facilitator could choose to distribute appliances "for free", or to price the appliances and provide players with an initial budget. The latter case would allow the community to explore the financial trade-offs of different appliances (e.g., incandescent vs. LED lighting).

Once the simulation clock starts, players explore consumption patterns by individually switching loads on or off. The main system window shows real-time total load and revenue collected. Currently, household energy costs are based on a minimum energy-based tariff set by the LCOE. Alternative tariff models, such as capacity-based payments or even dynamic pricing can be easily implemented (in future versions).

The current version allows the facilitator to set a varying generation output, but does not explicitly model the power generation process or include distribution losses or other network features. Warnings appear when total load is near capacity (90–100 %), at peak capacity, and over capacity (the system shuts down within a given duration).

The Simulation Gaming Process for Participatory Design

This participatory design process elicits system design parameters by alternating between simulation game playing and facilitated discussions. During discussion rounds, facilitators pose a set of questions aimed at obtaining the following system design parameters (Fig. 7.1):

- System capacity
- Demand side management options
- Organizational model
- Tariffs
- System policies and regulations

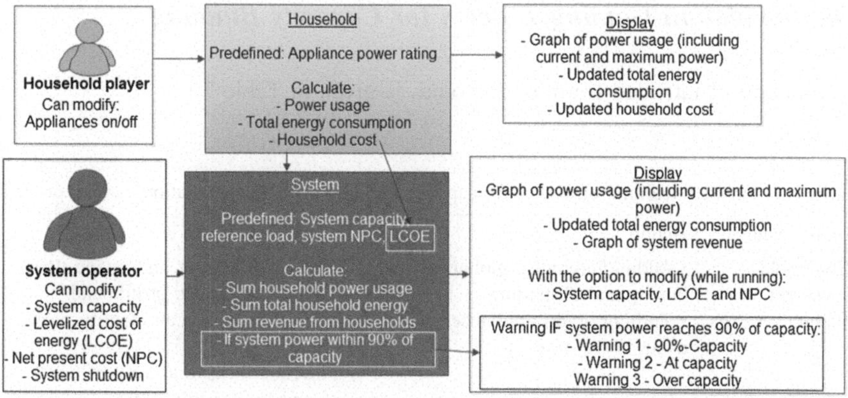

Fig. 7.1 Diagram explaining the simulation game methodology (Abdullah 2013)

The questions that lead to the above parameters are meant to represent an improved participatory design process over traditional planning methods such as surveys, questionnaires and techno-economic assessment tools. A representative set of questions is as follows:

- Identifying a reference load profile.
  - Based on observed system behavior in previous rounds, should the generation capacity be increased or decreased? → Installation costs and technology options will depend largely on system size.
  - What demand side management options are effective and acceptable? → Necessary especially when demand nears system capacity.
- Deciding organizational and management structures.
  - How would the community enforce demand side management strategies, system policies and regulations? → It is complex to enforce policies in rural communities. Design of such mechanisms must come from the community.
  - What type of organizational structure would work best for the micro-grid? Who will own and manage the micro-grid? → The community understands their requirements and social structures best.
- Deciding tariff structures.
  - Should the tariff be raised in order to generate revenue? → Ability to pay for micro-grid services is important to recover costs. However, rural users may have limited ability to pay and should decide on their own tariff structure.
  - What will revenue be used for, only cost recovery? What about an optional community savings fund?
- What will future load profiles look like? → Players can test potential future scenarios and load profiles to forecast future demand.

The Simulation Gaming Process for Capacity Building

A summary of intended learning outcomes is given in Table 7.1.

Table 7.1 Intended learning outcomes for capacity building using the simulation gaming process

Game feature	Action	Learning outcome
Household consumption playing	• Adding/removing appliances as if buying and discarding • Turning on/off appliances as if using in real-time	• Players will better understand their own household load profile and electricity use and can identify peak loads • Players will have a grasp of how different load profiles affect expenditure and be able to roughly estimate monthly charges and decide whether to lower consumption if costs are too high (assuming an energy-based tariff)
System-wide load profile and warnings	• Viewing system load profile in real-time against capacity	• Players will appreciate that capacity is limited and understand the need for both individual and collective load management • Players will learn about system load profiles, community peak loads and total electricity usage and behavior
Discussion rounds	• Collectively discuss findings, experiences and observations from the game • Discuss best way to manage consumption and micro-grid	• Players/community members will come together and work out issues with managing a shared resource • Community will apply existing governing structures to support decision-making process and strengthen local institutions • Community will apply existing (if any) resource sharing management to electricity use. E.g. agricultural communities that share water resources
Tariffs, revenue generation	• Viewing energy use against tariff, costs and revenue • Changing tariff in order to generate more revenue for the community micro-grid • Testing of future scenarios	• Players will understand the importance of cost recovery and how best to afford and use electricity for productive end use • Players will learn how different tariffs and corresponding consumption can contribute to greater revenue for a community savings fund. This could open up the opportunity to articulate desired services and capabilities for community development

Results: Simulation Game Screenshots

See Figs. 7.2 and 7.3.

Discussion: Case Study Observations

Development of the game is currently ongoing and has only been tested in mock-community settings. A beta version was trialed in August 2013 among university students and young professionals participating in an energy access workshop, and was used as a teaching tool for community participation and micro-grid planning. The participants were from various countries and originated from both rural and urban areas. They were divided into teams of 3–4 per group, each group representing households. The game was played on networked laptop computers, although future versions of the game will be played on mobile devices (i.e. tablet) for easier deployment.

Most of the case study participants have mainly experienced grid electrification and thus were new to the unique challenges in using and managing a micro-grid.

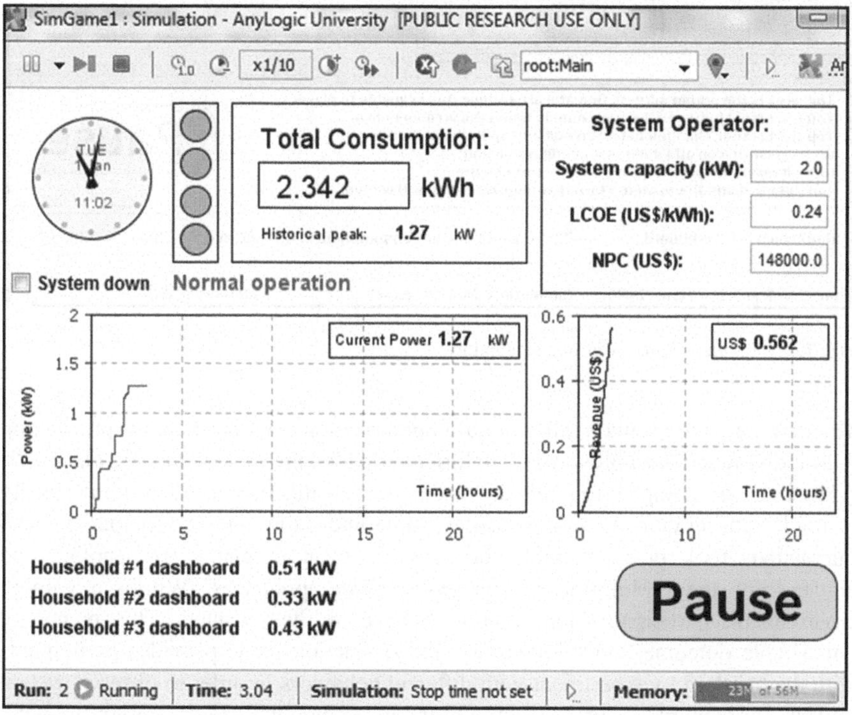

Fig. 7.2 System window showing micro-grid load profile

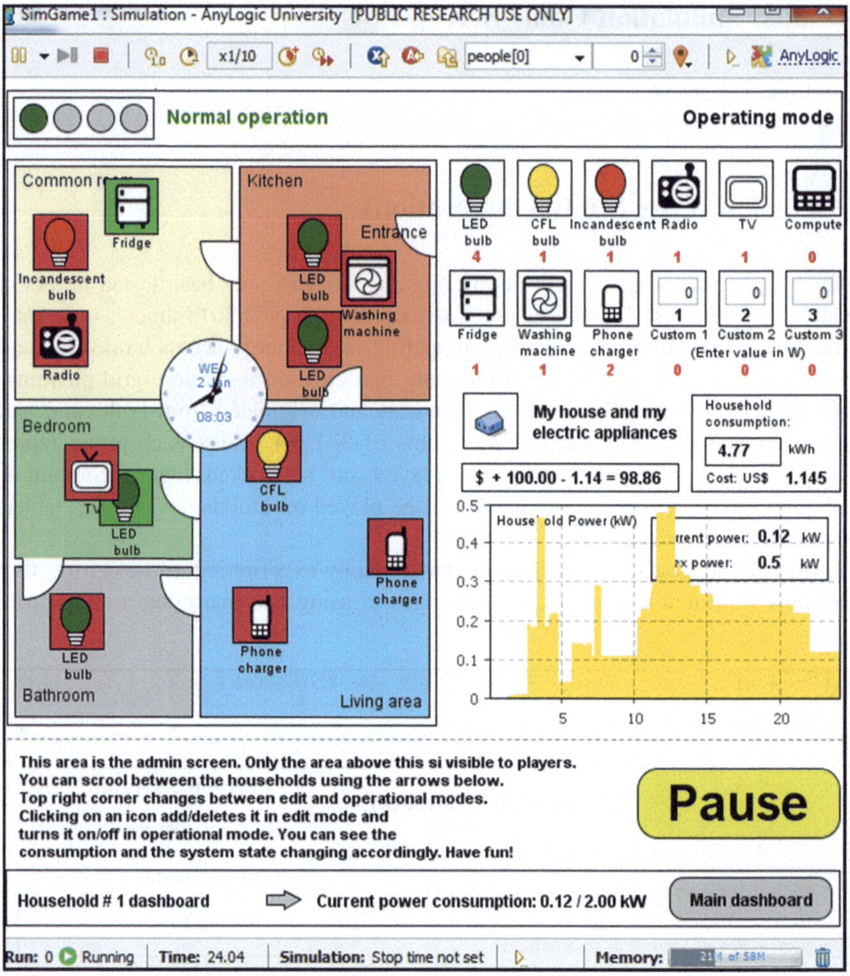

Fig. 7.3 Individual household player window

The first few game-rounds (after initial familiarization) saw participants prioritizing their individual households, which in turn overloaded the system causing blackouts. As facilitators attempted to guide the discussion towards increasing system capacity or implementing demand side measures, some interesting and unanticipated social interactions took place. Namely, households began to blame one another and demand for the prohibition of energy-intensive appliances. Within a gaming environment, participants felt freer to behave as they wanted. This is not an undesirable outcome. On the contrary, the simulation game provides participants with the freedom to experiment with different behaviors in order to observe system impact. For example, a future version of the game may include the option to 'steal' electricity, a prevalent problem with micro-grids. The game by itself is not meant to be biased in that it favors a desirable (or undesirable) behavior. It has been developed to remain as flexible as possible to enable participants to choose what normative actions and decisions work best for them.

It was observed that although the household teams could easily participate equally in discussions, the collective decision-making process still depended on group power dynamics or how comfortable participants felt voicing their opinions. However, the total system load profile displayed contributions of every household (an optional feature) and hence non-participating players could still be drawn out to discuss their contributing loads.

Like other participatory methods, the facilitator role is critical and requires an objective, unbiased (towards a particular solution) individual, knowledgeable about micro-grid delivery. However compared to conventional surveys and questionnaires, the simulation gaming design process has greater flexibility and can be applied in different local and social contexts without much customization. It is also dynamic in nature, providing immediate feedback and learning to players. In the trial, after three game-rounds, participants were collaborating well, had decided on a system capacity that suited their energy needs, and were no longer overloading the system.

The simulation game as a design tool was also useful in testing effectiveness of demand side management options. In the trial, participants were able to choose and test between limiting individual household capacity and voluntarily decreasing consumption when the game's warning signals displayed that the system was nearing capacity. During one game round, households were given the option of reducing their consumption within a specified time period when the system reached a critical state. Three out of the four households immediately reduced consumption when the warning appeared, while the remaining household waited until the others had reduced consumption. This provides interesting insight into human behavior and the game could help devise new demand side management strategies.

Ultimately, participants recognized that cooperation was necessary in order for everyone to have access to reliable electricity. The beta version trial was considered successful in providing an interesting and enjoyable learning experience for resource management and cooperative, participatory design. The process does have limitations, the most important one being that the simulation gaming process does require a certain level of community organization in order to successfully make design decisions. A hands-on, physical version of the game that works without devices is also being considered. The physical version will be for communities that have very little experience with technology and operating devices. Although the game would not be as dynamic nor will it provide immediate household load profiles, a physical version of the game will still be able to impart concepts such as cooperation in using a limited micro-grid capacity and making decisions as a community.

Acknowledgment To Dénes Csala (Masdar Institute) for his invaluable contribution in coding the beta version and developing its multi-user networking functionality in AnyLogic, and providing technical support during the trial.

References

Abdullah, M. A. (2013, June). *A participatory approach to rural energy systems: Design, planning and implementation* (Master of Science Thesis). Abu Dhabi, UAE: Masdar Institute.

Brandt, E. (2006). Designing exploratory design games: A framework for participation in participatory design? In *Proceedings of the ninth conference on participatory design: Expanding boundaries in design-Volume 1* (pp. 57–66).

Chua, A. Y. K. (2005). The design and implementation of a simulation game for teaching knowledge management. *Journal of the American Society for Information Science and Technology, 56*(11), 1207–1216. doi:10.1002/asi.20224.

Connolly, D., Lund, H., Mathiesen, B. V., & Leahy, M. (2010). A review of computer tools for analysing the integration of renewable energy into various energy systems. *Applied Energy, 87* (4), 1059–1082. doi:10.1016/j.apenergy.2009.09.026.

Cross, N., & Gaunt, C. T. (2003). Application of rural residential hourly load curves in energy modelling. In *Proceedings of power tech conference, 2003 IEEE Bologna* (Vol. 3, p. 4). doi:10.1109/PTC.2003.1304492.

Howells, M., Alfstad, T., Cross, N., Jeftha, L., & Goldstein, G. (2002). *Rural energy modeling*. Citeseer.

Chapter 8
The Energy Centre Model: An Approach to Village Scale Energy Supply

Charles Muchunku and Kirsten Ulsrud

Abstract A model for off-grid, village scale power supply has been developed through the Solar Transitions research project, led by the University of Oslo. Through a participatory approach the project developed an energy centre model which became operational on March 20th, 2012. The model creates affordable and accessible basic lighting and electricity services for off-grid households and businesses through a financially sustainable design. It is designed to be operated by local residents on commercial principles. Today, it provides lighting and other services to about 150 households. The paper describes and rationalizes the model and shares experiences from the first 1½ years of operation. The financial performance is documented and suggestions are provided on how the model could be improved.

Keywords Solar · Energy center model · Households

Introduction

Access to electricity has a positive impact on a wide range of factors influencing rural communities, from improved health, to access to communication and information, to better educational facilities, economic prosperity and improved standard of living. However, finding ways to expand energy services to marginalized households in developing countries is one of the most pressing challenges facing the world today.

C. Muchunku (✉)
Camco Advisory Services (K) Ltd, P. O. Box 76406-00508, Nairobi, Kenya
e-mail: charles.muchunku@camcocleanenergy.com

K. Ulsrud
Department of Sociology and Human Geography, University of Oslo,
P. O. Box 1096, Blindern 0317, Oslo, Norway
e-mail: kirsten.ulsrud@sosgeo.uio.no

S. Groh et al. (eds.), *Decentralized Solutions for Developing Economies*,
Springer Proceedings in Energy, DOI 10.1007/978-3-319-15964-5_8

Globally, there is no universal approach to provision of access to electricity in marginalized areas. The UN uses a three-step scale to denote various types of energy access[1]: Level 1—Basic human needs (electricity for light, education, health, communication and community services, and modern fuels/improved stoves for cooking with biomass); Level 2—Productive uses; and Level 3—Modern society needs.

Research Objectives

The action research project was aimed at identifying a sustainable approach for energy service delivery suited to local conditions of the selected location.

Methods

The project was initiated and carried out by a team of social scientists and practitioners from Kenya, India, Norway and Austria through the Solar Transitions research project, led by the University of Oslo.

Through a participatory approach with the local community, the project developed an energy centre model in Ikisaya, Kenya which became operational on March 20th, 2012. The process took into account local dynamics i.e. the type of settlement pattern, density of the population, ability to pay, electricity services prioritized by community members, gender aspects, available energy resources, proximity to the national grid and social inclusion.

The energy centre targets affordable, accessible provision of basic lighting and electricity services for off-grid households and businesses. Lighting needs are currently served through the use of kerosene and torches with expenditure from 1.4 to 3.6 €/month; which is likely to increase with increasing fuel costs. Mobile phone owners often have to travel long distances to charge their phones in addition to spending 18 €¢. Printing, photocopying, computer services and TV/video shows are also not readily accessible within the vicinity of most off-grid communities. All these services were pointed out as important by the community members. Furthermore, there is little or no competition for the provision of all these services under one roof as most entrepreneurs are unwilling to make large investments to provide energy services in off-grid areas.

The energy centre model is based on a 2.16 kW solar PV system which provides energy for a range of services i.e. lantern charging and renting, charging of mobile phones, IT-services (typing, printing and photo-copying) and television and video shows. Fees are charged for the provision of these services to cover operation and maintenance costs (e.g. battery replacement) and generate a surplus for energy centre improvement and expansion. The centre has the capacity to serve up to 200 households (1,000 people).

[1] http://www.un.org/millenniumgoals/pdf/AGECCsummaryreport[1].pdf (Accessed Jan 2014).

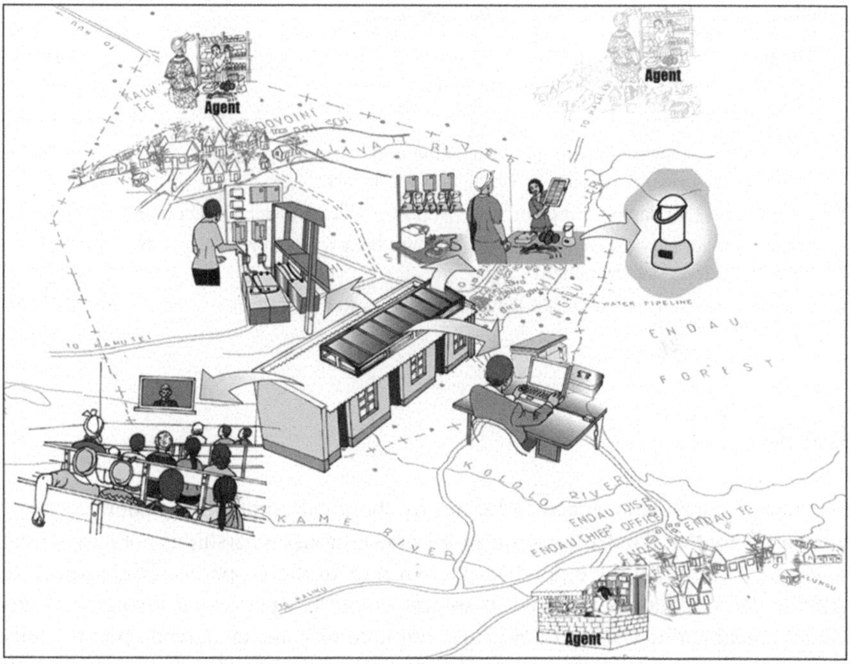

Fig. 8.1 Illustration of Ikisaya energy centre and its three first sub-centres, operated by agents

The energy centre building is designed to separately house the services offered. The area of the building is slightly below 70 ft^2. It has 4 main rooms as illustrated in Fig. 8.1:

- The charging station—where phones, lanterns and small batteries are received, charged and given out. It is also the room from which retail sales are made.
- The IT room—where photocopying, typing, printing and other IT related services are offered. It is envisioned that basic computer training could also be provided here.
- The TV room—where TV and video shows are screened. The room may also be hired out and used for meetings and training.
- The back office and store—where most of the installed equipment (batteries, inverters, charge controllers) is kept. The room is used as an office and a store.

The total investment cost for the centre was 42,527 €[2]; 10,509 € for the building structure which houses the energy centre. The capital costs of the system are broken down in the Table 8.1.

[2] Excludes costs related with developing the model, training the staff and follow ups.

Table 8.1 Summary of capital costs for the energy centre

	Item	Cost (€)
1	Building, construction and oversight costs	10,509
2	Furniture and fittings	818
3	Solar PV system equipment and installation costs (including lanterns and other accessories) at the energy centre and for agents	25,700
4	Appliances (e.g. laptop computer, laser-jet printer, TV, decoder, DVD player)	1,782
5	Start-up stationary and equipment (e.g. receipt books, counter books, cash boxes, folders, stapler, punch)	982
6	Retail outlet start up stock (40 lanterns and 40 power packs)	2,736
	Total (€)	42,527

Results

The energy centre is designed to be run by the local community under a concessional arrangement. Under the terms of the concession, ownership is not transferred; the energy centre and the equipment therein remain the property of the project. In addition, the operator is required to submit quarterly operational financial reports and an annual audit. The operator is also required to make an agreed upon monthly deposit to a joint bank account for the battery replacement and maintenance fund. If the operator fails to adhere to these terms and no suitable remedial action is identified and effected, then the concession is terminated and the system reassigned or relocated to another community.

The services provided by the centre generate an average revenue of 460 €/month. Operation and maintenance costs average 400 €/month. This includes staff salaries, petty cash, consumables, general maintenance and contribution to a battery replacement fund i.e. to raise 4,600 € every 2–3 years (130–190 €/month).

The revenue and expenditure trends from 18 months of operation are summarized in Tables 8.2 and 8.3 below. As revenues and expenditures vary from month to month, 3 month averages are provided to give typical values. The different periods are also selected to show how revenues and expenditures are evolving. The revenue trends show that the lantern renting and mobile phone charging services are by far the highest and most consistent source of revenue; they represent 70 % of all revenue generated by the energy centre and about 50 % of the costs. They essentially subsidize the other services provided at the centre i.e. the IT services and TV/video shows. Whereas, these are viewed as important services by the local community they are not sufficiently subscribed to cover their operational costs.

The lantern charging and renting service has evolved from being a purely centralized model where the lanterns and lantern charging system were all housed at the energy centre to a decentralized model where lantern renting services are also provided in neighbouring villages through agents. The reason for this change was demand for lantern renting in these other villages around Ikisaya. The Ikisaya energy centre currently has 5 agents located in other trading centres that are 10 km

Table 8.2 Average monthly revenues Oct–Dec 2012 and Apr–Jun 2013

Business section	Services	Average daily users (Oct–Dec 12)	Average monthly revenue (Oct–Dec 12) (€)	Monthly revenue/ section (€)	Average daily users (Apr–Jun 13)	Average monthly revenue (Apr–Jun 13) (€)	Monthly revenue/ section (€)
Charging services	Lantern charging	26	72	123	40	109	194
	Mobile phone charging	9	50		31	86	
	Battery charging		1				
Agents	Lantern charging	55	149	149	52	143	152
	Mobile phone charging				3	9	
IT services	Photocopying	15	42	84	16	44	55
	Typing and printing	2	13		4	11	
	Laptop charging		30				
Retail outlet	Lanterns		6	19	2	6	6
	Power packs		13				
Multipurpose room	TV and video shows	7	20	22	13	36	37
	Room hire		2			2	
Other services	Hair cutting	3	8	8	7	18	18
Totals (€)		117		406	169		462

or more from the energy centre. These agents are in the trading centres of Endau, Malalani, Kalwa, Kathua and Yiuku.

The charging units are easily set up by installing one or more lantern charging units per agent. Each charging unit, is made up of a 50 Wp module, a junction box/ charging unit and 10 lanterns. In some areas agents have also been provided with additional equipment to enable them to also offer phone charging services. The lantern charging capacity of the energy centre and of the different agents is currently: Ikisaya–56, Endau–56, Malalani–22, Kalwa–20, Kathua–20 and Yuiku–20.

The agents are typically existing shop-owners in these trading centres interested in generating additional income. These agents earn a commission of 30 % on the lantern renting and phone charging revenue they collect.

The energy centre has also been experimenting with an alternative kind of agent. In one trading centre the energy centre has employed a local person to manage the lantern charging services there. This person is paid a fixed monthly salary (irrespective of the revenue collected). The centre also directly covers the rental costs of the premises where the services are housed.

The major challenge with the agent approach is the inability to determine the actual revenue collected by the agents. A number of options are being tried out to address this:

- A fixed monthly amount payable by an agent to the centre be determined and any revenue generated over and above this amount be retained by the agent as his/her income. It is anticipated that this approach would not only guarantee the centre a predictable monthly income but that it could also incentivize the agent generate as much revenue as possible.
- Implementing a system that could remotely monitor an agent's performance and that could also remotely control the agent's system. Such a system would ensure timely monthly payments by enabling the centre to remotely disconnect an agents system if payment is not remitted. In Kathua and Yuiku, the energy centre is piloting lantern and mobile phone charging systems with the capacity for remote monitoring and control.

It is important to consider that the logistics of remittance or collection of payments from agents is difficult in areas with limited inter-village public transport and poor or limited mobile network coverage.

The total number of services used and paid for in a given day averages about 169; an increase of 44 % over 5 months. Lantern renting and mobile phone charging are the most used services; they together represent 77 % of all services used at the centre. The relative usage of other services compared to the total number of services used at the centre is: IT services at 13 %; TV/Video at 7 %; and hair cutting at 3 %.

The projected and actual average monthly expenditure for the energy centre are shown in Table 8.3 below. On average the energy centre's expenditure has not varied much over the course of its operation. Staff salaries and the battery replacement fund represent the largest operational expenditures.

Table 8.3 Average monthly expenditure Nov 12–Jan 2013 and Apr–Jun 2013

Energy centre expenses	Details	Projected expenditure per month	Average monthly expenditure (Nov 12–Jan 13)	Average monthly expenditure (Apr–Jun 13)
Salaries	Manager	73		
	IT clerk	59	59	65
	Centre technician	59	59	
	Evening attendant	45	45	58
	Part time accountant	36	36	55
Other staff payments	Overtime (evening attendant)		9	7
Agents commission	Commissions for lantern renting agents		30	17
Consumables	Printing paper	4	0	7
	Cartridges black	14	20	25
	DSTV monthly subscription	39	20	9
Petty cash		36	21	29
Transport	Monitoring of agents		8	14
Business permit	County council payments			4
Maintenance fund contribution	Estimated 500,000 needed after 2 years for battery replacement and other emergency maintenance requirements	194	124	112
Total expenditure, monthly (€)		560	425	400

The battery replacement and maintenance fund is based on raising an estimated 4,600 € every 2–3 years. This amount is to cover future battery replacement costs and unexpected system component repair or replacement costs. The centre target is therefore to set aside 194 € every month towards this fund. The actual amount raised is typically the difference between the monthly revenue generated and the recurrent monthly operational expenditure; as at the end of August 2013, the centre had managed to set aside a total of 1,532 € for the battery replacement and maintenance fund.

Discussion

With a typical solar PV system life span of 15 years, the project is low-cost compared to other options for provision of basic energy services to low density off-grid communities. Considering the average population density in the general area is 9 households/km[3]; electrification through a mini-grid would not be a viable option.

The significant upfront investment required for the energy centre and small margins make the model uninteresting for private sector investment. In its current form, the model would not attract businesses that could easily generate quicker and more substantial returns elsewhere. A capital subsidy would therefore be required if an identical model is to be replicated.

Nevertheless, components of the model could be interesting for private sector investment. Revenue and expenditure trends indicate that the lantern renting and mobile phone charging services are the highest and most consistent source of revenue; they represent 70 % of all revenue generated by the energy centre and cover about 50 % of the operation and maintenance costs. An analysis of investment costs also indicates that a model that targets only the provision of lantern renting and mobile phone charging services would require less than one third of the investment costs used for the Ikisaya energy centre (see Table 8.1; considering that the lantern and phone charging represent 30–40 % of the solar PV equipment and installation costs).

A key lesson is that location is a key consideration when selecting the services to be provided and best way to deliver them (i.e. whether through an energy centre or simply through an agent). In areas where household incomes are low it may be necessary to focus on the provision of the most basic services i.e. lighting services (lantern renting) and phone charging only. In economically active areas, the demand for TV and IT services would be higher, as would be the ability to purchase lanterns or small solar home systems. For purposes of standardization and replication, it may therefore be necessary to develop criteria for assessing and categorizing sub-locations based on socio-economic activity and then develop services best suited for each category.

Reference

The Kenya Bureau of Statistics. (2009). Population Census.

[3] Kenya Population Census data 2009 (*Source* Kenya Bureau of Statistics).

Chapter 9
Lessons from the Edge

Peter Newell, David Kerner and Scott Thomas

Abstract Lessons learned by the Army in changing how it meets energy needs on the battlefield can well inform how energy solutions are brought to rural areas around the world. Just as bringing conventional, grid-based electricity to remote villages often imposes a high economic cost, delivering energy on the battlefield incurs a high human cost, one that has led the Army to reexamine its assumptions about how energy is provided and used. Changing the way they framed the problem led to better energy solutions, saved lives, and made the missions more resilient. The resilience paradigm is similarly well suited to rural energy applications. Based on the Army's lessons learned, the authors describe the attributes of a resilient system and discuss how these terms can be applied to instill energy resilience in rural communities.

Keywords Army energy · Resilience

Introduction

Late in 2010, the US Army and other forces in Afghanistan were faced with a growing trend in Improvised Explosive Device attacks that no amount of technological advantage seemed to abate. With no clear path to a technical solution to the problem, experts began looking for ways to reduce the number of Soldiers exposed

P. Newell (✉)
BMNT Partners, Palo Alto, CA, USA
e-mail: pnewell@bmntpartners.com

D. Kerner
The Tauri Group, LLC, Alexandria, VA, USA
e-mail: david.kerner@taurigroup.com

S. Thomas
Stetson Engineers Inc., Diamondhead, MS, USA
e-mail: scottt@stetsonengineers.com

S. Groh et al. (eds.), *Decentralized Solutions for Developing Economies*,
Springer Proceedings in Energy, DOI 10.1007/978-3-319-15964-5_9

to potential attacks. This fresh look at a decade-old problem also coincided with a growing desire to find a way to curb the growing costs of providing fuel to fixed sites throughout the country by inserting renewable energy systems as a source of power.

These weren't the first attempts to insert renewables onto the battlefield; there had in fact been several others dating back to early-2006. Unfortunately, the driver in previous attempts was a desire to save dollars, something that is a tough sell to Soldiers who are most concerned with just making it back to base in one piece every day.

Real progress did not come until the Army targeted a reduction in resupply efforts at the far reaches of the battlefield, where resupplying fuel was largely a human capital-intensive endeavor.

In these locations, for these units on the "tactical edge"—the burden of providing power and energy relies primarily on the force on the ground where the operational costs to deliver power and water are the greatest. At the tactical edge, sustainment operations are tactical operations that consume combat power, and at the tactical edge, savings will likely be measured in lives rather than gallons of fuel saved.

As we learned to ask the right questions, leaders across the country discussed the challenges of resupplying ever more distant Combat Outposts and their associated Observation Posts. The problem was most telling in one Brigade's area along the Afghanistan-Pakistan border, where logistics resupply operations between a battalion-level Forward Operating Base and a Combat Outpost was a 10-h trip—one way—all on routes that had to be cleared of improvised explosive devices and then secured lest they be repopulated with more. From that Combat Outpost it was a 5-h walk up a mountain to reach the closest Observation Post (OP). At the OP there was no landing zone; at best, there was only a single touchdown point with barely enough room for a helicopter to get two wheels on the ground. That Observation Post was within 2.5 km of the Pakistani border and was in near daily contact with insurgents.

Two squads of Infantrymen who rotated into the site every 7–10 days manned the Observation Post. They lived and worked in open bunkers and had only minimal power requirements—just enough to power their radios and Intelligence, Surveillance and Reconnaissance (ISR) systems and recharge their batteries. Their sole source of power was a single 5-kW generator that they had somehow managed to get to the Observation Post. Unfortunately, the generator was unreliable and difficult to repair, no surprise given that the power draw on it was less than 20 % of its capacity.

The Observation Post relied on air resupply for fuel and water, something that could not always be counted on in the winter. Close by were three other Observation Posts along the border in similar conditions.

Our observations indicated that "spot power" was a reality for small units who moved often and changed missions and operating bases as often as they change socks. At these locations there were no grids, save those installed on the fly by very industrious Soldiers.

At one small Combat Outpost representative of the problem, we found power provided by five generator sets. The first generator ran the Combat Outpost's Command and Control node with 15 computers, three large television screens, a number of computer monitors and six radios. That 60-kW generator was running at 17 % of its capacity. The second generator we found was powering the Combat Outpost's living areas and dining facility—this 60-kW generator was operating at 36 % capacity. The third generator, also a 60-kW unit, operated a single pump at the shower point that had a power draw of just 750 W. The fourth generator our team found was the best utilized, with 53 % of its capacity being used to power four refrigerator vans that supported the dining facility. The final generator set we found was a 15-kW set that was used to provide the 1.2-kW of power needed to run the Combat Outpost's Rapid Aerostat Initial Deployment (RAID) tower. Moreover, in our review of human capital interaction with energy systems at the outpost, we found no one on the outpost was trained in energy system management and only a few outside the lone generator mechanic had any basic knowledge of energy systems maintenance. Ultimately leaders knew things did not work right but were exasperated in the efforts to explain why, much less improve them.

Our visits to the forward edges of the battlefield, our site surveys, and our efforts to curb their energy problems, taught us many lessons.

First, energy solutions must be developed in open, transparent, and collaborative environments, so that new strategies and technical solutions are informed by user (i.e., Soldier) and technology developer insights and modified per actual field experience. Several of our early efforts lacked this coordination. For instance, in the early days of foaming tents for insulation, we failed to load-balance the generators that powered the air conditioning for the tents. While we were successful in reducing the temperature inside the tents, the reduced load on the generators caused them to be more inefficient and resulted in more fuel burned. Despite our best intentions, our lack of engagement with the entire community of interest involved in providing power to the bases actually led us to increase, rather than improve, fuel consumption on the bases. In quite similar fashion, it is not hard to imagine that very well intentioned, but one-dimensional, solutions applied in rural settings would have similar effects.

Through subsequent efforts, we recognized the real power of our organization was in our ability to bring together teams of subject matter experts from other organizations in order to more rapidly understand and solve complex problems. We became heavily reliant on partnerships and collaboration with other government organizations, academia, and industry, all of whom had the skill-sets necessary to support rapid requirements analysis, unit coordination, accelerated procurement, contracting, training, assessments, and sustainment.

Likewise, at the tactical edge the operators are Infantrymen, cooks, and medics, not electricians. Clearly, materiel solutions delivered for these Soldiers' use had to be robust yet functionally easy to integrate, operate and maintain. Additionally, training had to go beyond simple maintenance procedures and include the principles of power management, to ensure the optimal performance and benefit was realized from any energy solutions.

Third, our Highly Qualified Experts and assessment teams had to be 'solution-agnostic.' These teams had to be useful from the day they arrived onsite. Their initial duties included helping to optimize what was already present while also gathering the data required to help us determine the right solution. We also found that we had to gather data not just from one part of the system but had to check downstream to insure the solution didn't cause a bigger problem someplace else. We certainly could not assume that the improved effectiveness of a single device in a system would mean that the entire system would be better optimized.

We came to understand that fuel reliance is just part of the sustainment problem for tactical units; access to clean water and waste removal account for a significant portion of our sustainment convoy requirements and must be addressed as well.

Our most significant lesson was that despite our tremendous resources, we truly lacked resiliency—the ability to ensure a mission can function despite shocks and challenges—when it came to sustaining ourselves in remote areas. While our method of brute force accommodation made up for that lack of resiliency in most cases, when it failed, the results were catastrophic. And those same accommodations—massive resupply convoys—became assailable weaknesses for others to exploit.

In the future, we have much more to learn about solving challenges in austere environments, be they remote combat outposts or rural villages.

Understanding the Challenge

How we characterize the problem, a step that explicitly or implicitly embodies our assumptions and expectations, is a central concern. This steers how we define 'success,' what terms we choose for analysis, our metrics for measuring progress towards goals, and the technologies and strategies we select to achieve those goals.

If our focus is exclusively on economics, we inherently assume certain things about what will be good for the affected stakeholder. In the case of energy delivery in Afghanistan, the original assumptions included that energy supplies would focus almost exclusively on the use of military-grade JP-8 fuel, an assumption that carried mortal implications for those who delivered, retrieved, and relied upon that singular approach. Once we reframed the narrative, leaders were able to understand the full implications of logistics choices; the calculus no longer focused on economics, but on Soldiers' lives and mission effectiveness.

In a similar manner, efforts to meet rural energy needs often focus on narrow economic considerations, such as the cost per unit of energy delivered or the overall energy capacity installed for a given amount of money. Job creation and second-order effects may be included in the calculation, but the emphasis tends to be on the energy delivered. We argue that, just as the military energy supply problem was refocused from just supplying energy to the broader goal of ensuring the mission—which carries with it the need to protect troops, reduce reliance on singular supplies of a resource, ensure mission capabilities in the face of disruptions—so too the rural energy initiatives should focus on the community's existential needs to function and carry forward its character and identity.

A Resilience Perspective

The focus, then, must be on resilience. While several definitions exist for the term, the one of greatest use for open systems such as a rural economy derives from what natural resource managers have observed about ecosystems that adapt, survive, and thrive despite a wide range of stresses and disruptions. Systems characterized by uncertainty and unpredictability appear more tractable when examined from this ecological systems perspective. In this context, resilience has been defined as the capacity of a linked social-ecological system (SES)—very much a description of a rural society—to experience shocks while retaining essentially the same functions, structure, feedbacks, and therefore identity (Holling 1973; Walker et al. 2006; Walker and Salt 2006). How much disturbance can the system—people, infrastructure, resources—accommodate while still maintaining its basic structure, capabilities, and capacity to function? How far can it bend and adapt without breaking?

Systems often fail in unpredictable ways, but resilient systems continue to function despite the challenges. Planning assumptions do not always hold true, and planners often fail to ask how well the system will function in the face of large, unexpected challenges, or the accumulation of many smaller stresses; these include market failures; geopolitical and demographic shifts; resource shortages (e.g., water, fuel, fertilizer, minerals); epidemics; climate change; severe weather events; technology disruptions (NAS 2012); and myriad other shocks and perturbations.

Many systems appear to share attributes by which resilience can be characterized and assessed. Kerner and Thomas have investigated which system attributes relate specifically to the ecological definition of resilience, asking, "What attributes reflect whether a system will be able to continue to function and retain its identity in the face of existential challenges?" They considered attributes for all types of systems, including natural and manmade, physical and institutional, small and large, simple and complex. Building on others' efforts (e.g., Holling 1973; Lovins and Lovins 1982; Walker et al. 2006), they have delineated and defined common resilience attributes. These attributes, the full collection of which is presented with definitions in the Appendix, fall into three overarching categories that help us to understand the resilience 'posture' of a given social-ecological system. These categories include:

Stability: The degree to which a system can continue to function if inputs, controls, or conditions are disrupted. It is a reflection of how minor a perturbation is capable of rendering the system inoperable or degraded; the types of perturbation to which the system is especially vulnerable; whether the system can 'ignore' certain stresses; and the degree to which the system can be altered by surprise. Stability entails a number of attributes including controllable degradation, resistance, dispersion, and others.

Adaptive Capacity: The ability of a system to reorganize and reconfigure as needed to cope with disturbances without losing functional capacity and system identity. It reflects an array of response options and the ability to learn, collaborate, adapt, and create new strategies to ensure continued functionality. Adaptive Capacity entails a number of attributes including response diversity, connectivity, learning capacity, and others.

Readiness: How quickly a system can respond to changing conditions. It is affected by the physical, organizational, social, psychological, or other barriers, internal or external, that might impede timely response. Readiness is a measure of responsiveness; its converse is entanglement, a measure of the forces impeding responsiveness. Readiness entails a number of attributes including situational awareness, preparedness, simplicity, and others.

Put simply, we want to know if the system can survive a challenge as things currently stand (Stability), have the ability and options to respond if necessary (Adaptive Capacity), and understand if there are factors that help or hinder that response (Readiness). These attributes can be used to develop system-specific resilience metrics to guide assessment, as was done for a notional assemblage of Combat Outposts in Afghanistan, following up on the field evaluations described above (Army REF 2013).

Energy Resilience

In meeting the needs of a rural community, we seek energy resilience as one of the key enablers of the community. We define energy resilience as the ability to maintain the community's identity and function—to keep it alive and thriving—in the face of energy supply perturbations. This definition refocuses energy policy from assuring supply to assuring preparedness, thus setting the framework for stakeholders to improvise, adapt and overcome the effects of potential supply interruptions (Thomas and Kerner 2010). This shift in the narrative moves us to value "inefficiencies" such as redundancy and back-up capabilities over optimal solutions; it moves us to consider *all* of the factors that keep the community viable, and how those factors are affected by different energy strategies.

So how does a rural community measure success? Some goals include local employment, enduring social and cultural integrity, self-determination, safety and security, self-reliance, and dependable sources of energy, water, food, and other resources. Changing the focus to energy resilience leads to solutions that ensure societal functions such as farming, transportation, education, public safety, waste disposal, etc., can endure energy supply disruptions; adapt to accomplish those functions in new and different ways; and respond in a timely manner, unencumbered by dependencies and constraints. A resilience perspective considers the totality of the energy-water-food-livelihood nexus.

Examining energy resilience at multiple scales is important because the controlling variables for different systems may operate at different speeds or in different physical dimensions (Holling 1973; Walker and Salt 2006; Thomas and Kerner 2010). The Combat Outpost resilience assessment included one scale "higher" (interaction with the Forward Operating Base) and one scale "lower" (operations of the Observation Post as well as the individual components of the Combat Outpost) (Army REF 2013). Assessing the resilience of a rural community might focus one level higher (i.e., resilience within the region) and one level lower (resilience of family farms and businesses) in order to capture information at these different scales.

Solutions for rural energy concerns will be more varied than just a short list of grid and alternative power (photovoltaics, wind, hydro, etc.) options. Appropriate technologies—those of a size, simplicity, and sustainability commensurate with the particular community—should resonate with the inherent goals of that community and may fit well here. It should be noted, though, that solutions will not only be technological in nature, but will likely require organizational, regulatory, and procedural changes. Moreover, leadership and initiative, highly important traits in the Army, will play a crucial role in realizing any resilience goals. Finally, it should also be noted that efficiency measures, while important, should not be equated with resilience; depending on less of something is not the same thing as being less dependent on it.

Conclusions

Military forces are designed to be highly resilient in a number of ways, but they have been dependent upon reliable delivery of large quantities of specific, high-quality energy resources. Defending the vulnerable supply lines to which forces have been tethered skews the "tooth to tail" ratio and places additional service members in harm's way. This energy dependency has created significant vulnerabilities and limits force resilience. By reframing the planning considerations to a resilience perspective, new solutions emerge that both ensure the mission and protect the lives of Soldiers.

The narrative we choose steers outcomes and affects whether change is palatable. Framing rural community energy decisions in terms of resilience leads to outcomes that favor a community's existential needs; a focus on a monetary return on investment skews decisions to favor cost factors. As such, a resilience "return on investment" is not defined by dollar metrics, but by metrics associated with the continued identity and function of the community. Resilience can be assessed, developed, and maintained within a social-ecological system—e.g., a rural community—to avoid tipping points and ensure its existence for tomorrow. Focusing on system resilience provides a broad framework for selecting between options, prioritizing investments and initiatives, and making decisions about vital issues affecting the fate of rural communities.

Appendix

Definitions of resilience attributes

Stability	
Single points of failure	Singular features or aspects of the system, the absence or failure of which will cause the entire system to fail
Pathways for controlled reductions in function	Whether the functionality of a system, operation, or capability can be reduced in a manner that avoids the overwhelming effects of an unconstrained failure
Resistance	The insensitivity of the system to stresses of a given size, duration, or character
Balance	The degree to which a system is not skewed toward one strength at the expense of others
Dispersion	The degree to which the system is distributed over space and time
Adaptive capacity	
Response diversity	The variety and disparity of steps, measures, and functions by which an operation can carry out a task or achieve a mission
Collaborative capacity	The capacity to act through coordinated engagement
Connectivity	How readily resources and information can be exchanged to ensure continued functionality
Abundance/reserves	The on-hand resource stores upon which a system can rely when responding to stress
Learning capacity	The ability to acquire, through training, experience, or observation, the knowledge, skills, and capabilities needed to ensure system functionality
Readiness	
Situational awareness	How well system, component, and functional capabilities are monitored. How readily emerging stresses or failures can be detected
Simplicity/understandability	How well system functions and capabilities can be understood
Preparedness	The level of preparation in plans, procedures, personnel, and equipment for responding to system perturbations
False subsidies	Whether inputs, outputs, or internal processes receive incentives disproportionate or unrelated to their value
Autonomy	A system stakeholder's authority to select and employ alternate actions, configurations, and components in response to stress

Source Kerner and Thomas (2014)

References

Holling, C. S. (1973). Resilience and stability of ecological systems. *Annual Review of Ecology and Systematics, 4*, 1–23.

Kerner, D., & Thomas, S. (2014). Resilience attributes of social-ecological systems: Framing metrics for management. *Resources* , *3*(4), 672–702.

Lovins, A., & Lovins, L. H. (1982). *Brittle power: Energy strategy for national security*. Andover, MA: Brick House Publishing Co. Inc.

National Academy of Sciences (NAS). (2012). *Disaster resilience: A national imperative*. Washington, DC: National Academy Press.

Thomas, S., & Kerner, D. (2010). *Defense energy resilience: Lessons from ecology*. Carlisle, PA: Strategic Studies Institute. United States Army War College.

U.S. Army Rapid Equipping Force (Army REF). (2013). *Final report energy resilience study: Resilience assessment of notional combat outpost* (Contract SP0700-D-0301). Alexandria, VA.

Walker, B., Gunderson, L., Kinzig, A., Folke, C., Carpenter, S., & Shultz, L. (2006). A handful of heuristics and some propositions for understanding resilience in social-ecological systems. *Ecology and Society, 11*(1), 13.

Walker, B., & Salt, D. (2006). *Resilience thinking: Sustaining ecosystems and people in a changing world*. Washington, DC: Island Press. 165pp.

Part II
Innovations in Value Chains and Financing Schemes

Chapter 10
Value Chain Thinking and Energy Projects—A Problem-Centered Value Chain Approach to Energy Based Upgrading of Rice Farmers in the Philippines

Henrik Beermann, Utz Dornberger, Ben Sebitosi, Sebastian Groh and Jonas van der Straeten

Abstract This paper conceptualizes the link between value chain theory and productive use (PU) focused energy projects based on microfinance mechanisms. Its main argument is that all PU of energy projects focusing on micro, small, and medium enterprises (MSME) development can be interpreted as value chain upgrading attempts. It is argued that successful upgrading greatly depends on the MSMEs embeddedness in specific market contexts. For that reason, the context must be assessed to derive energy based intervention points that cause additional income for MSMEs and consequently development. Based on this rationale, a problem-centred value chain approach is proposed. A case study of the Philippine

H. Beermann (✉)
International Joint Master Programme in Sustainable Development,
Leipzig University Grimmaische, Straße 12, 04109 Leipzig, Germany
e-mail: beermann.henrik@gmail.com

U. Dornberger
International SEPT Programme, Leipzig University Beethovenstrasse 15,
04107 Leipzig, Germany
e-mail: dornberger@uni-leipzig.de

B. Sebitosi
Centre for Renewable and Sustainable Energy Studies, University of Stellenbosch
Matieland, Matieland 7600, South Africa
e-mail: sebitosi@sun.ac.za

S. Groh · J. van der Straeten
Research Group Microenergy Systems, Technische Universität Berlin,
Secr. HBS 1, Hardenbergstr. 16–18, 10623 Berlin, Germany
e-mail: groh@ztg.tu-berlin.de

J. van der Straeten
e-mail: straeten@ztg.tu-berlin.de

S. Groh et al. (eds.), *Decentralized Solutions for Developing Economies*,
Springer Proceedings in Energy, DOI 10.1007/978-3-319-15964-5_10

rice market illustrates the usability of the method by outlining some risks and opportunities the rice value chain context poses to energy based upgrading attempts of rice farmers.

Keywords Productive use of energy · Value chain theory · Systems theory · Microfinance · Development

Introduction

Be it ecosystem degradation, global warming, persistent inequality and poverty, or food insecurity—most of the challenges making up the poly-crisis of today's world are in essence system failures (Swilling and Annecke 2012). Solutions to these problems must be based on holistic thinking rather than separating the problem into small, isolated pieces and solving them stepwise (Batie 2008). Fueled by advanced communication technology, trade agreements and globalized cross-border capital markets, cross-border trade and production has linked the state and future prospects of evolving countries micro, small- and medium-sized enterprises (MSMEs) to the (global) market system they participate in Kula et al. (2006). Energy related research in agricultural markets needs to "change intellectually and operationally from a narrow focus on agriculture and technological research to a better understanding of rural societies and their needs. There is a need to seek greater understanding of alternative pathways for rural economic development, placing the role of agriculture (and energy) in perspective, and redefining the role, mission, and strategy of agricultural institutions as agents as facilitators for rural economic growth" (Anandajayasekeram and Gebremedhin 2009: 8). The application of value chain theory is one result of the paradigm shift that has occurred in agricultural research during the last decades. Value chain thinking anticipates these challenges by assessing MSMEs development potential from the viewpoint of the market system they are part of. Energy projects fostering the utilization of energy services by MSMEs, but develop interventions solely at the firm's level of energy need and use patterns, risk overlooking key external drivers' growth and competitiveness (Wolfe and Page 2008). Therefore, the development practitioners' perspective needs to go beyond the energy technology and farm system to ensure that the energy based intervention gains economic significance.

Research Objectives

The power of energy technology for MSME development is based on the multifacetted opportunities energy services pose for altering the way these firms interact with market systems. Unsurprisingly, Fakira (1994) states energy is a critical resource to liberate MSMEs from low value, low productivity and low income activities. Following Boardman and Kumani (2012: 152), "there is often a two-way relationship between the lack of access to adequate and affordable energy services and poverty. The relationship is, in many respects, a vicious cycle in which people who lack access to cleaner and affordable energy are often trapped in a re-enforcing cycle of deprivation, lower incomes and the means to improve their living conditions while at the same time using significant amounts of their very limited income on expensive and unhealthy norms of energy that provide poor and/or unsafe services". Hence, a general consensus exists amongst development practitioners in regard to the high relative potential energy technology has for MSME development. Groh (2014) argues that based on the existence of an energy poverty penalty, it is likely that households' and micro-businesses' development path is inhibited or at least delayed. According to Kirubi (2006), energy is a necessity, though not a sufficient means for MSME development. This implies that, even if energy is considered as a barrier to development, removing this barrier does not necessarily cause additional income and economic growth. To cause profound changes, complementary factors, such as infrastructure, access to capital, the availability of information, skills or social services must be integrated in the design of energy based development interventions (UNDP 2011). Despite those basic insights, literature that systematically assesses the importance of context for energy based MSME development projects remains scarce. Against this backdrop, this paper aims at answering the following research question: Which role can value chain theory play to support energy projects aiming at MSME development? The question is addressed by conceptualizing PU focused energy projects as value chain upgrading attempts. On this basis, a problem-centred value chain approach is proposed. The practical relevance of the approach is elucidated by assessing some of the risks and opportunities the rice value chain in the Philippines poses to energy based upgrading of rice farmers by means of solar based drying technology.

The Systemic Value Chain Approach

Agricultural goods usually pass through many hands as they move from farm to fork. The goods move along a value chain, defined as "the full range of activities and services required to bring a product or service from its conception to sale in its final markets" (Kula et al. 2006). The value chain approach centres on the "interrelatedness of those actors gradually adding value to product or service as they pass it from one link of the chain to the next" (UNIDO 2011: 1). These different actors undertaking value adding activities are linked by the flow of products, finance, information and services (KIT and IIRR 2010). Value chain assessments analyse key market actors, the relationships between them, and other factors influencing the performance of an industry. The assessment is centred on the chains metabolism:

Flows of information, finance, knowledge as well as the formal and informal relationships determining these flows. Limiting factors to increased efficiency, productivity and competitiveness are identified and strategies to overcome these barriers are developed (Fries 2007; Miller and Jones 2010). These strategies are based on value chain interventions—concerned activities that facilitate a systemic change of the value chain in regard to an intended goal, such as increased competitiveness of the chain or single firms.

The advantage of such an approach for development projects is that interventions can be tailored according to the context they are embedded in. Its disadvantage, however, is that researchers are endangered by losing sight of the bigger picture because one gets easily caught in particular value chain details (UNIDO 2011). A systematic and systemic analysis of the factors affecting the performance of the firms in a value chain is needed: Systematic in a sense that the process of data gathering must conducted according to an organized method guided by the assessments purpose, and systemic in a sense that that the gathered data must be analysed from structuralist viewpoint. As there is "no single instrument or a defined 'recipe' to follow" (Miller and Jones 2010), it's up to the researcher to decide on how to do that. The approach applied herein is based on the value chain framework propose by USAID.[1] It applies a "market system perspective to analyse microenterprises needs and opportunities to [...] prioritize programming options available" (Wolfe and Page 2008). According to the framework, "value chains have both structural and dynamic components. The structure of the value chain influences the dynamics of firm behaviour and these dynamics influence how well the value chain performs" (Kula et al. 2006). The systemic nature of this causal model becomes obvious when comparing Kula's statement with Sterman's (2000) explanation of the basic rationale of systems theory: "The behaviour of a system arises from its structure. The structure consists of feedback loops, stocks and flows, and nonlinearities created by interaction of the physical and institutional structure with the decision-making processes of the agents acting within it". A solid understanding of the systems structural patterns is necessary to understand how they cause behavioural patterns. This, in turn, is a prerequisite to indentify places to intervene in (market) systems, and to develop interventions in order to change the system's behaviour according to a given goal (Senge 1990; Meadows 2008). As such, the value chain framework is an application of this fundamental rationale and is therefore subsumed herein under the systems theory umbrella. Following this rationale, systems theory and value chain theory form a nested hierarchy. Just like a matryoshka doll, systems theory represents the highest level of analytical abstraction and the value chain approach an application of basic insights of systems theory (see Fig. 10.1). On a further subordinated level, PU projects can be subsumed as value chain upgrading attempts, which is discussed in the following section.

[1] www.microlinks.org/good-practice-center/value-chain-wiki.

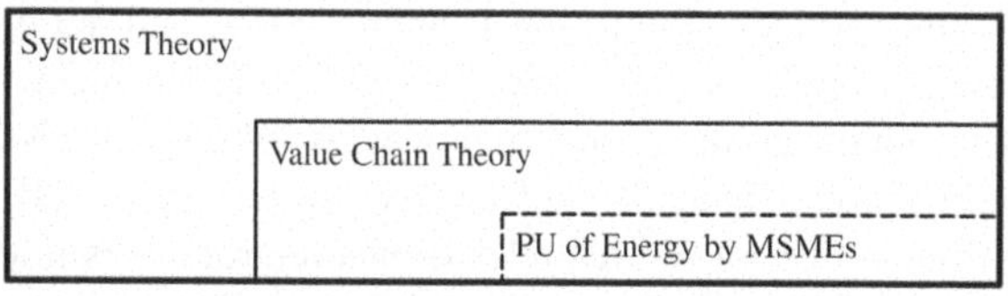

Fig. 10.1 Nested hierarchy of systems theory, value chain theory, and PU of energy

Value Chain Theory and PU of Energy

From a value chain perspective, a firm's decision to invest and thereby facilitate an intended economic change is a dynamic response to an opportunity to do so. This opportunity is posed by the interplay of a value chain's structural and dynamic elements. In value chain theory this process is understood as upgrading, defined as the process of implementing an innovation that increases the value a firm adds to a good or service (Pietrobelli and Rabelotti 2005). Following Schumpeter (1939), an innovation is an economic decision to adopt a particular intervention in order to cause an intended economic change. Value chain literature distinguishes between five upgrading types, namely process upgrading (improving production efficiency), product upgrading (improving a products quality), functional upgrading (doing things different, performing higher level stages of the chain), channel upgrading (tackling different end markets), and sectoral upgrading (applying skills gained in one value chain to participate in another) (e.g. Humphrey and Schmitz 2002; Miller and Jones 2010). The nature of a successful innovation process, in other words the application of a single or combination of these upgrading strategies, depends on the market structure an MSME is embedded in.

Productive use (PU) of energy is defined as the utilization of energy "either directly or indirectly for the production of income or value" (White 2002). The definition is based on a contemporary understanding of the term development as goes beyond the sole increase of financial income (see Sen 1999). However, for the sake of the discussion herein, PU of energy is understood as the utilization of an energy service in a way that the financial income of an MSME in increased. The term "energy service" is used herein to apply an end-users-perspective to energy projects. Energy itself does not make a difference in poor people's lives'; it is rather the service the energy provides such as cooling, heating, or communication (Allderdice et al. 2007). The potential of energy technologies to increase the income of MSMEs provides an opportunity to finance these technologies on a loan-basis. This provides an opportunity to break the poverty cycle many MSMEs are trapped in, despite the fact that the MSME might not be creditworthy from the viewpoint of conventional banking as they lack bankable collaterals. Designing energy based loan projects in a way that the potential income effect energy technology holds is maximized is a necessity to realize the "synergy potential of financial and energy inclusion" (Groh 2013). Following this logic, microfinance institutions (MFIs) are showing an increased interest in diversifying their portfolio by including energy related products and services (Kebir and Heipertz 2010) and applying innovative finance mechanisms to finance these offers (e.g. value chain finance). PU focused interventions cause an economic change in a way that the value a MSME adds to a good or service increases. Hence, PU of energy can be interpreted as value chain upgrading based on an energy service related intervention. This implies that every energy project focusing on PU of energy by MSMEs can be interpreted as one or a combination of the 5 upgrading types process-, product-, functionals-, channel- and sectoral upgrading. In theory, access to energy services can generate income in three major ways: First, a currently used energy source can be substituted by a more cost efficient alternative (substitution effect). Second, access to energy can offer new business opportunities and thereby promote the emergence of new firms (entrepreneurial affect). Third, access to energy services can offer the possibility to alter a firms production process in a way that the value added to a good or service is improved (development effect) (see Fig. 10.2).

It is this income effect that constitutes the link between energy projects and value chain theory. By means of four examples, this linkage is exemplified in Table 10.1. In case of process-, product-, functional-, and channel upgrading, the discussed logic applied. Taking the example of seafood value chains, an investment in an energy efficient refrigerator can improve the energy service cooling in a way that the relate expenses reduced. This example is a process upgrading approach, as the efficiency of the value adding activity cooling is optimized. This income effect can be a basis for a micro-loan based finance approach. The case of energy-based sectoral upgrading is different though. Following the example used in Table 10.1, a biomass gasification power plant that is embedded in an agricultural community offers new value adding activities to local farmers. If the power plant is based on energy crops, providing the opportunity to farm and sell these crops means

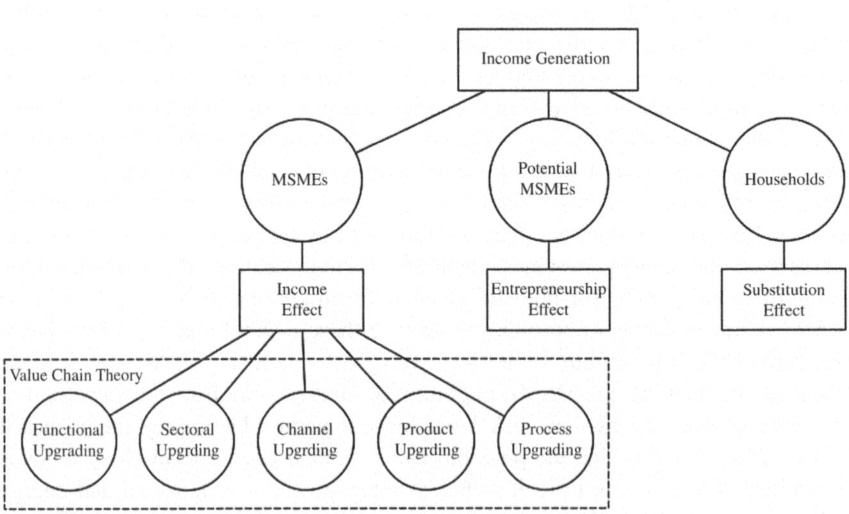

Fig. 10.2 Energy-based income effects and value chain upgrading

Table 10.1 Linking energy technology and value chain upgrading

Energy technology	Energy service	Improvement	Upgrading type
Energy efficient refrigerator	Cooling	Reduced cooling expenses by	Process upgrading
Solar tunnel dryer	Drying	Improved quality of agricultural goods	Product upgrading
Energy efficient electric vehicle	Mobility	Ability to bridge exploitive middlemen	Functional upgrading
Solar home system (SHS) to provide energy for internet access	Information	Ability to respond to changing market conditions	Channel upgrading
Biomass gasification power plant	Does not apply	Offering new value chain based on energy crop needs of biomass power plant	Sectoral upgrading

exposing these farmers to a new value chain. The major difference of this channel upgrading example to the other upgrading types is that the income effect is not based on an energy services implemented on the level of the MSME, but it is a the provision of a new value chain based on the biomass needs of the power plant. The second major difference is that the investment costs for projects' like this usually exceeds the financial capabilities of MFIs.

Proposal of a Problem-Centred Value Chain Approach

What are the implications of the previous discussion for energy based development projects? The basic rationale of the applied value chain approach is that an investment in energy technology must be seen as a dynamic response of a market actor in terms of an opportunity the value chain context poses. The context consists of different structural elements and dynamics—with each of these either supporting an investment opportunity, posing a risk to an investment opportunity, or being neutral. It is the sum of these influences that determines whether or not an opportunity to invest exists [for a generic discussion of these causal linkages see Dunn et al. (2006)]. Value chain assessments are undertaken ex-ante to the development of a context tailored intervention. However, despite the importance of market based information for the design of energy based MSME development projects, conventional value chain theory seems to be incommensurate with the reality of energy based development work. New projects are often started with a given, idea of where and how to intervene in a firm system, as well as an assumption in regard to the financial benefits of such an intervention. Undertaking an open-ended (in terms of the intervention) value chain assessment is resource intensive and therefore out of realm of most energy projects. Hence, the question is how to bridge both approaches—how to utilize "intervention-open" value chain thinking in order to support “intervention-closed” PU of energy project? As a first attempt to answer this question, this paper proposes to reverse the value chain rationale and undertake problem- centred value chain assessments. In this regard, problem centered means that all gathered information is related to an ex-ante defined upgrading strategy. The data collection process is limited to the influence (supportive, hindering, neutral) the value chains structural and dynamic elements have on a given energy based upgrading strategy. Thereby, market-related risks and opportunities to the given upgrading strategy are derived, which enables decision-makers to design a project in a way that the risk/benefit ratio of the project is optimized and the income generation effect maximized.

Case Study: Energy Based Upgrading of Rice Farmers in the Philippines

The problem-centred value chain approach is applied herein to the case of energy based upgrading of rice farmers in the Philippines. Ex-ante to the assessment, a financially promising intervention has been proposed, which is the alteration of the drying process of rice farmers by means of solar based drying technology. Hence, the energy service this upgrading strategy is based on is “drying” of rice. The focus on this intervention is justified on the basis of the projects baseline-scenario, which is the current palay-drying practice of agricultural smallholders: Farmers usually dry their palay on public roads, a practice with the consequence that parts of it is consumed by free-range livestock, grains are contaminated by livestock and cracked by vehicles, the drying process depends on climatic conditions, etc. The total assumed income generation effect of the intervention must be seen in relation to business-as-usual case. The assumed financial benefits of improving the farmers drying process are: (1) Increased market value of a higher quality produce, (2) prolonged quality preservation of dried product due to lesser contamination, and (3) more efficient drying in a high humidity environment. It is assumed that all of these alterations are increasing the farm-gate price of palay. The problem-centred value chain approach contextualizes these assumptions by relating them to the local rice market. It discards, enhances, or adjusts the assumed benefits and serves as a basis for a realistic calculation of the financial viability of investing in the upgrading strategy. The gathered information serves as a basis for deriving the project’s feasible design space—potential project settings that are in line with the projects goals function while taking into consideration the projects constraints and local market conditions. On this basis, the project can be designed in a way that the income generation effect for palay farmers is maximized, which, in turn, minimizes the lending risks MFIs.

Method

The data collection process was explorative and mainly based on qualitative interviews with value chain internal market actors and value chain external experts. The major problem of attempts to analyze the rice market in the Philippines is the unwillingness of certain market actors to talk with strangers about their business practices, a condition that defies an approach based on a large sample survey over a wide area with standardized questionnaires (Hayami et al. 1999). As discussed in the previous section, the data gathering process has been guided by challenge of designing a project in a way that the financial benefits of the drying based intervention are maximized. Structural and dynamic parameters potentially impacting on this goal have been detected during the research process—which is only possible in an explorative manner. Based on this rationale, 48 open-ended explorative interviews with key value chain actors and local rice market experts have been undertaken between August 13th and October 10th 2013 on the Philippines main island Luzon.

Analysis

Following the rationale of the problem-centred value chain approach, and given the limitations of this paper, most conventional features of value chain assessments (e.g. a map of the market, discussion of value chain actors) are excluded from this case study. Instead, a few selected aspects of importance are explained and their relation to the projects goal function discussed.

End market conditions: The value chain approach is driven by the principle of demand-driven supply. End-markets play a central role in the value chain study, as they determine demand characteristics terms of quality, quantity, timing, and pricing (see Kula et al. 2006). The rice value chain must be subdivided into the palay chain (un-milled rice), and the rice chain. Both are connected by the miller. The end-market of interest for farmers is defined by the palay chain, with its different market segments defined by the specific needs of palay buyers. According to the assumptions, quality increase caused by upgrading the drying process leads to income increase on the level of farmers. However, quality is a multidimensional concept, with its subjective assessment being related to a specific end and the resources to achieving it (Allaire 2012). Whether, and to which extent an income generation effect is caused depends on the end-markets notion of quality, which determines their willingness to pay for quality alterations. Three end-markets segments have been identified: (1) The governmental National Food Authority (NFA) applies a very differentiated price mechanism based on a matrix incorporating several quality related attributes, (2) Local buyers who apply a straight payment scheme mainly based on the type of rice and the moisture content (often reduced to "wet" and "dry", as well as the type of palay), and (3) buyers connected to institutional markets, who are in the need to supply large quantities of high quality rice. The crux of benefit maximization is to link farmers to those of these end-market segments that (financially) acknowledge the quality alteration, which is in this case segment 1 or 3. However, quality is not the only determinant of the rice price. Figure 10.3 illustrates the dynamics governing the price setting process, which is interpreted herein as a dynamic process based on the interplay of a set of cultural, technical, market-based and political variables (Fig. 10.4).

Enabling Environment: Policies, institutions, climatic conditions, and other attributes collectively creating the external business setting in which value adding activities take place are subsumed under the term enabling environment (Christy et al. 2009). Various variables of this structural element are of importance. One brief example is representing all not mentioned findings: During the last decades, the developmental policy of the Philippines has been characterized by free "dole-outs" of financial and technical means, a practice that caused several unintended consequences: (1) A "dole-out mentality" has been caused that makes farmers wait for governmental interventions rather than proactively causing change. (2) Governmental presence in development projects undermines the peoples willingness to

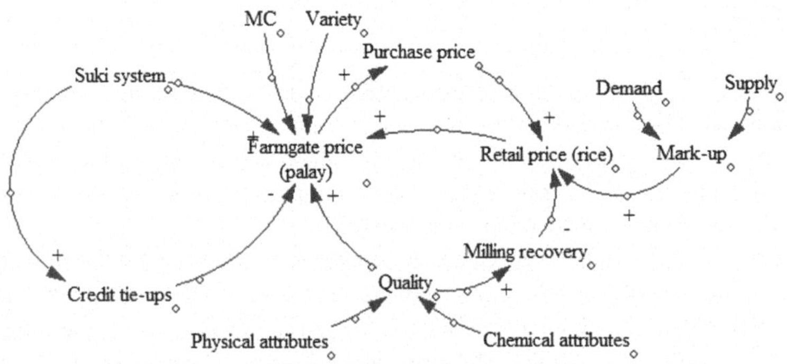

Fig. 10.3 Causal-loop diagram of the price setting mechanisms in the Philippine rice value chain

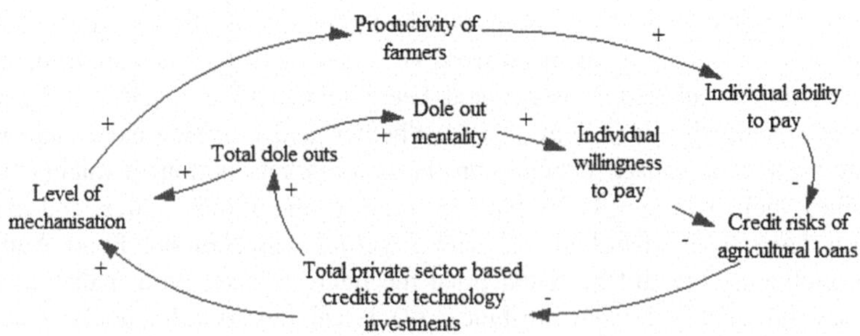

Fig. 10.4 Causal loop diagram illustrating the dynamics caused by governmental dole-out programs

pay (WTP) for micro-loans. (3) Governmental dole outs of technologies are wrecking market-based efforts to provide these technologies. The dynamic interplay of these aspects is depicted in Fig. 10.2. Governmental institutions are providing a whole range of supporting services for development projects like this. However, it must be assumed that cooperation could negatively impact on the WTP of farmers for loans. This could negatively impact on the repayment rates. For that reason, a potential cooperation with governmental institutions must be seen as very critical.

Vertical and horizontal linkages: According to the United Nations Industrial Development Organization (UNIDO 2011: 8), "networks and linkages are the building blocks of collective efficiency". Vertical linkages are the relationships between market actors at different value chain nodes. Horizontal linkages refer to the relationship between market actors undertaking the same activity in a value chain. Of critical importance herein are the linkages between palay farmers and segments of potential buyers. As mentioned earlier, the NFA is applying a price setting mechanism based on several quality determinants. By upgrading the drying process of palay farmers in the intended way, the farmer's ability to meet these determinants is improved. For this reason, farmers could theoretically realize higher palay prices by selling to the NFA. However, some aspects are questioning this approach: (1) The NFA is a governmental entity. For that reason, cooperation could cause a dole out mentality amongst farmers. (2) The complicated bureaucratic application process constitutes transaction costs. These must be seen in relation to the paid premium. (3) Research indicates that the NFA does not necessarily pay farmers immediately. More than the fact that farmers need cash immediately after harvest, the uncertainty whether or not the NFA is capable to pay entails high planning risks for the project. In sum: Although the NFA appears to be a market segment worth tackling, the assessment of the farmer-NFA linkage revealed some of the obstacles such an attempt is prone to.

Supporting Services: Supporting services can be subdivided into either formal or informal (1) financial services (e.g. lending), cross cutting services (e.g. legal advice), and (3) sector-specific services (e.g. the availability of a certain technology) (Campbell 2008). Agricultural smallholders are usually excluded from the formal financial system. However, agriculture is an investment-intensive activity, with returns only realized at the end of the cropping season (MCPI 2010). Palay buyers are usually stepping into fill this finance gap. By providing farmers with the financial means or inputs necessary to initiate the next harvest, trader-credits are crucial for the functioning of the rice market and for ensuring food supply. However, the public perception of these credit tie-ups is rather negative; as they are often characterized by excessive interest rates (e.g. one interviewed trader charged 28 % interest per month). Farmers usually pay back in kind after harvest by accepting prices dictated by the traders. For that reason, many farmers are depended on new loans and trapped in a state of constant indebtedness. This aspect must be from the viewpoint of *utang na loob*, which is the Philippine concept of moral indebtedness. Once indebted, farmers are obliged to show gratitude even when the the financial debt is settled. From the viewpoint of designing a project that upgrades the drying process, the discussed issues have two major implications: (1) Local traders usually apply a straight buying scheme that reduces the quality determinants to class A (bad quality), class b (average quality), class C (good quality), as well as the moisture content (either "wet" or "dry"). Upgrading the drying has a positive impact on various other quality determinants (e.g. whiteness, aroma, foreign matter, milling recovery, etc.). For maximizing the financial benefit of the upgrading attempt, farmers have to be linked to an end-market that is willing to financially acknowledge these alterations, which most probably isn't the local trader. However, *utang na loob* might force farmers to stick to their traders, even though their financial debts are settled and another market segment offers higher prices. (2) The risk management rationale of the project is based on the income generation effect of improving the energy service "drying"—Maximizing the income generation effect means minimizing loan failures. But what if farmers are indebted by traders or even by some other suspicious external parties like 5, 6 Bombay lenders[2]? It can be assumed that the party applying highest social or even physical pressure is paid first—which won't be the MFI providing energy loans.

Discussion and Concluding Remarks

The conceptualization of PU of energy projects with the aim of MSME promotion as upgrading attempts provides a new viewpoint on PU projects. By connecting these previously unconnected dots, a new, intrinsically trans-disciplinary discursive way to think about energy projects is offered. From the viewpoint of the systemic value chain approach, it is the value chain system that poses an opportunity to invest in energy technology. Energy projects neglecting the importance of the market context will a priori limit their space of interventions to those located within the borders of the individual firm. Such a self-imposed restriction does not meet the opportunities offered by the multifaceted applicability of energy services, as the state and future development paths of MSMEs is often determined by forces located outside the firms' borders, and access to energy services offer the opportunity to tackled intervention points beyond fuel substitution on a firm level. Furthermore, especially if an intervention is developed without proper knowledge of the local market conditions, the external market context poses risks to a given upgrading strategy that can't be overseen and managed in advance. It might seem to be worthwhile to increase the productivity of an agricultural smallholder, but how does that make sense if the end-market is already saturated? How does it make sense to improve the productivity of Philippine banana farmers producing for the European market, if the European standard for bananas will be changed soon, challenging the farmer to alter the quality of their produce, not the quantity? Whereas traditional micro-lending neglected questions like this, there is a growing recognition amongst MFIs that a broader,

[2] Their name is a play on their lending scheme and origin: For every 5 pesos, 6 have to be paid back after one month.

market-based approach to the design and utilization of their services is needed. One of the leading Philippine MFIs in this regard is the CARD Bank. Based on their past experiences in MSME lending, the institution came to the conclusion that "most of its clients, after attaining commercial-level status through its financial services, are now in need of essential non-financial services to fully develop their businesses. This range of non-financial services, known as business development services (BDS), represent the entire spectrum of services a business requires to attain sustainability when analyzed within the context of value-chain analysis. Within this context, assessing what specific types of BDS enterprises require is the first vital step in addressing the goal of helping enterprises fully realize their business potentials" (Alip et al. 2009). Energy services offer a huge development potential, but only if their implementation and application is contextualized. The problem-centred value chain approach is a first attempt to utilize the power of value chain thinking for overseeing and managing the risks and opportunities a particular context poses to a pre-defined energy based intervention. By means of a case study on upgrading of Philippine rice farmers, the approach has been tested. The discussed aspects are only a fraction of the insights gathered during the research process. However, they are sufficient to illustrate how external, value chain related factors are limiting the feasible design space of energy projects. Neglecting the local market context, and assuming an energy project can be designed in a way that the risk/benefit ratio is optimized, leaves interventions with the risk of causing adverse effects which can otherwise not only be mitigated but development effects be strengthened.

Acknowledgments The case study was partly conducted in the framework of the Energy Inclusion Initiative, an initiative by ADA and MicroEnergy International. We thank both institutions for their support.

References

Alip, J., Guarin, R., & Magampon, M. (2009). *Business development service. The case of card MRI*. Retrieved from http://cardbankph.com/.

Allaire, G. (2012). The multidimensional definition of quality, In A, Louis., H. Ilbert., & N. Saavedra-Rivano (Eds.), *Geographical indications and international agricultural trade: The challenge for Asia*. New York: Palgrave Macmillan.

Allderdice, A., Wienicki, J., & Morris, E. (2007). *Using microfinance to expand access to energy services. Summary of findings*. Washington DC: The SEEP Network.

Anandajayasekeram, P., & Gebremedhin, E. (2009). *Integrating innovation systems perspective and value chain analysis in agricultural research for development: Implications and challenges*. Nairobi: International Livestock Research Institute.

Batie, S. (2008). Wicked problems and applied economics. *American Journal of Agricultural Economics, 90*(5), 1176–2119.

Campbell, R. (2008). *Value chain famework briefing paper*. Washington, DC: USAID.

Christy, R., Mabaya, E., Wilson, N., Mutambatsere, E., & Mhlanga, N. (2009). Enabling environments for competitive agro-industries. In C. Da Silva, D. Baker, A. W. Shepherd, C. Jenane, & S. Miranda da Cruz (Eds.), *Agro-industries for development* (pp. 136–185). Wallingford, UK: CABI, with FAO and UNIDO.

Dunn, E., Sebstad, J., & Batzdorf, L. (2006). *Lessons learned on MSE upgrading in value chains, AMAP BDS knowledge and practice microREPORT #71, USAID/G/EGAT/MD*. Washington, DC: USAID.

Fakira, H. (1994). *Energy for micro-enterprises. Energy policy research and training project.* Energy and Development Research Centre: University of Cape Town. (Paper No. 12: 16–25).

Fries, B. (2007). *The value chain framework, rural finance, and lessons for TA providers and donors*. Presentation at the Asia International Conference Agri Revolution: Financing the Agricultural Value Chain, Mumbai, India.

Groh, S. (2013). *The synergy potential of financial and energy inclusion*. Center for Financial Inclusion blog post. August 28, 2013. Website: http://cfi-blog.org/2013/08/28/the-synergy-potential-of-financial-and-energyinclusion/#more-12227.

Groh, S. (2014). The role of energy in development processes—The energy poverty penalty: Case study of Arequipa (Peru). *Energy for Sustainable Development*, *18*, 83–99 (2014).

Hayami, Y., Kikuchi, M., & Marciano, E. (1999). Middlemen and peasants in rice marketing in the Philippines. *Agricultural Economics, 20*(2), 79–93.

Humphrey, J., & Schmitz, H. (2002). How does insertion in global value chains affect upgrading industrial clusters? *Regional Studies*, *36*(9), 1017–1027.

Kebir, N., & Heipertz, J. (2010). *Financing energy. The role of microfinance institutions*. UMM Workshop report. E-MFP Action Group. Bergamo: Planet Finance.

Kirubi, C. (2006). *How important is modern energy for micro-enterprises? Evidence from rural Kenya*. Berkerly: University of California.

KIT & IIRR. (2010). *Value chain finance: Beyond microfinance for rural entrepreneurs*. Nairobi: Royal Tropical Institute Amsterdam and International Institute of Rural Reconstruction.

Kula, O., Downing, J., & Field, M. (2006). *Globalization and the small firm. A value chain approach to economic growth and poverty reduction* (AMAP BDS Knowledge and Practice microREPORT No. 42). Retrieved from Microenterprise Learning Information & Knowledge Sharing website: http://www.microlinks.org/sites/microlinks/files/resource/files/ML5023_mr_42_strategy_paper.pdf.

Meadows, D. (2008). *Thinking in systems. A primer*. Chelsea Green Publishing: Vermont.

Microfinance Council of the Philippines (MCPI). (2010). AGRI—MF. *Agriculture microfinance. Guidebook*. Pasig City: MCPI.

Miller, C., & Jones, L. (2010). *Agricultural value chain finance. Tools and lessons*. Rome: FAO.

Pietrobelli, G., & Rabelotti, R. (2005). Upgrading in global value chains. Lessons from Latin American clusters. *World Development, 33*(4), 549–573.

Schumpeter, J. (1939). *Business cycles. A theoretical, historical, and statistical analysis of the capitalist process*. New York: McGraw-Hill.

Sen, A. (1999). *Development as freedom*. Oxford: Oxford University Press.

Senge, P. (1990). *The fifth discipline. The art & practice of the learning organization*. New York: Currency Doubleday.

Sterman, J. (2000). *Business dynamics. Systems thinking and modelling for a complex world*. Boston: Irwin/McGraw-Hill.

Swilling, M., & Annecke, E. (2012). *Just transition. Exploration of sustainability in an unfair world*. Cape Town: Juta.

United Nations Development Program (UNDP). (2011). *Towards an energy plus approach for the poor*. New York: UNDP.

United Nations Industrial Development Organization (UNIDO). (2011). *Pro-poor value chain development. 25 guiding questions for designing and implementing agroindustry projects*. Vienna: UNIDO.

White, D. (2002). *Productive uses of renewable energy. Experience, strategies, and project development*. Workshop synthesis and report. Rome: FAO.

Wolfe, J. & Page, C. (2008). *Nepal inclusive economic growth assessment*. Microenterprise Development. Washington, DC.: USAID.

Chapter 11
Advanced Solar-Irrigation Scheduling for Sustainable Rural Development: A Case of India

Hari Dilip Kumar, N. Tejas Kumar, K.R. Suresh, H. Mitavachan and Ganesh Shankar

Abstract In the past few years, there have been large migrations from the rural to the urban areas of India as the country has developed. This has led to shortage of agricultural manpower in rural India, a region which is essential for the country's food security. Also, groundwater depletion in rural areas is affecting agriculture. Automating some agricultural tasks in a sustainable, scientific way could address this, leading to decreased water and energy use and increased crop yields. We develop a research platform to implement proof-of-concept of a system for precision agriculture incorporating renewable energy sources and information technology. Data collected by this system will be used to design a low-cost, commercial version of this technology for the rural Indian farmer.

Keywords Precision agriculture · Renewable energy · Solar irrigation · Indian farmer · Information technology · Water management

H.D. Kumar (✉) · N. Tejas Kumar · H. Mitavachan · G. Shankar
FluxGen Engineering Technologies Pvt. Ltd, Bangalore, India
e-mail: hari.coding@gmail.com

N. Tejas Kumar
e-mail: tejas@fluxgentech.com

H. Mitavachan
e-mail: mitavachan@gmail.com

G. Shankar
e-mail: ganesh@fluxgentech.com

H. Mitavachan
Postgraduate Program Renewable Energy, Oldenburg, Germany

K.R. Suresh
BMS College of Engineering, Bangalore, India
e-mail: suri.civ@bmsce.ac.in

S. Groh et al. (eds.), *Decentralized Solutions for Developing Economies*,
Springer Proceedings in Energy, DOI 10.1007/978-3-319-15964-5_11

Introduction

India is basically an agrarian society wherein around 60 % of land area is cultivated for agriculture and about 50 % of the population directly or indirectly depends on agriculture. Agriculture in India is mainly dependent on the monsoon. Due to crop failures from monsoon variability, water scarcity, increasing cost of cultivation and many other unsustainable agricultural and trade related practices, the number of farmer suicides in rural areas has assumed alarming proportions. An estimated 250,000 Indian farmers have committed suicide since the mid-1990s (Mishra 2008). Moreover, recent climate change studies indicate that fluctuations in temperature and precipitation on the Indian sub-continent (especially in the monsoon) are very likely to increase in the near future (Turner and Annamalai 2012). This will lead to increasing occurrences of droughts and floods, which in turn will increase conflict over scarce resources, especially land and water, impacting Indian agriculture negatively as a direct consequence (Jamir et al. 2013; Pullenkav 2013).

The Kharif crop is mainly rain fed (monsoon), whereas the Rabi (winter) and summer crops require irrigation. The increasing unreliability of the monsoon and the scarcity of electricity in rural areas are forcing Indian farmers to depend on diesel fuel operated tube wells for irrigation. The high operational cost of diesel based irrigation pumps is forcing farmers to practice deficit irrigation, and as a result crop yields are much lower than their potential (Pullenkav 2013).

Given that India has more than 300 sunny days per year in most regions, the alternative option of solar water pumping has been seen as an obviously promising one. Although the growth of the solar water pumping market has been insignificant so far due to various regulatory, market and technological barriers (Pullenkav 2013), India's recent national solar mission (JNNSM) is expected to bring a significant change in this direction. It is estimated that around 70 million solar PV pumps can be installed in India by 2020 (Purohit and Michaelowa 2008).

As of 2010, there were around 9 million diesel irrigation pump sets installed in India. If solar photo-voltaic (PV) pumps replace at least half of these, then about 225 billion liters of diesel can be saved per year. The cost of solar PV water pumping without any subsidy comes to around 64.2 % of the cost of a diesel pump over ten years, and has a payback time of around 4 years (from diesel savings). The levelized cost of solar PV pumping works out to be Rs. 8.6 per kWh compared to Rs. 13.9 per kWh for diesel pumping for a lifetime of 25 years and without taking into account any subsidies for solar PV (Raghavan et al. 2010). Hence, it is becoming obvious, especially with the increasing diesel fuel prices and the significantly falling PV prices, that solar PV water pumping for irrigation is a promising and cost-effective solution for India.

However, just going solar for agricultural irrigation is not enough. Rodell et al. (2009) in their study based on satellite observations provide evidence that unsustainable consumption of ground water for irrigation and other anthropogenic uses is leading to significant ground water depletion in various Indian states. The study further highlights that, as a consequence of this, these states might experience reduction in agricultural output, shortages of potable water and serious socioeconomic stresses, if proper measures are not taken very soon. Hence, irrigation techniques and water harvesting methods have to be very efficient.

All these facts make the following evident. First, Indian agricultural fields inevitably need better irrigation; second, there is an urgent necessity to take proper safeguards so that the adopted irrigation techniques are climate friendly and really sustainable ones. In this paper, the authors try to address this dual problem by developing a novel technological method of integrating solar PV pumping technology with information technology (IT) and environmental parameter sensing.

Research Objectives

This research project[1] aims to develop and deploy advanced sustainable technology for efficient irrigation of agricultural land in rural India. The technology developed will be based on sound principles from scientific/precision agriculture and incorporates renewable energy sources as far as possible. This is to make the technology sustainable in places where grid electricity access is limited or absent. In this paper we document the development of a proof-of-concept system based on these ideas.

Methods

The aim of our studies is to design and test a system that uses computer-controlled solar water pumping and closed-loop irrigation scheduling of a target agricultural field containing a given crop. "Closed-loop" irrigation scheduling means that the irrigation to the field is controlled on the basis of the soil moisture content in the field, apart from other environmental parameters.

The primary research methodology is to quantify the amount of electricity and water consumed, and the yield obtained after the crop life-cycle, using this method of irrigation scheduling. The next set of experiments will compare these parameters with the existing system of irrigation prevalent in rural India (flood irrigation), which does not use any form of feedback control for scheduling of irrigation in agricultural fields. The intent of these experiments is to come up with good engineering design decisions when developing and commercializing this technology at the lowest possible unit cost for the rural Indian farmer.

We now present background material that will allow an understanding of the technology we have developed.

[1] This work is a joint effort between BMS College of Engineering, Bangalore and FluxGen Engineering Technologies pvt. ltd., Bangalore.

Soil Moisture Sensing

The moisture content in a sample of soil is the ratio of the volume of water in the sample to the total volume of the soil sample (Black 1965) and is generally abbreviated "VWC" (Volumetric Water Content). Several techniques have been devised to allow automated and non-destructive measurement of soil moisture, ranging from neutron scattering to soil dielectric constant based instrumentation (Walker et al. 2004). In our research, we use capacitive soil moisture sensors of the "soil dielectric constant" type. These sensors return a voltage from which the VWC of the soil surrounding the sensor probe may be inferred.

Soil Parameters

We now explain the main parameters related to soil moisture (Doorenbos and Pruitt 1977) that are used by the irrigation scheduling algorithm. The Field Capacity (FC) of a soil is the moisture held in the soil after excess water has drained away and the rate of downward movement has decreased. This usually takes places 2–3 days after rain or irrigation in pervious soils of uniform structure and texture. The Permanent Wilting Point (PWP) for a given crop is the minimal point of soil moisture that the plant requires not to wilt. If the moisture falls below the PWP, the plant will be damaged. The Readily Available Moisture (RAM) for a particular crop in a soil is the range of available water that can be stored in the soil and is available for crop growth. The RAM is computed as the difference between the FC and the PWP. The Maximum Allowable Depletion (MAD) is the portion of the RAM that is allowed for crop use prior to irrigation. It is expressed as a percentage of the RAM, usually between 50 and 70 % based on the crop type.

The Irrigation Scheduling Algorithm

The algorithm works as follows. The system contains a database of soil parameters across twelve different types of soil (e.g. clay, loam) and 13 crop types (e.g. maize, sugarcane). The parameters maintained in this database are the FC, the PWP, the RAM and the MAD, as explained before. The lower threshold for activation of irrigation scheduling is computed using the RAM and the MAD. When the soil moisture level falls below the lower threshold, the irrigation valve is opened. Once the soil moisture increases to cross the FC, the irrigation valve is closed. This is a form of closed-loop control with hysteresis. We emphasize that this is the basic algorithm; one of the aims of our research is to come up with more efficient variations on this theme.

Weather Parameters

A precision weather station relays real-time environmental parameters including ambient temperature, daily rainfall, solar irradiation, wind speed and wind velocity to the software, which logs all these parameters continuously. There is a two-fold utility in logging these parameters. Firstly, it may allow a "weather forecast" to be computed. If a computed weather forecast indicates that it will rain within some period, making irrigation superfluous, water savings may be increased by appropriately pre-empting the next irrigation.

Logging weather data also allows correlation studies to be performed between the weather conditions and the irrigation schedule. In the final version of this product, some components (e.g. the weather station) may be removed in order to deliver it to the end customer (an Indian farmer) at the lowest possible unit cost. The research system with full weather instrumentation allows us to infer a logical rule-set which may be implemented when localizing the product (without the weather station) to a particular region with a known climate through the year.

Renewable Energy

The pumping of water into an overhead storage tank is performed under computer control. The pump is a commonly available "solar pump" of efficient design that is powered by (in this case) a 50 W-peak solar PV module. In the field version of our system, all hardware, including the weather station and the embedded controller, run from renewable energy sources. We note that the technology we present is a naturally compatible "application" for a rural renewable energy microgrid.

The Role of Information Technology

IT is central to our approach to problem-solving in this domain. The system we have developed acquires and logs several parameters of interest, including the pumping and irrigation schedules, weather parameters and energy usage. This data allows for extensive analysis of the performance of this type of system under various environmental conditions. Having these logs allows us to reuse the data for other academic and industry projects related to renewable energy, rural electrification systems and water management.

Since our system is built using versatile tools from National Instruments (http://www.ni.com), we are able to easily develop, test and deploy improved irrigation scheduling algorithms. Specifically, we utilize the LabVIEW platform and RIO-based reconfigurable hardware from National Instruments in this system. For further technical details of our system, please see Hari et al. (2013).

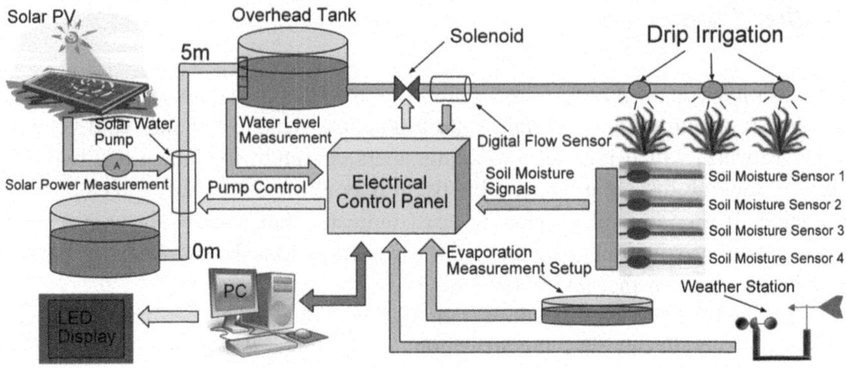

Fig. 11.1 Block diagram of the proof-of-concept system

Results

As a concrete first step towards this technology, we have developed a proof-of-concept system[2] for research and development (R&D) in this area incorporating all the techniques outlined previously. As shown in Fig. 11.1, irrigation is provided by drip irrigation sprinklers placed in a crop field. The irrigation is controlled (switched on and off) using the solenoid, which is under digital control of the computer (PC) running custom software. The computer also controls the solar water pump, which is powered by solar PV panels or a renewable energy microgrid. The power delivered from the solar panels for pumping over the course of the day is measured and logged.

The soil moisture sensors return a voltage related to the moisture content in the surrounding soil. The precision weather station provides real-time data on the weather parameters as mentioned previously. Since this is a research system, other parameters useful for problem-solving in this area are measured—including the water level in the overhead tank to which water is pumped, and daily evaporation (measured in a special setup.) The latter allows for fine-tuned corrections to the irrigation scheduling algorithm based on moisture loss due to evaporation. The electrical control panel contains all the field wiring housed in a weatherproof enclosure, and transmits signals over a bus to a computer in the lab.

The system continuously acquires and logs all parameters. It displays them in graphical format on the Display (see Fig. 11.2). The most important feature implemented is the closed-loop irrigation scheduling, as explained previously.

A major use of the system we have developed is to research and test innovative irrigation scheduling algorithms in-field. For example, would it make more sense to irrigate only at night, when the water loss due to evaporation is less? What soil

[2] Patent pending.

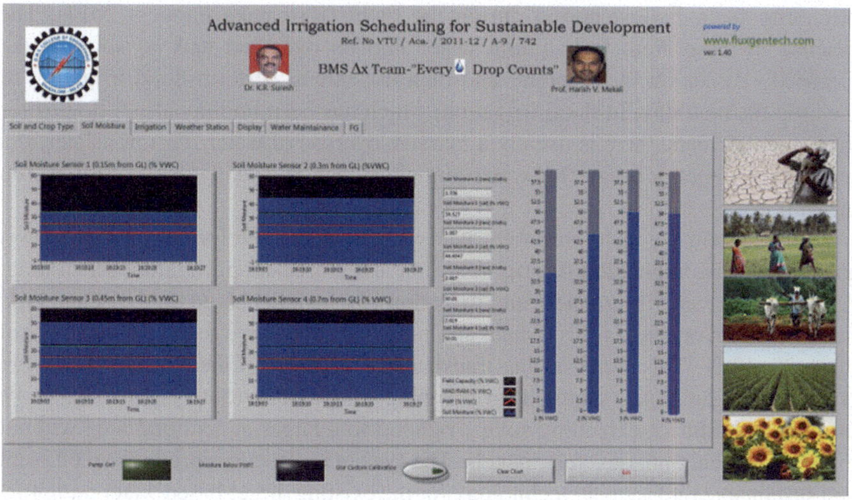

Fig. 11.2 Display screen, showing soil moisture levels

moisture sensor placement is required for a robust reading? Similarly, questions may also be posed and answered regarding the optimum solar water pumping schedule, efficiency of renewable sources, etc.

The use of software to perform the irrigation scheduling allows us to test the hypothesis that we can use the same basic algorithm across any crop or soil type, allowing for an economy of effort in developing and commercializing this technology for rural India. Our technology currently supports 13 crops (e.g. maize, sugarcane) across 12 soil types (e.g. loam, clay).

The system has been running autonomously since July 2013 at a field site at BMS College of Engineering, Bangalore. The crop being grown is maize in a clayey soil. We currently use four soil moisture sensors located one below the other at sensing depths of 15, 30, 45 and 70 cm under the ground. The irrigation scheduling algorithm currently being tested contains a simple linear root water uptake model (Prasad 1988) so that, in effect, the soil moisture level is determined across the current depth of the plant's root system.

We have also developed a completely portable version of the research system based around the RIO embedded technology from National Instruments (Hari et al. 2013). This version contains all the functionality of the proof-of-concept system without need of a PC.

Discussion

The technology described in this paper will find application in areas where there is land suitable for agriculture, but inadequate supply of water and/or electricity. In grid-connected rural areas of Karnataka, the government currently provides free electricity for a limited period during the night (Karnataka Farmers 2014). The practise in these areas is to pump water into an overhead tank as soon as the electricity comes on in the night, and then use this to perform flood irrigation of the crop the next day. In flood irrigation, which has been traditionally followed for thousands of years, there is usually a massive overuse of water. This translates to a much higher energy usage (for water pumping) than is actually necessary for crop growth.

In the system we have proposed, water saving (and therefore energy saving) is achieved by a combination of drip irrigation and closed-loop irrigation scheduling. Every drop of water is counted, and irrigation is scheduled based on the actual crop demand. From our field trials, which are in progress, we have seen water savings of up to 30–40 %. This water saving translates to lowering of the size, and therefore cost, of the solar panels required for pumping water. Hence our system can also be used as a retrofit for extending solar irrigation to a larger agricultural area without extra investment on larger solar panels. The system is also appropriate as an efficient and sustainable replacement technology for the existing diesel-based water pumping.

A major factor suggesting adoption of this technology is the dropping groundwater levels in India, as mentioned previously. Dropping groundwater levels in rural areas makes agriculture unsustainable and forces migration of people from the rural to urban areas, resulting in loss of skilled manpower. Hence adoption of this system may increase food security in India even though labour costs are increasing and the size of the agricultural workforce is decreasing. If this technology is successful and leads, for example, to increased yields and therefore increased incomes for Indian farmers, their livelihood may be improved.

As groundwater levels are dropping even in urban areas, the technology may find application, after suitable modification, in urban spaces like apartments, parks and botanical gardens. The final product we are developing can be used without specialist knowledge of agriculture. It may therefore become a useful tool for those entering the farming profession from a non-agricultural background.

A challenge in moving this technology out of the lab and into rural India is making it cost-effective for the end user. The most expensive components in the field version will likely be the solar panels, batteries if present and the various sensors. Indigenization of some components for the Indian scenario may be required to make this technology affordable to the Indian farmer.

Our future pathway for developing this technology is to continue research at BMS College of Engineering and field sites in rural Karnataka. There are many steps to achieving the full potential of this technology, including solving technical problems, performing rigorous field trials, making it economically viable, educating farmers on scientific agriculture, exploring financing options and building the supply chain, market and distributor network for the final product. We hope to involve many students and faculty across disciplines in this effort, which will give them first-hand experience of the challenging real-world problems that must be solved in order to bring this technology to life.

Acknowledgments We sincerely thank ADA for sponsoring the visit of Dr. Suresh K.R. at the MES-BREG 2014 Symposium held at UC Berkeley in April 2014.

References

Black, C. A. (1965). *Methods of soil analysis: part I—Physical and mineralogical properties*. Madison, USA: American Society of Agronomy.

Doorenbos, J., & Pruitt, W. O. (1977). Guidelines for predicting crop water requirements. *FAO Irrigation and Drainage Paper, 24*, 15–29.

Hari, D. K., Shankar, G., Kumar, T., & Voleti, L. (2013). *Performing advanced irrigation scheduling for sustainable agriculture using LabVIEW and NI single-board RIO*. Retrieved from National Instruments Web Site: http://sine.ni.com/cs/app/doc/p/id/cs-15713.

Jamir, C., Sharma, N., Sengupta, A., & Ravindranath, N. (2013). Farmers' vulnerability to climate variability in Dimapur district of Nagaland, India. *Regional Environmental Change, 13*, 153–164.

Karnataka farmers to get 7 hours three-phase power supply. (2014). *The Hindu*, Retrieved on Jan 29, 2014, from http://www.thehindu.com.

Mishra, S. (2008). Risks, farmers' suicides and agrarian crisis in India: Is there a way out? *Indian Journal of Agricultural Economics, 63*(1), 38–54.

Prasad, R. (1988). A linear root water uptake model. *Journal of Hydrology, 99*(3), 297–306.

Pullenkav, T. (2013). Solar water pumping for irrigation: Opportunities in Bihar, India. GIZ (Indo-German Energy Program—IGEN). Retrieved from http://www.igen-re.in/files/giz__2013__report_solar_water_pumping_for_irrigation_in_bihar.pdf.

Purohit, P., & Michaelowa, A. (2008). CDM potential of SPV pumps in India. *Renewable and Sustainable Energy Reviews, 12*, 181–199.

Raghavan, S., Bharadwaj, A., Thatte, A. A., Harish, S., Iychettira, K. K., & Perumal, R. (2010). Harnessing solar energy: Options for India. Center for Study of Science, Technology and Policy: Bangalore.

Rodell, M., Velicogna, I., & Famiglietti, J. S. (2009). Satellite-based estimates of groundwater depletion in India. *Nature, 460*, 999–1002.

Turner, A. G., & Annamalai, H. (2012). Climate change and the South Asian summer monsoon. *Nature Climate Change, 2*, 587–595.

Walker, J. P., Willgoose, G. R., & Kalma, J. D. (2004). In situ measurement of soil moisture: a comparison of techniques. *Journal of Hydrology, 293*(1), 85–99.

Chapter 12
Towards a Waste Management System for Solar Home Systems in Bangladesh

Alexander Batteiger

Abstract One of the most recognized rural electrification programs in the world is the SHS-Program in Bangladesh. Based on the steadily growing high installation rates of SHS future waste generation was estimated on a country level. In 2012 at least 200,000 lead-acid batteries (LABs) from the SHS-Program were disposed. Lifespan of LABs is the crucial factor for the estimation of future waste generation. Depending on the average lifespan, between 800,000 and 1.2 Million LABs equaling an amount of 6,000 to almost 10,000 metric tons of lead per year is estimated to enter the waste management system of Bangladesh in 2016. Further research on regional levels and mass flows should be conducted to prepare the waste management system of Bangladesh.

Keywords Solar home system · Lead-acid batteries · Waste management system · Waste generation estimation · Bangladesh

Introduction

The main objective of the United Nations sustainable energy for all initiative is to ensure universal access to modern energy services until 2030 (Ki-moon 2011). Off-grid areas of developing countries can be electrified by using decentralized energy systems based on renewable energies (OECD/IEA 2011). Especially Solar Home Systems (SHS)[1] have already proven to be a possible solution for rural

[1] A SHS consists in general of (a) Solar modul (b) Lead-acid-battery (c) Charge controller, (d) light bulbs or LED (e) Mounting structure, (f) Installation kits and (g) Cables and connecting devices (UNFCCC 2013).

A. Batteiger (✉)
Postgraduate Program Microenergy Systems, Center for Technology and Society,
Technische Universität Berlin, Sekr. HBS 1 Hardenbergstr. 16–18,
D-10623 Berlin, Germany
e-mail: batteiger@ztg.tu-berlin.de

S. Groh et al. (eds.), *Decentralized Solutions for Developing Economies*,
Springer Proceedings in Energy, DOI 10.1007/978-3-319-15964-5_12

off-grid electrification. Bangladesh is worldwide known for the successful implementation of SHS.[2] The yearly installation rate of SHS grew from about 100,000 in 2008 to more than 600,000 in 2012 (WB 2013a). By the end of 2013 more than 2.7 Million SHS were installed and SHS are close to become economical viable without subsidies. To fulfill the goal of 6 Million SHS by the end of 2016 (IDCOL 2014), yearly installation rates have to exceed one million SHS.

In the first round of the "Rural Electrification and Renewable Energy Development Project" no environmental impacts other than CO_2-Mitigation were assessed. Nevertheless, in the final report was stated that the project helped to increase standards in battery recycling (WB 2013a). In 2012 the second round of the project was launched, since than regular assessments of environmental impacts have to be reported (WB 2013b). In the first environmental and social impact report of IDCOL, two major environmental impacts related to the SHS-Project were reported: Improper disposal of lead-acid batteries (LABs) and solar panels (IDCOL 2013).

In 2010 the Blacksmith Institute (2010) in cooperation with the Green Cross published a report about the six worst pollution problems of the world, with lead being one of it. About three quarter of all lead is used for batteries (Roberts 2003). In low- and middle-income countries car battery recycling is one of the main sources of lead pollution (BSI 2010). According to a recent research photovoltaic energy systems are expected to become a significant new source of lead pollution in China and India (Gottesfeld and Cherry 2011). Lead losses over a lifecycle of a battery in countries with advanced infrastructure are around 5 %; and up to 30 % in developing countries. In the informal sector losses only in the recycling phase can be as high as 50 % (Hoffmann and Wilson 2000).

So far assessments of SHS have mainly focused on social impacts and CO_2-Mitigation. The goal of this paper is to identify future waste-flows on a country level to integrate them into the existing waste management system.

Research Objectives

Since high installation rates of SHS are a comparatively new development, there is little empirical data dealing with end of life of solar home systems. The main focus of most of the studies on environmental impacts is greenhouse gas and indoor emissions. The market for SHS is steadily growing and lower prices for solar panels will make SHS economical viable without subsidies soon. Figure 12.1 illustrates the expected growth of installation of SHS in Bangladesh.

[2] The Solar Home Systems (SHS) program of Bangladesh is supported by the World Bank (WB), Asian Development Bank (ADB), Islamic Development Bank, Japan International Co-operation Agency (JICA), GEF, GIZ, KfW, GPOBA, USAID and DFID. SHS are being installed under the ongoing Renewable Energy Program of Infrastructure Development Company Limited (ICDOL) in Bangladesh.

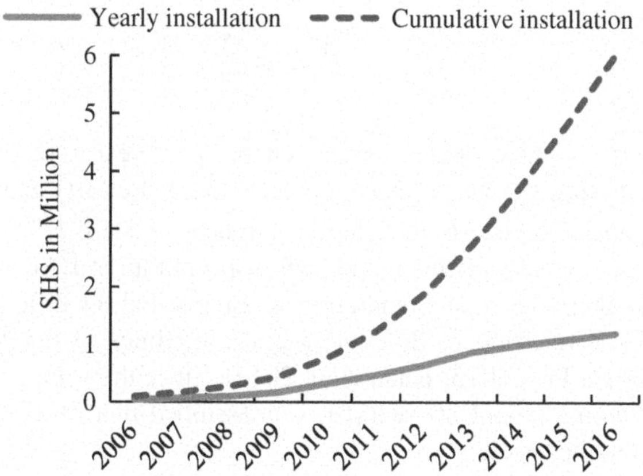

Fig. 12.1 SHS in Bangladesh (WB 2013a; IDCOL 2014; own calculations based on IDCOL 2014)

The main objective of the research is an estimation of the future waste generation in particular of LABs due to the SHS-Program in Bangladesh. The total number of LABs and the resulting amount of lead are the estimated parameters. The research is a starting point for a deeper inventory of waste-flows over the whole life-cycle of a SHS. It is necessary to close material cycles, optimize the waste management system and minimize environmental impacts.

Methods

To prepare and adapt a waste management system to future waste generation information on the number of expected devices and the total mass of the waste flows on a country level is necessary. The model considers only national waste streams. Potential cross-national waste streams are not taken into account. The research is based on literature review. No field research was conducted. Excel was used as a modelling tool.

Estimation of Disposal of Lead-Acid Batteries

Depending on the data availability and quality different standard models for quantifying future waste generation of electronic devices can be used (Wang et al. 2013). Due to data scarcity, especially for the distribution of the life-time of the LABs, the market supply model—a comparatively simple model—was used. It was already used for assessing waste-processing and generation in the informal sector in Delhi (Streicher-Porte et al. 2005). Future waste generation is estimated from historic product sales with their respective disposal rates in the evaluation year. The model is represented by the following Eq. (12.1):

$$W(n) = \sum_{t=t_0}^{n} POM(t) \cdot L^{(p)}(t,n) \quad (12.1)$$

W(n) is the estimated waste generation in year n, representing the number of LABs entering the waste management system. POM(t) equals the historic sales, in this paper the historic installation numbers of SHS. $L^{(p)}(t,n)$ is the discard-based lifespan, which reflects the annual disposal rate of LAB used in SHS in the evaluation year n. The waste generation was modelled for different constant disposal rates of LAB referring to different average lifetimes of the batteries (Laufer and Schäfer 2011; Palit 2013; Khan et al. 2012). Since the solar panel has a significantly higher lifetime (about 20 years) it was assumed that obsolete batteries were replaced by new batteries.

Estimation of Lead Flows on a Country Level

Based on the total number of LABs the amount of lead entering the waste management system was estimated. The following Eq. (12.2) was used:

$$M(n) = W(n) \cdot m(Ah) \cdot P \quad (12.2)$$

M(n) represents the estimated amount of lead entering the waste management system. W(n) is the estimated number of LABs calculated with Eq. (12.1). W(n) is multiplied with m(Ah), the average total weight of a LAB depending on the capacity. P is the percentage of lead of the total mass of a LAB.

$$Ah = W_p \cdot F \quad (12.3)$$

The average capacity of the battery (Ah) depends on the nominal power (W_p) of the solar panel. Data for the average nominal power, due to historically installed and estimated future installations SHS, were analyzed (WB 2013a, 2014; Brossmann 2013). The factor F (Ah/W_p) defines the size of the battery (GSHAKTI 2014; Khan et al. 2012).

Results

Figure 12.2 shows the number of disposed LABs for several constant annual disposal rates from 2006 to 2016 based on historical data and linearly extrapolated into the future. As installation rates began to grow strongly in 2008 the estimated waste flows follow with a time delay of about two to three years. In 2006 the model estimates that around 15,000 LABs are entering the waste management system. In

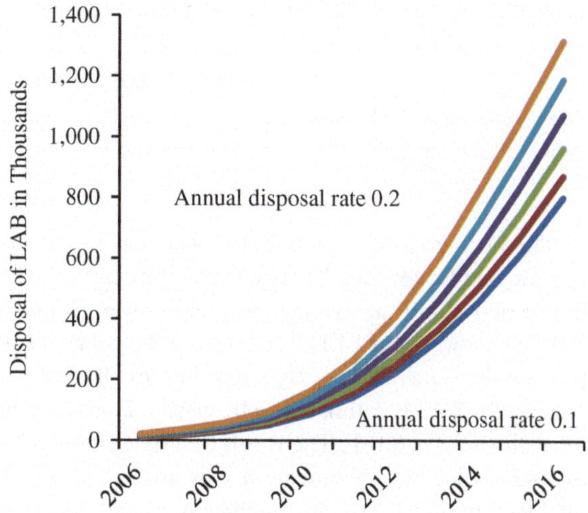

Fig. 12.2 Estimated disposal of LABs

2012 a minimum of 200,000 LABs were disposed. Obsolete batteries are in general replaced by new ones. This strengthens the difference between the disposal-rates. If the average lifetime was only 2.5 years, which is not unusual (Khan et al. 2012), the annual disposal rate would be 0.2 resulting in about 400,000 LABs being disposed in 2012. Due to the high installation rates between 800,000 and 1.2 Million LABs have to be recycled in 2016.

Estimation of Lead Flows

For estimating the resulting lead flows the average capacity of LAB needs to be assessed. Grameen Shakti has the highest market share in the SHS-Market of Bangladesh and Table 12.1 shows their SHS-Specification for off-grid systems. The Factor (Ah/Watt) varies between 1.1 and 1.6. For the estimation of the total amount of lead the factor 1.5 was used.

Table 12.1 SHS-specification of Grameen Shakti (GSHAKTI 2014, own calculations)

System nominal power (Watt)	Battery capacity (Ah)	Factor (Ah/Watt)	Price in USD	Specific invest. cost (USD/W)
10	15	1.5	97	9.7
20	30	1.5	178	8.9
30	40	1.3	229	7.6
30	40	1.3	255	8.5
40	44	1.1	332	8.3
50	80	1.6	422	8.4
65	100	1.5	474	7.3
85	130	1.5	579	6.8

Table 12.2 Estimation of the total amount of lead in metric tons depending on the annual disposal rate

	2006	2008	2010	2012	2014	2016
Mass of lead in t ($L^{(p)} = 0.1$)	81	233	666	1676	3469	6015
Mass of lead in t ($L^{(p)} = 0.2$)	162	448	1235	3068	6221	9888

The average system nominal power was in the first round of the Rural Electrification and Renewable Energy Development Project around $50W_p$ (WB 2013a). In the second round the average nominal power of a new installed SHS is $35W_p$ (WB 2014). Due to better LED Technology the solar panels need less nominal power to provide the same service. By summing up all SHS until 2016, the average nominal power of a SHS is estimated to be $40W_p$. Therefore the average battery capacity for the estimation is 60Ah. The result of a quick research on alibaba.com on the weight of LABs with this specification was around 15 kg. The percentage of lead of the total weight of a LAB varies between 40 and 60 % (ANL 2010; Pavlov 2011). The estimation is calculated with 50 % lead.

Table 12.2 shows the mass of lead disposed as a function of time for the years 2006 to 2016, based on the estimated disposal of LAB as shown in Fig. 12.2. The minimum mass of lead represents an annual disposal rate of 0.1 and the maximum an annual disposal rate of 0.2. In 2012 at least 1,676 metric tons of lead entered the waste management system due to the SHS-Program. The number could grow to almost 10,000 metric tons of lead annually depending on the annual disposal rate of LABs. A market study conducted on the lead acid battery recycling in Bangladesh stated that the formal recycling sector recovered about 3,400 tons of lead in 2006 annually (Waste Concern 2006). In 2006 the SHS-Program had little impact on the local battery recycling system. From 2006 to 2012 the amount of estimated lead already grew by 20 times.

Discussion

The objective of the research was a first estimation of future waste generation due to the SHS-Program in Bangladesh. Since a comparatively simple model was used and real distributions of the lifespan of LABs were not available, further research should be conducted on that topic. The lifespan of LABs is the main driver of the estimated future waste generation and especially on environmental impacts. Lead losses are very low during the use phase, but are about 30 % during the recycling phase, even in the formal recycling sector of Bangladesh (Waste Concern 2009).

In 2005 IDCOL developed a policy guideline on the disposal of warranty expired batteries. Costumers should be notified three months before the warranty of five years expires to change their batteries. But, none of the battery recyclers in Bangladesh collects the batteries with the electrolyte, according to IDCOL "the electrolyte is poured here and there" (IDCOL 2013). Depending on the number and local distribution of the disposal of the electrolytes, toxic-emission to the soil and water are expected. Furthermore almost 60 % of the SHS-System not having changed their battery during warranty time stated in a survey, that they plan to sell it or hand it to a local battery store, if it is not working anymore (Brossmann 2013, 62).

In 2010 the national 3R-Strategy[3] of Bangladesh already took into account that LABs used for solar units will increase the number of LABs entering the waste management system (DOE 2010). In the same report it is stated, that one of the biggest battery producer in Bangladesh recently opened a new battery factory with the capacity for recycling 3,000 tons of lead per year. Nevertheless, for integrating the estimated LABs into the waste management system several new factories have to be installed.

Therefore a working collection system needs to be designed, due to significantly lower lead losses in the regular recycling sector. Since the lifetime of the solar panels is significantly higher (20 years or more) than the average lifetime of batteries, there is still time left to adapt the waste management system to that need.

Acknowledgements The authors gratefully acknowledge the Hans-Böckler Foundation for funding the postgraduate program microenergy systems at Technische Universität Berlin and this doctoral thesis.

References

ANL. (2010). *A review of battery life-cycle analysis: State of knowledge and critical needs.* Department of Energy (USA): Argonne National Labaratory.

Brossmann, M. (2013). *Off-Grid rural electrification and fighting poverty.* Master thesis, University of Tübingen.

BSI. (2010). World's worst pollution problems report 2010. Blacksmith Institute, www.worstpolluted.org/files/FileUpload/files/2010/WWPP-2010-Report-Web.pdf. Accessed 8 Jan 2014.

DOE. (2010). National 3R strategy for waste management. Department of Environment. Minstery of Environment and Forests. Government of the People's Republic of Bangladesh, Dhaka.

Gottesfeld, P., & Cherry, C. (2011). Lead emissions from solar photovoltaic energy systems in China and India. *Energy Policy, 39*(9), 4939–4946. doi:10.1016/j.enpol.2011.06.021.

GSHAKTI. (2014). Grameen shakti bank—price list for SHS, http://www.gshakti.org/index.php?option=com_content&view=article&id=115&Itemid=124. Accessed 20 Jan 2014.

Hoffmann, U., & Wilson, B. (2000). Requirements for, and benefits of, environmentally sound and economically viable management of battery recycling in the Philippines in the wake of basel convention trade restrictions. *Journal of Power Sources, 88*(1), 115–123. doi:10.1016/S0378-7753(99)00519-4.

IDCOL. (2013). Environmental and social management framework. Infrastructure Development Company Limited, Dhaka. http://www.idcol.org/Download/Harmonised%20ESMF%202013.pdf. Accessed 14 Jan 2014.

IDCOL. (2014). Renewable energy projects. http://idcol.org/energyProject.php. Accessed 20 Jan 2014.

[3] 3R: Reduce, Reuse, Recycle.

Khan et al. (2012). Reduced battery sizing in a solar home system with respect to the night load and solar panel size. In *International Conference on the Developments in Renewable Energy Technology (ICDRET), 2012 2nd*, pp. 1–4.

Ki-moon, B. (2011). Sustainable energy for all: A vision statement. United Nations Organization. New York. Available www.sustainableenergyforall.org. Accessed 31 July, 2014.

Laufer, D., & Schäfer, M. (2011). The implementation of solar home systems as a poverty reduction strategy—A case study in Sri Lanka. *Energy for Sustainable Development, 15*(3), 330–336. doi:10.1016/j.esd.2011.07.002.

OECD/IEA. (2011). Energy for all. Financing access for the poor. Special early excerpt of the world energy outlook 2011, Paris.

Palit, D. (2013). Solar energy programs for rural electrification: Experiences and lessons from South Asia. *Energy for Sustainable Development, 17*(3), 270–279. doi:10.1016/j.esd.2013.01.002.

Pavlov, D. (2011). *Lead-acid batteries: Science and technology: A handbook of lead-acid battery technology and its influence on the product*. Amsterdam UA: Elsevier.

Roberts, H. (2003). Changing patterns in global lead supply and demand. *Journal of Power Sources, 116*(1–2), 23–31. doi:10.1016/S0378-7753(02)00701-2.

Streicher-Porte, M., et al. (2005). Key drivers of the E-waste recycling system: Assessing and modelling E-waste processing in the informal sector in Delhi. *Environmental Impact Assessment Review, 25*, 472–491.

UNFCCC. (2013). *Solar home systems monitoring report*. http://cdm.unfccc.int/filestorage/k/2/AP8S6U5KHY7Z2XJ9BECFW0N4LI1V3M.pdf/SHSMonitoringReport.pdf?t=cjF8bjBiZXptfDDqwkVkqWFI57lw1QhKAkF3. Accessed 14 Jan 2014.

Wang, F., et al. (2013). Enhancing E-waste estimates: Improving data quality by multivariate input-output analysis. *Waste Management, 33*(2013), 2397–2407.

Waste Concern. (2006). *Lead acid battery recycling in Bangladesh*. http://www.wasteconcern.org/Publication/BatteryRecyclingBroFinal.pdf. Accessed 14 Jan 2014.

Waste Concern. (2009). *Waste data base 2009*. http://www.wasteconcern.org/documents/Waste%20Data%20Base_2009.pdf. Accessed 14 Jan 2014.

WB. (2013a). *Implementation completion and results report. Rural electrification and renewable energy development project*. The World Bank. http://www-wds.worldbank.org/external/default/WDSContentServer/WDSP/IB/2013/07/15/000333037_20130715125106/Rendered/PDF/ICR26090ICR0Ba00Box377378B00PUBLIC0.pdf. Accessed 20 Jan 2014.

WB. (2013b). *Project appraisal document: Rural electrification and renewable energy development II (REREDII) project*. The World Bank http://www-wds.worldbank.org/external/default/WDSContentServer/WDSP/IB/2012/09/03/000356161_20120903020519/Rendered/PDF/717000PAD0IDA00se0Only090Box370115B.pdf. Accessed 18 Jan 2014.

WB. (2014). *Bangladesh—rural electrification and renewable energy development II (RERED II) project : P131263—implementation status results report : Sequence 25. ISR13274*. The World Bank. http://documents.worldbank.org/curated/en/2014/01/18819309/bangladesh-rural-electrification-renewable-energy-development-ii-rered-ii-project-p131263-implementation-status-results-report-sequence-25. Accessed 24 Jan 2014.

Chapter 13
How to Scale up Green Microfinance? A Comparative Study of Energy Lending in Peru

Natalia Realpe Carrillo

Abstract Among the green microfinance initiatives, energy lending still appears to lack long-term sustainability, and therefore remain difficult to up-scale. In the rollout of these programs, implementing a two-hand model approach—where microfinance institutions (MFIs) partner with energy suppliers—becomes a challenge for both parties. Obstacles entail the access to technical assistance, and the development of efficient supply chains and profitable business models. In this paper, a comparison of the energy diversification process—from a pilot phase towards a large commercialization—of the Peruvian MFIs *Fondesurco* and *CMAC Huancayo* is presented, highlighting their experiences, challenges and discusses opportunities to build from their lessons learned in order to innovate green microfinance.

Keywords Energy lending · Two-hand model · Up-scaling

Introduction

Green microfinance, microfinance services addressing the triple bottom line—impacting at the economic, social and environmental level, entails a variety of internal or external actions that MFIs undertake with the common denominator of environmental preservation. In particular, *energy lending* aims at enhancing access to clean energy by offering credit for modern energy technologies. The portfolio diversification into energy represents an opportunity to extend the market and to reach vulnerable communities in need of accesses to finance and/or energy. However, profitable business models have not yet been validated on a large scale.

Energy lending programs worldwide vary in engagement strategies, product offerings and service delivery models. Parkerson (2005) differentiates three credit sale forms: the lease purchase model, the dealer credit model '*one-hand model*' and

N.R. Carrillo (✉)
MicroEnergy International GmbH Potsdamer, Street 143, 10783 Berlin, Germany
e-mail: natalia.realpe@microenergy-consulting.com

S. Groh et al. (eds.), *Decentralized Solutions for Developing Economies*, Springer Proceedings in Energy, DOI 10.1007/978-3-319-15964-5_13

the end-user credit model '*two-hand model*'. While in the first two models one organization assumes the delivery, financing and maintenance of the systems, in the two-hand model, through strategic partnerships, the responsibilities between the MFIs and the energy suppliers are clearly divided. For the initiation of such energy lending programs, MFIs are rather attracted to develop strategic partnerships with energy companies, instead of incurring heavy organizational changes or assuming the entire supply chain of the technologies by themselves (MicroEnergy International and PlaNet Finance 2010).

In such partnerships, MFIs require the motivation and the abilities to channel capital into loans for energy, as well as a high capacity to assume the largest risk, as green loans are built upon the linkage between the microfinance and energy sectors (Rao et al. 2009). Therefore, engaging in these initiatives requires strategic decisions from the MFIs' management and operational capacities, as well as full support from the energy companies (Morris et al. 2007). Indeed, given the right stakeholder constellation, capital, interests, and favorable external conditions, this approach has potential to succeed.

A comparative analysis of two energy-lending programs implementing a two-hand model approach is presented in this paper, displaying the variety of challenges both institutions face in reaching scale. Since 2011, the Peruvian NGO *Fondo de Desarrollo Regional* (Fondesurco) and the *CMAC Municipal de Ahorro y Crédito*, (CMAC Huancayo) have diversified their portfolio by incorporating energy-lending among their financial services. Both green microfinance programs have been supported with financial and specialized technical assistance. Through partnerships with local energy companies, both MFIs are offering tailored "green loans" in the south and center of Peru. Driven by their social commitment and their aim to differentiate themselves from their competitors, after the conclusion of the pilot projects, *Fondesurco* and *CMAC Huancayo* are scaling up at varying paces.

Objectives

Why is it difficult to scale-up? The ability of the MFIs to foresee and tackle market obstacles is required in order to overcome the challenges of disbursing green loans. Financially sustainable returns, and institutional and operational capacity to manage energy programs, both at the head office and branch levels are needed in order to achieve viable business models (Levaï et al. 2011). Moreover, one of the most constraining factors for MFIs engaging in green initiatives is the access to technical expertise and knowledge in order to establish new management and operational procedures (Allet 2011).

On the other hand, local energy companies have limited access to funding and technical assistance (Kebir et al. 2013). In particular, challenges entail the set-up of the distribution networks and the adaptation to a long-lasting relationship with the client. In order to guarantee a reliable supply of products, the local expansion of energy services into the working areas of the MFIs requires well-built distribution capabilities (Levaï et al. 2011). Given the differences between their infrastructures and business models, achieving balanced partnerships is crucial to success of this two-hand model.

The comparative analysis of *Fondesurco's* and *CMAC Huancayo's* energy programs aims at understanding, from a practitioners point of view, how to address these obstacles—e.g. acquiring technical assistance, developing right supply chains and developing profitable business models for both parties, identify the challenges involved, and highlight opportunities moving forward to scale-up the two-hand approach.

Background

In 2010, the Luxembourgish NGO *Appui au Développement Autonome* (ADA) and the German consulting company MicroEnergy International (MEI), agreed to jointly develop and implement a green microfinance pilot, supporting MFIs to integrate energy services or products into their portfolio. As a first step, a *Microenergy Atlas*[1] was developed, in which Peru was identified as one of the countries with the greatest potential. An increasingly competitive microfinance market and a well-developed market for clean energy technologies made Peru an ideal environment to initiate an energy and financial inclusion program. Moreover, the Atlas highlighted Peru as one of the countries with extended energy needs, potential energy products and service partners, and a mature microfinance sector (Realpe and Palomares 2012). Among the shortlisted Peruvian MFIs, *Fondesurco* and *CMAC Huancayo* were finally selected in October 2010. Moreover, the local support of the project Energy, Development and Life (*Energía, Desarrollo y Vida*) from the GIZ (EnDev/GIZ Peru) was identified as local technical partner in order to support implementation and evaluation of the programs. The role of EnDev/GIZ entailed the pre-selection and evaluation process of the products and suppliers, monitoring of the supply chain and support of the technical activities.

Engaging in Green Microfinance

The MFIs' size, culture, experiences and structures largely differ from each other, ultimately influencing the energy program development.

Fondesurco, operating in southern Peru, was recognized as one of the most innovative and customer-oriented MFIs in the sector and was ranked in first position due to its high motivation to launch a structured large-scale pilot project.

[1] A tool to assess the potential of microfinanced modern energy services based on renewable energy and energy efficiency evaluating the microfinance and energy sectors.

By 2010, besides serving a rural client base with high-energy needs, *Fondesurco* had already experience in financing a solar home system for a household and a solar thermal system for an eco-tourism hostel, assuming high risks with informal agreements with suppliers.

The portfolio diversification became an opportunity to attract new and potentially cheap sources of funding (Casal Ribeiro 2012).[2] Despite the fact that donor pressure to achieve financial sustainability might limit MFIs' ability to innovate and experiment with new products (Hall et al. 2008), *Fondesurco*, following its intrinsic social mission, conducted the energy lending program, and perceived it as a way to offer a differentiated service in an increasingly competitive market (von Wolff and Phalpher 2014).

CMAC Huancayo, covering almost one fourth of the country, demonstrated substantial experience with product loans for hardware products such as housing appliances and cars valued up to 20,000 USD. Its commercial operations consisted of contracting a network of suppliers and facilitating purchases from its clients through product loans. The analysis highlighted the remarkable enthusiasm of the management, the systematic approach for the introduction of new products and the sound experience with suppliers. *CMAC Huancayo* recognized the first-mover advantage in adopting a triple bottom line approach, while the microfinance sector tended to prioritize profit-oriented types of portfolio diversification (Lentz 2011).[3] Yet, the political embeddings of the MFI at the management level was recognized as a potential factor that could negatively affect its performance.[4]

Fondesurco perceived micro-energy lending as a way to fulfill its social mission, as well as a way to offer a differentiated service in an increasingly competitive market (Casal Ribeiro 2012; von Wolff and Phalpher 2014).[5] Moreover, the portfolio diversification provided Fondesurco with an opportunity to attract new and potentially cheap sources of funding (Casal Ribeiro 2012). Despite the fact that donor pressure to achieve financial sustainability leads to a further limitation of the MFI's ability to innovate and experiment with new products (Hall et al. 2008), Fondesurco followed its intrinsic social mission and wagered for micro-energy lending.

---

[2] Through a partnership between ADA and the ULB (*Université Libre de Bruxelles*), internships for students of the Master European Microfinance Program at the local MFIs were sponsored. Pierre Casal was awarded a scholarship from May to July in 2012, supporting *Fondesurco* in its initiation of the small commercialization phase.

[3] Ibid. Caroline Lentz assisted *CMAC Huancayo* in the implementation of the pilot project from May to July 2011.

[4] As a public institution, stability of the management positions largely depends on the municipal elections and political agreements, which influence personnel rotations.

[5] In the framework of the partnership between ADA and the *Université Libre de Bruxelles* (ULB), internships for students of the Master European Microfinance Program at the local MFIs have been sponsored for a period of 4–6 months, for their master thesis projects. Pierre Casal was awarded a scholarship from May to July in 2012, supporting Fondesurco in its initiation of the small commercialization phase.

Results from Allet (2012) showed that drivers to go green mostly come from strong leaders in an organization, who are motivated to find the necessary resources to roll out the initiative. However, MFIs without sufficient commitment to the environment at the management level tend to identify a trade-off between their financial and environmental bottom lines. Caja Huancayo's former Finance and Administration Manager attended the awareness raising workshops conducted by ADA and MEI and followed up the selection process until the technical assistance and co-financing of the pilot project was confirmed. Caja Huancayo recognized the first-mover advantage in adopting a triple bottom line approach, while the microfinance sector tended to prioritize profit-oriented types of portfolio diversification (Lentz 2011).[6]

Greening the MFI

A field study conducted in order to assess the energy needs, expenses and habits of the MFIs' current and potential clients, identified a potential market for solar thermal systems (STS) replacing electric heaters, improved cooking ovens (ICO) reducing in 50 % firewood consumption than traditional ovens, and solar crop dryers (SCD) substituting traditional methods for coffee drying.[7]

Technologies were selected based on their potential productive use, aiming to meet the micro-entrepreneurs' needs, enabling them to either increase their income or reduce energy expenses (electricity or firewood, with the solar thermal system and improved cooking oven respectively) in the long run. Indeed, investments in modern energy systems by MFI clients and by energy companies can become more attractive if these are coupled with increased economic productivity (Morris et al. 2007). Furthermore, the increase in productivity in rural areas is expected to result in economic growth, a rise in rural employment and a reduction in the migration of the rural poor to urban areas (Cabraal et al. 2005). Specifically, the solar thermal systems targeted hostels and rural eco-tourism households; the improved cooking ovens were promoted for restaurants, and the solar crop dryers were selected for both independent coffee farmers and cooperatives.

Both MFIs enabled access to the technologies through the creation of a specific green loan with preferential interest rates: *Fondenergía* (*Fondesurco*) and *Credi-ecológico* (*CMAC Huancayo*), first offered in few selected pilot branches. *Fondesurco* acknowledged its program as a re-launch of the previous pilot, while *CMAC Huancayo* offered its first green credit product. By targeting the BoP, the credit specifications offered privileged conditions to the customers, not only through special prices, but also with the provision of guaranteed after-sales services. At first,

[6] Ibid. Caroline Lentz assisted Caja Huancayo in the implementation of the pilot project from May to July 2011.

[7] Only for *CMAC Huancayo*.

the green financial products were offered in few selected pilot branches. Fondesurco acknowledged its program as a re-launch of the previous one-shot pilot, while Caja Huancayo offered its first environment-linked credit product. The coordination of the program was under the umbrella of the Research and Development department in Fondesurco and of the Credit Department in Caja Huancayo.

Model and Set-up

The energy programs were implemented as *two-hand models*, fostering energy-microfinance partnerships, with the guidance and support of the technical assistance that identified, evaluated and selected the energy service suppliers for each MFI. Initially, after having validated the technologies with specific product testing,[8] one supplier per technology was considered, establishing a list of prices and procedures for the piloting regions.

ADA together with MEI supported the MFIs in the elaboration of internal procedures of the two-hand model. High investment costs in the learning process were delegated to both MFIs during the set up of the pilot projects. In order to properly transfer and integrate the methodologies and tools provided by the technical assistance, the MFIs were advised to create a new position of *Energy Technical Advisor* (ETA) in addition to the assigned responsible person of each Department in charge, aimed at enabling the MFI to manage the supply chains. *Fondesurco* hired an ETA after two months, at *CMAC Huancayo*'s an ETA was enrolled after more than a year, close to the end of the pilot phase.

The ETA coordinated the credit disbursement process as well as the aftersales services provided such as customer satisfaction follow-up, programmed visits and complaints handling, as well as the planning, project monitoring and knowledge transfer from the technical assistance. The ETA reduced the amount of tasks traditionally allocated to the suppliers, including the technical assessment of potential clients, the organization of order placements and the logistics coordination with the supplier for the delivery/installation of the technology (Lentz 2011; Casal Ribeiro 2012). The extended role of the MFIs in the supply chain management surpassed the expected responsibilities in a two-hand model approach. The engagement of the MFIs to assume a larger role in the supply chain was crucial to the functionality of the programs.

The pilot phase began in June 2011 and concluded in April 2012 for *Fondesurco* and September 2012 for *CMAC Huancayo,* turning into a small-scale commercialization phase. The scaling-up phases aim at including additional branches and

[8] Laboratory tests, contracted by EnDev/GIZ, were conducted at the *National University of Engineering of Lima* and the *University of San Agustin* to validate the ICOs and STS, while a technical study conducted by the *National Institute of Agricultural Innovation* compared the performance of the SCD with traditional drying methods.

new technologies. In this second stage, technical assistance focused particularly on the supply chain reinforcement and on the selection, evaluation and validation of new technologies.

Challenges

Energy finance is not a core competence of most MFIs and entry barriers limit MFI participation in the sector (Levaï et al. 2011). The Peruvian energy market and the willingness to participate among players have catalyzed innovation in green microfinance, despite the relatively high costs and limited revenues for the MFIs. However, both MFIs faced a series of challenges in the scaling-up process, related to their structure and their capabilities to innovate and adapt their processes to client needs.

Green Loan Disbursement

Fondesurco. The disbursement of the green loans involves a significant amount of work for low credit amounts (Casal Ribeiro 2012). Compared to the average loan size of 1.811 USD, the *Fondenergía* ranges from 300 to 1,000 USD, with limited portfolio profitability. The lack of a proper incentive structure hinders the motivation of loan officers to promote the *Fondenergía.*[9] Nevertheless, the value of differentiation in a competitive market has prevailed and promotional activities have been widely undertaken.

CMAC Huancayo. Cash sales at the exhibitions affected the motivation of loan officers to promote the green loans, which focused on the sole add-ons of the credit, such as the level of after-sales service, the validation process of the technologies and guarantees offered. This ensures that customers wishing to acquire micro-energy devices will opt for the *Crediecológico* rather than financing it through a traditional credit line (Lentz 2011).

Institutional Commitment

Fondesurco. The integration of energy lending represented a high-risk diversification. Given the correlation between the technology and the MFI reputation risk, the quality of the technologies and of the supply chain play a major role. This institutional awareness motivates the organization to adapt the green loan disbursement in

[9] A project evaluation—contracted by ADA—was conducted in 2012 by *COPEME*, a Peruvian microfinance network and consultancy company.

a learning-by-doing process. However, more efficient and systematic processes are needed in order to scale-up. A further challenge includes the deployment of the clean energy technologies in the 18 branches, whilst continuing to improve communication with financial and technical partners (Convergences 2013).

CMAC Huancayo. Despite the incentive structure designed for the loan officers—initially planned only for senior officers, and offered for the first time for a specific product at the institution—the awareness raising campaign for loan officers required as much effort as the one directed for clients. Following the recommendations of the pilot evaluation,[10] the complexity of the loan disbursement has been reduced, and a systematic calendar of trainings and exhibitions has been established in half of the agencies selected for the expansion of the program. Nonetheless, the MFI requires larger commitment at the operational level and better coordination skills from the energy suppliers.

Supply Chain Design

Fondesurco. Each pilot branch was organizing its own supply chains, conducting technical assessments of interested clients. Observations from Kebir et al. (2013), affirm that some MFIs go beyond the financing of products to selling them as well, contracting local partners to handle the sourcing of parts, installation and servicing of systems. However, the considerable involvement of *Fondesurco* might not be sustainable on a larger scale. A close collaboration with suppliers is needed in order to manage the complexity of the value chain and to establish sustainable partnerships (Casal Ribeiro 2012). Furthermore, Kebir et al. (2013) explain how MFIs tend to be the only formalized institutions able to lend to the customers of energy SMEs or small informal local distributors. The functioning of the supply chain has been stabilized through measures such as *Fondesurco*'s financing of ICO manufacturers[11] and EnDev/GIZ's technical support to maintain a minimum stock.

CMAC Huancayo. The main challenge regarding STS has been the need for local installers to support the selling process. Concerning the ICOs, a relatively new technology in the market, the manufacturers faced serious challenges in the expected quality of the systems. Moreover, due to the absence of ETA during the pilot phase, despite the rigid structure and restricted flexibility of *CMAC Huancayo*, loan officers had to coordinate the delivery and installation of the devices. Considering that nascent energy companies are unable to provide sufficient quality (Levaï et al. 2011),

---

[10] COPEME Evaluation (2012).

[11] Kebir et al. (2013) explain how MFIs tend to be the only formalized institutions able to lend to the customers of energy SMEs or small informal local distributors. In this case, being *Fondesurco* the financer of the supplier, the risks of its energy portfolio is intensified.

the ICO manufacturers faced serious challenges in the expected quality of the systems. Finally, external risk factors, such as the coffee disease "*roya*", affected the business of the SCD and the overall portfolio of this market segment.

Client Response

Fondesurco. Despite the great customer acceptance and satisfaction, and the excellent performance of the portfolio (zero default rate), energy lending was uncommon for both the clients and the partner energy enterprises. Hence, major efforts have been required from the MFI side in order to raise awareness of the benefits of the technologies.

CMAC Huancayo. Due to a less mature market for STS, larger efforts were needed for promotional campaigns. Moreover, a change of suppliers of the ICOs was needed due to quality issues, and the higher price set by the new supplier did not match the willingness to pay of the customers, requiring further promotional activities.

Empowering Green MFIs

Characterized by sourcing equipment and maintenance services from local suppliers, two-hand model projects have a strong focus on local market development and capacity-building measures, prioritizing productive energy uses to promote local business activity. Indeed, energy lending enables clients to improve their profitability by providing better services to their own customers, while also generating positive social effects such as improved working conditions and improved hygiene (von Wolff and Phalpher 2014). Through the systematic implementation of their green microfinance programs, *Fondesurco* and *CMAC Huancayo* continue to adapt their businesses and operational models to their required conditions and capabilities. Achieving a triple bottom line allows them to stand out in a competitive microfinance environment, and at the same time, contribute to the mitigation of climate change effects. This case study is limited to the specificities of the support provided by the financial and technical assistance to both MFIs. Further research on two-hand models underlying the role of energy partners shall be conducted to counterbalance these results.

The outcome of this study identifies the following ways forward towards a successful scale-up for policy makers and practitioners in the field of green microfinance.

- External support for MFIs and Energy Partners
  Considering the costs and the risks entailed in the experimentation of new business models, the necessity of external support to launch green microfinance programs is apparent, at least in the start-up phase. The financial assistance from ADA since 2010 and REEEP in 2013, as well as the technical assistance from

ADA, MEI and EnDev/GIZ, have been essential for the MFIs from the pilot until the small-commercialization phase, in the implementation of the scale-up structures, particularly in the reinforcement of the supply chain, the validation of new suppliers for the existing products and the inclusion of new technologies in their portfolios. Furthermore, a special follow-up of the energy partner on its growing business solidifies not only the emerging energy market, but also paves the way for the success of the green microfinance businesses.

- Consolidation of green programs within MFI internal structure and governance
  Formalized internal structures facilitate decision-making processes, a strict execution and monitoring of project plans. This allows, on the one hand, the internalization of the know-how transferred from the technical assistance and the building of internal capabilities, and on the other hand, the ownership to react with strategic decisions in the business development. The ETAs enrolled at *Fondesurco* have been promoted within the institution, strengthening the role of the R&D department in the green activities. This shows how *Fondesurco* has been able to capitalize on the established structure of the green program. At *CMAC Huancayo*, the only ETA enrolled since 2012, despite several changes of champions, has gradually assumed further responsibilities for the coordination of the energy program. Within its internal structures, *CMAC Huancayo* adapted its financial product to 'credit destination', allowing loan flexibility to access energy technologies, as well as getting a lower interest rate depending on the chosen and fit loan category.
- Capacity building and empowerment process
  In the transition towards large-scale commercialization, the technical assistance shall focus on capacity building and the 'autonomy' process for both MFIs. This entails support in introducing new technologies and providing the necessary technical tools to develop the respective business plans, assess energy needs and market size, and design efficient supply chains. Furthermore, the MFIs shall be supported to enable them to design and implement their tailored energy business. This support includes assessing their capabilities and business opportunities to develop separate energy business lines or enhance win-win business models with selected suppliers through profitable operational structures.

References

Allet, M. (2011). *Measuring the environmental performance of microfinance*. CEB Working paper N. 11/045.

Allet, M. (2012). Microfinance and Environment: Why do Microfinance Institutions Go Green? CEB Working paper N. 12/015.

Casal Ribeiro, P. (2012). *Can micro-financing green energy be sustainable? A case study of Fondesurco, Peru*. Master thesis, European Microfinance Programme, ULB.

Cabraal, A., Douglas, F., & Agarwal, S. (2005). Productive Uses of Energy for Rural Development. *Annul. Rev. Environ. Resour.*, *30*, 117–144.

Convergences. (2013). *Microfinance partnerships to help access basic services*. Convergences World Forum. September 17–19, 2013, Paris.

Hall, J. C., Collins, L., Israel, E., & Wenner, M. D. (2008). *The missing bottom line: Microfinance and the environment*. Social Performance MAP: Green Microfinance. SEEP Network Social Performance Working Group.

Kebir, N., Spiegel, N., Schrecker, T., Groh, S., Scott, C., & Aliaga, G. (2013). *Exploring energy SME financing in emerging and developing countries*. Oestrich-Winckel: Sustainable Business Institute.

Lentz, C. (2011). *Case Study on the integration of micro-energy credits into the MFI's loan portfolio*. Master thesis, European Microfinance Programme, ULB.

Levaï D., Rippey, P., Rhyne E., & Allderdice, A. (2011). *Microfinance and energy poverty: Findings from the energy links project*. Final Report to AFD (now FHI 360) and USAID under the FIELD Project. Center for Financial Inclusion. Publication No. 13.

MicroEnergy International and PlaNet Finance. (2010). Fact sheet the potential of linking microfinance and energy supply. 4th Microfinance-Breakfast. UMM, Up Micro Loans. March 4th 2010, Berlin.

Morris, E. J., Winiecki, S. C., & Cortiglia, D. (2007). *Using microfinance to expand access to energy services: Summary of findings*. SEEP Network.

Parkerson, D. J. (2005). *Assessing the potential for microfinance institutions to finance solar photovolatic systems in the Dominican Republic*. Working Paper (14). SEEP Network.

Rao, P. S. C., Miller, J. B., Wang, Y. D., & Byrne, J. B. (2009). Energy-Microfinance Intervention for Below Poverty Line Households in India. *Energy Policy*, *37*, 1694–1712.

Realpe Carrillo, N., & Palomares, C. (2012). *Microfinance and energy: A profitable partnership*. Lima.: Amaray Energy and Development for rural regions.

Von Wolff, S. & Phalpher, K. (2014). Green Finance. Successes and Challenges—A Landscape Overview. Finance in Motion. pp 40.

Chapter 14
Microfinancing Decentralized Solar Energy Systems in India: Innovative Products Through Group Approach

Satish Pillarisetti

Abstract The apex development bank in India-NABARD (National Bank for Agriculture and Rural Development) first facilitated the microfinancing scheme for solar home lighting system by rural banks. This turned out to be a success and the government introduced a subsidy linked bank credit programme to upscale and mainstream the programme. Initially the programme met with only a limited success as it had too many loose ends. Later, with a number of modifications, the program gradually took off and is now making steady progress. However, NABARD realized that there are many areas and communities which may not be able to benefit by this scheme and may need a different financial product. NABARD developed group based products for such communities. These products are based on a partnership with NGOs, and at times involving retail banks. NABARD utilised the funding available under the rural innovations fund and the umbrella programme for natural resource management to support these innovative products. First product was for lighting girls' hostels and the second was for financing solar micro grids through community participation. Both these products were implemented in the Sunderbans region of West Bengal state in joint liability group mode and combined the elements of loan and grant. The products were designed and rolled out with the combined involvement of NABARD and an NGO-Mlinda Foundation. The third product was in collaboration with another NGO-ASSEFA in the states of Bihar and Jharkhand. This product was through the self help group mode. In respect of this product the end users were dissatisfied with the product and the NGO had to modify it to suit the needs of the users. The modified product was accepted by the users and was rolled out to their satisfaction. The successful take off of three such products and failure of one product to take off and its subsequent modification are discussed here. This paper uses the sectoral innovation systems (SIS) to analyze as to how successful new products and processes emerge from a favorable combination of technologies, knowledge, institutions, networks and actors.

S. Pillarisetti (✉)
National Bank for Agriculture and Rural Development (NABARD),
C-24, Bandra-Kurla Complex, Mumbai 400051, India
e-mail: satishp99@usa.net
URL: http://www.nabard.org

S. Groh et al. (eds.), *Decentralized Solutions for Developing Economies*,
Springer Proceedings in Energy, DOI 10.1007/978-3-319-15964-5_14

Keywords Microfinancing · Solar home lighting · Joint liability groups · Self-help groups

Introduction and Background

Worldwide it is estimated that 2.6 billion people in rural areas do not have electricity (Terrado et al. 2008). A large proportion of these people live in India. To address this problem of access to energy, one of the missions started by the government of India was the Jawaharlal Nehru national solar mission (JNNSM). The mission envisages deploying 20,000 MW of solar power by 2022 of which 2000 MW would be through off-grid applications. A major part of the off-grid applications would be through solar lighting and heating systems supported by bank finance. Modeled on the initiative taken by a regional rural bank in northern India and facilitated by the apex bank for agriculture and rural development-the national bank for agriculture and rural development (NABARD), the government launched a capital subsidy-cum-refinance scheme for installation of solar off-grid (photo-voltaic and thermal) and decentralized applications. The scheme had undergone several changes to improve its implementation which included changes in the pattern of subsidy, financing modalities etc. Later the central bank of the country—Reserve Bank of India was persuaded to include financing of solar home lightning systems also as a part of priority sector advances. Further on the suggestion of NABARD, the government has accepted the proposal to include cooperative banks also in the scheme. The main product in focus has been the solar home lighting system. Though further action is needed in terms of training and sensitizing the bankers as well as users to upscale it, the programme is now set to take off on an even keel, with nearly 200,000 units financed by banks so far, and a large number in the pipeline.

Literature Review

While there is extensive literature on microfinance per se, literature dealing with microfinance for energy and more specifically for renewable energy is limited. A few of the writings on the subject are discussed here. Rao et al. propose an energy microfinance framework that caters to the energy (lighting and cooking) needs of low-income household population by engaging a microfinance institution. The model/framework proposed encompasses two independent entities. One has an energy expertise (non-profit organization) and the other possesses finance management skills (microfinance institution). The funds from a larger corpus or an individual flow to a smaller microfinance institution (MFI) and/or self-help group committee. The primary responsibility of the MFI is to supply credit to its members and/or enterprise identified by the non-profit. The non-profit organization identifies the energy service company which is responsible for designing, installing and servicing the clean energy technology products. It also identifies the entrepreneur among the member beneficiaries and trains him/her in daily energy services and maintenance of the system. The member-beneficiaries are the existing clients of the MFI. They are offered 'energy' loans by the MFI to obtain reliable, renewable and energy efficient lighting and cooking technologies. The non-profit and the MFI/SHG network can be treated as a single entity or two independent organizations under one large entity (Rao et al. 2009).

A briefing note of the foundation for development cooperation (FDC) states that it has been widely recognized that public sector provision of centralized energy is insufficient to meet the growing demands of developing countries, and that private sector involvement is a necessity. This is also reflected in the microfinance for energy lending sector. As a consequence, understanding, identifying and realizing the economic opportunities of modern energy systems for low income clients is essential for attracting financing for energy lending. MFIs must demonstrate that the cost of energy systems can be recovered through end-user payments, and consequently attract private sector investments in energy enterprises. Energy lending by MFIs is in formative stages. In order for it to become a mainstream product within the microfinance industry, MFIs and their energy industry partners must continue to identify and implement least-cost technologies that adds value for clients. At the same time, gaps in financing models specific to energy sector must be identified and mechanisms established to address them (Bedson 2007).

Kapadia and Hande are of the view that meeting the needs of the energy consumers calls for partnerships between energy service providers and microfinance agencies. Energy service providers can greatly increase access to poor and/or rural consumers if these consumers could also access credit facilities. Microfinance agencies could greatly increase the reach of their financial services, and augment the economic development opportunities they enable, by lending for energy. Although lending for energy service is not widespread, the experience has consistently been that once people gain access to modern energy, they are determined to keep that access, and are, for the most part, responsible borrowers. Lending for energy services can also open up new markets for microfinance agencies that have little to do with energy (Kapadia and Hande 2004).

A study on energy access in South Asia emphasizes that micro-lending for energy consumers opens up great potential for MFIs to expand business beyond their existing client base. However, proficiency in energy technology and interactions with energy sector players are fundamental elements for designing effective energy lending programs with good loan portfolio quality. The global initiatives to utilize micro-lending for small scale, decentralized or individual energy products as strategic means of expanding access to modern energy services also provide opportunities for energy market players to tap into a critical customer base. The four MFIs in the study show that initiatives to foster development of micro-lending for energy consumers can come from any stakeholder as long as collaborations with other key stakeholders are established (SEEP 2007).

Microfinance can facilitate both the ‘push’ and ‘pull’ of renewable energy technologies in developing country markets. It can enable a push mechanism by financing installation costs of renewable energy technologies, as well as generate a pull mechanism by financing energy distribution to consumers and industry. Microfinance has also the potential to expand renewable energy distribution by making the energy systems more affordable. MFIs are well positioned to do so because they are able to offer flexible loans and structure their lending schemes and financial products in accordance with the needs of the poor. They are able to tie loan repayment rates to seasonal energy needs, or seasonal variations in rural household and industry income. Innovative partnerships have already emerged between renewable energy generators and MFIs to support distribution to rural populations (Mohiuddin 2006).

The fact sheet of Micro Energy International and PlaNet Finance, Deutschland is of the view that there exist ways to effectively combine microfinance with renewable energies, especially in rural areas where access to energy remains limited. Out of about 10,000 MFIs serving worldwide more than 155 million clients, 30–40 MFIs today offer energy loans. Each of these MFIs has its own strategy, but their three main objectives are: to finance the access to energy and to renewable energies for micro and small enterprises (MSEs) and low-income households through adapted financial services; to enhance the productivity and the business opportunities of MSEs through energy efficient solutions; to raise awareness among MSEs and low income populations, especially in rural areas, on the opportunities offered by renewable energies. Although a few MFIs have been successful, most of them are still in the learning phase and face important challenges on extending energy loans (Micro Energy International and PlaNet Finance, Deutschland 2010).

The role of the Alternate Energy Promotion Centre (AEPC) in developing the renewable energy technologies (RET) sector in Nepal and the opportunities for MFIs to become key players in financing this sector is discussed by Wegstein. MFIs play a role by lending to households for installation of biogas plants and solar home lighting systems. Entering this sector helps the institutions increase their business volume, organizational growth, outreach, sustainability and profit. If demand for RET exceeds the financial capacity of an MFI, AEPC steps into create MFIs’ partnerships with commercial banks. It also arranges to provide training and capacity building for MFIs in financing RET (Wegstein 2010).

The financing models discussed in this paper are novel and innovative, and have evolved based on the grassroots level needs as expressed by the users. Documentation of such models, combined with a theoretical framework, provides us with insights on the methods to be used to overcome field level issues. The analysis and documentation of innovative financing models for micro energy systems introduced in India in the present paper strengthens the literature on microfinance for renewable energy, specifically in the rural and decentralized contexts, as the existing literature has few references to group based products implemented in a combined manner by financial and non-financial institutions.

Methodology

As the focus of this paper is on renewable energy systems the concept of sectoral innovation systems (SIS) is a relevant and appealing framework. SIS has been described as a set of new and established products for specific uses and the set of agents carrying out market and non-market interactions for the creation, production and sale of these products (Malerba 2002). It is a framework that helps in analyzing complex dynamics, because it takes into account multiple actors, policy makers and institutions at different levels (Al-Saleh 2010). The main structural elements of SIS are: knowledge and technologies, actors (i.e. firms and other organizations) and networks, as well as institutions (e.g. standards, laws and regulations). Thus in order to understand the dynamics and the innovation processes of a given sector, one needs to give due consideration to these five key elements. By focusing on these key characteristics of a sector, one might be able to unfold the 'dominant design', 'paradigm', and the 'trajectory' of that sector. SIS is a theoretically rich concept because it presents a rather well articulated, multi-dimensional, integrated, dynamic and systemic view of innovation in globalized sectors; as such it could provide a framework for the understanding and analysis of a sector as well as allowing comparative analysis across range of sectors (Malerba 2005).

This framework is used to analyze the innovations introduced by NABARD. In the presented case studies, technologies were a given element, while the actors i.e. NGOs were playing an isolated role. The networks were basically at various levels connecting different players to facilitate their interaction, some of these set up formally by governments and regulators and others are informal understandings. NABARD is a part of the building block of both 'knowledge' and 'institutions' as it leverages its knowledge of innovative product development as well as standards, laws and regulations. In the case studies discussed, the framework of SIS was thus used to analyze the parts played by the users (borrowers), the NGOs, the banks, the technology providers and NABARD in evolving a new product at each stage to overcome some glitch thrown up during the course of implementing the programme.

When an existing or progressing arrangement encounters problems in implementation a set of innovative systems have to be brought in. While the solar home lighting scheme was picking up steady progress, NABARD was receiving constant feedback from NGOs and other field based organizations and its own district level officials that there are a large number of communities and groups which would not be in a position to make of use of this scheme and would require a special dispensation. NABARD utilized two funding arrangements available with it to innovate products for group based financing. The implementation of these new products was studied on the basis of secondary data and progress reports received at NABARD from implementing agencies (NABARD 2013). The basic research questions which this paper attempts to address are the need for a group approach to financing, the necessary conditions for a group approach to be successful and the reasons for clients preferring one type of group approach to another. The criteria on which the success of these projects is measured are complete utilization of the loan amounts, installation of the systems and the regular repayment of the dues as per the schedule fixed.

Discussion and Results

The rural innovations fund (RIF) was a fund created in NABARD with the support of the Swiss agency for development and cooperation (SDC) and was being used to support innovations in rural areas through grant, loan or soft loan. The Umbrella Programme on Natural Resource Management (UPNRM) is a programme of NABARD launched with the funding support of KfW, the German development bank. This programme supports any type of activity which contributes to the conservation of natural resources. Innovative schemes for financing solar lighting systems were launched through joint liability groups (JLGs) in the inaccessible estuarine islands of Sunderbans in West Bengal state and through self help groups (SHGs) in the remote tribal communities of Jharkhand and Bihar states, utilizing the funds available under RIF and UPNRM.

1. West Bengal Regional Office of NABARD has covered new ground by financing joint liability groups (JLGs) for solar electrification of school hostels in the remote Patharpratima block which falls in the off grid region of Sunderbans area in the South 24 Paraganas district in the year 2012–2013. As this region is in the estuarine area of the Ganges River, many island areas are cut off from the grid. Grid solutions require a minimum threshold level of electricity demand and certain level densities to achieve the economies of scale (Sorensen 2004). The villages did not have electricity connection and the girl students/hostellers of the schools, majority of them belonging to backward strata of the community, pursued their studies using kerosene lamps. The families of the hostellers made a contribution of Rs. 2 per day for use of kerosene lamps. A local NGO working in the region in the area of clean environment-Mlinda Foundation motivated the school hostel committee to replace the kerosene lamps with solar LED lights. However the foundation realized that none of the existing schemes could finance such an activity. This predicament of the NGO came to the notice of NABARD's district level official. After several rounds of exchange of views and discussions between the NGO, parents and NABARD officials at various levels, the conclusion was that a group approach could work and the financing could be done through joint liability groups (JLGs). A joint liability group (JLG) is an informal group comprising of 4–10 individuals coming together for the purpose of availing bank loan either singly or through the group mechanism against mutual guarantee. JLG is primarily intended to be a credit group, but group savings may be an optional activity. The quantum of credit need not be linked to savings and no collaterals need be insisted upon by banks against their loans to JLGs. JLGs are formed by facilitators like NGOs or other individual rural volunteers.

The Mlinda Foundation organized the committee members involved in running the school hostels into JLGs. These JLGs then approached NABARD for support for providing solar lights in the school hostels under the RIF. This innovative project was sanctioned to three JLGs of parents and school teachers. The three JLGs covered a total of 172 hostel children and a total of 157 solar powered LED lamps were installed. The total financial cost of the installation in the three hostels was Rs. 581,128 of which 50 % was loan and 30 % was grant under RIF from NABARD and 10 % each was the contribution of Mlinda Foundation and the JLGs themselves. With the help of this financial assistance the kerosene lamps being used by school children for studies during evening hours were replaced with solar lights. The lights have been installed with the condition of repayment of loan through contribution from the school children. The responsibility of maintenance of solar lighting system as well as collection and repayment of loan assistance lies with the respective JLGs.

The project is being implemented successfully at ground level in association with the NGO and the loan repayment has already started coming in from the JLGs. For the timely repayment of loan assistance, the JLGs opened their bank account in a nearby bank branch. They used this account to deposit the monthly collection from the school students. The per monthly collection from students was decided on the basis of the number of students and the capacity of solar panel installed for electrification. The total number of students covered under these project in the 3 JLGs was 172 and the per LED unit collection per day from the students was Rs. 2.00 in two JLGs and Rs. 3.00 in another JLG. This was the amount the students were paying to the school hostel authorities for provision of kerosene lamps. Per student per month collection for repayment was related to the number of units installed in each of the hostels. It was Rs. 62 in one hostel, Rs. 129 in the second and Rs. 106 in the last one. The loan repayment fixed was 2 years for two JLGs and 5 years for one JLG. The project is expected to improve the health of the girls residing in the hostels, reduce absenteeism and provide better environment for students and also save the environment from carbon dioxide generated from the kerosene lit lanterns.

2. In the same region of Sunderbans another project focused on household solar micro grids through cluster mode of JLGs. To address this problem of non-availability of electricity, solar micro grids were financed through community participation in association with the NGO, Mlinda Foundation. The project aimed at replacement of the use of kerosene for lighting purpose by solar energy. The repayment of loan was to be made from savings accrued from non-usage of kerosene for domestic lighting. The business model involved adoption of solar system by community in a financially sustainable manner with huge potential for replication across the rural geographies. This model was adopted as depending on the technology employed, energy can be provided on a household basis as with solar home systems or on the level of villages and communities (Saghir 2005). Adoption of JLG model in the project aims to incorporate the element of both individual and collective ownership and accountability and also

reduction of risk of non-repayment of default. The loan cum grant assistance of Rs. 71,500 was sanctioned to 13 JLGs for installation of solar panels in 13 micro cluster units comprising 7–10 households to provide 3 LED lights points and one mobile charging point to each household of the cluster. The loan component was Rs. 36,000 and grant component was Rs. 35,500. Both the loan and grant support for the project was sourced from the RIF of NABARD. The total loan component for 13 JLGs was Rs. 468,000 and grant component was Rs. 461,500. Mlinda Foundation spent Rs. 8000 per JLG for formation and nurturing.

The repayment of loan was to be through monthly collection of savings from the non-usage of kerosene by JLG members for lighting purposes. The repayment period was fixed for 4 years. The JLG members proposed to save an additional amount for replacement of battery and maintenance of solar panel. The maintenance of the panels would be the responsibility of each of the JLGs. The project is based on the concept of innovative business model in which end users are enabled for payment of energy services through utilization of amount presently spent on the purchase of kerosene for home lighting purpose. The element of both individual and collective ownership and accountability is inherent in the JLG model and the risk of non-repayment and default is greatly reduced.

3. NABARD is currently implementing the Umbrella Programme on Natural Resource Management (URNRM) supported by the German development bank KfW. The programme finances a wide range of activities which encourage the sustainable use of natural resources. Financing of a variety of renewable energy projects was taken up under this programme. One such project taken up by NABARD in 2010–2011 is situated in Jamui, Chakai and Khaira blocks of Jamui district of Bihar state and Deoghar district of Jharkhand state. The project area is rich in natural resources, but inhabited by poor and marginalized tribal communities. Livelihoods are based on subsistence farming and forest produce. Energy consumption is based on firewood from forest. For the lighting of the homes communities are solely dependent on kerosene lamps. In these communities the solar lanterns alternative was proposed as these communities are extremely backward and would not be able to maintain even if the lower end of solar home lighting systems are financed and supplied. As the capital cost of solar home lighting system depended on the size, site and technology of the PV system components (Patel 2006) the model propagated under lighting a billion lives (LaBL) programme of The Energy Resources Institute (TERI), New Delhi was adopted. The project was taken up for implementation by a well-known NGO-Association of Sarva Sewa Farms (ASSEFA). TERI, based on its wide expertise and rich on-field experience in energy sector, has initiated the (LaBL) scheme with the mission of enabling a billion lives to access light from solar technologies. The major players in the LaBL scheme are TERI, who has the technology; LaBL Associate who is interested in promoting clean lighting (in the present case, ASSEFA) and LaBL Franchisee (the identified SHGs who will run the units) who will be authorized to set up and manage the solar lantern charging stations.

LaBL model operates on fee-for-service or rental model where centralized solar lantern charging stations (SLCS) are set-up in villages for charging the solar lanterns and providing the lanterns daily on rent to households and enterprises. A typical SLCS consists of 50 solar lanterns with 5 solar panels and junction boxes. The solar lantern will provide light for 5–6 h daily using LED lamps on full charge of the battery providing light equivalent of a 40 W incandescent bulb. If operated on dimming option, it will work for 8 h. The SLCS will be operated and managed by entrepreneurs and are provided hand holding support by LaBL Associate (in this case, ASSEFA). Rent collected by the LaBL franchisee will be used for operation and management of the charging station and a part of the rent will be kept apart for replacement of battery after 18–24 months of operation. TERI provides the requisite training support to both the LaBL Associate and entrepreneurs.

Under this project, 40 charging stations were to be set up across these districts and per charging station 50 solar lamps can be charged. Thus a total of 2000 families are covered through this project. The cost per charging station was estimated at Rs. 175,000 and the total cost for 40 stations was at Rs. 7 millions. Additionally a grant component of Rs. 0.86 million was envisaged for accompanying measures of training and skill building. The contribution by participants was Rs. 0.48 millions. The loan and grant were both advanced by NABARD under the UPNRM programme. In each block one project coordinator was appointed to oversee the project activities. The project activities were to be implemented and monitored through women self help groups which have been promoted by ASSEFA in the identified villages. Each charging station would be under the supervision of SHGs and the loans for all members are overseen by the SHG which also maintains a regular record of repayment of loan by members as well as the servicing and maintenance of the charging station. From each SHG, 1–2 women are identified for implementing the project and made responsible for the implementation.

The outcome of the project was to install sufficient lighting in evening and night hours, empowering the women in handling innovative technology, reduce health hazards by preventing in-house pollution, lessen the cost of consumption on kerosene and to provide congenial environment for students and workers to pursue their studies and occupations during the night. Last but not the least, the solar lamps are portable and ensure safety from fire hazards.

However the project did not proceed as planned. ASSEFA has indicated certain issues faced by them during implementation of the project, viz. the project did not gain popularity among the beneficiaries due to low illumination of the lamps and need for regular visits to charging stations. Community was not managing the asset properly—there was no belongingness and ownership and the maintenance was poor. Each charging station was to be looked after by 3 SHGs covering 50 members. This led to diffusion of belongingness and ownership. The villagers were preferring CFL solar lantern with individual charging panel. In view of this, ASSEFA has proposed to introduce CFL solar lantern

with individual charging panel model in place of LED type solar charging model (LaBL model). ASSEFA has also indicated that the change of activity is possible within the budgeted unit cost approved in the sanctioned UPNRM project. ASSEFA has already covered 250 beneficiaries with LED type solar charging station and CFL lamps will be distributed to the remaining 1750 participants. Thus, the initially envisaged 2000 number of participants will be covered.
The per unit cost of CFL lantern with panel is Rs. 2550 compared to LED type (Rs. 3500). So the total loan requirement would be Rs. 4.46 million as against the balance available at Rs. 5.78 million. ASSEFA proposed to utilise the entire grant assistance sanctioned for training the remaining 1750 participants. The physical units covered will be—250 participants under LED charging station mode and the remaining 1750 participants to be assisted for CFL lanterns with individual solar panels. The total UPRNM loan was revised to Rs. 5.34 million. Considering that the coverage of participants has remained the same, the total grant assistance for accompanying measures sanctioned to ASSEFA remains at Rs. 0.86 million. In the modified project also the loan administration, monitoring and recovery is to be through self help groups. The maintenance work is also to be overseen by the SHGs.
This project of solar home lighting through CFL lantern with individual charging panel has found a wider acceptance than LED lantern with a link to charging station in the remote areas of these backward states, as these are direct replicas of the kerosene lanterns hitherto being used.

Conclusions

The above analysis of the innovative financial products introduced in India for solar lighting systems indicates a favorable combination of all segments as indicated in the SIS theory-technologies, knowledge, institutions actors and networks. The paper underlines the fact that successful product evolution has emerged from a combination of all these factors in all the case studies. The lessons from these projects can be evaluated only on the basis of MIS being received for both these programme, as there is no other data source. The data received indicates that 93 % of the units financed are in working condition. In both the JLG projects of West Bengal the entire loan amount has been utilized and all the units have been installed. By the time of the study 10 monthly installments of school hostel JLG project and 8 monthly installments of solar micro grid JLG project were due and all these installments were fully paid (NABARD 2013). Thus all the loans advanced are being serviced regularly in terms both principal and interest, and there are no defaults. These two parameters are sufficient to underscore the fact of the success of these innovative group based finance products.

To achieve success in financing this scheme a multi-pronged involvement is required from the side of all stakeholders. The scheme can be grounded only with a public-private partnership where a partner organization takes a major share of motivating and convincing people to ensure their participation. The group based financing in West Bengal was successful because, firstly the groups themselves had a common aim of educating their girl children and secondly because the NGO expended considerable energy in building up of the social capital in terms of nurturing and capacity building. Similar social mobilization has taken place in the other project of micro-grids through JLG approach.

The two JLG based projects in West Bengal demonstrate effective community participation for seeking clean environment and also opened avenues for banks for providing loan to JLGs for non-agricultural purposes. These projects proved helpful in the development of sustainable business models for solar lighting system and showcasing the same to mainstream banks for providing bank finance.

Another measure of success of these pilots is the fact that the lead bank of the area, United Bank of India has adopted the same model for financing solar lighting systems in the off-grid areas through JLGs. The main attraction for a mainstream bank was the cent percent loan recovery which was taking place in this project.

Though there was a mid-project change in the product in case of Bihar-Jharkhand project, there too all the units planned for under the project were installed. The mid-course correction in the Bihar-Jharkhand project reflects a poor pre-project community mobilization as well as neglect of peoples' participation aspect by the NGO. This indicates that either before or during the project formulation phase the NGO has not involved the people and elicited their views. The project was thought out for the people by the NGO and its personnel, not by the people themselves. This is an error which many peoples' organizations commit in their over enthusiasm of knowing everything the people require. But to its credit the NGO ASSEFA realized its mistake when the project clientele brought the problems to it, reconfigured the project and switched over to the model desired by the people.

Further this project reveals to how the people are prepared to work in a group for using and servicing the financing product but are not willing to do so for the physical product. This is a reflection of deeply ingrained social mobilization issues and a clear indication as to why in certain objective conditions only certain group based activities work.

Despite the mid-course product change in this project, on one measure of success-loan repayment-it has not faltered. As at the time of the study, 14 loan installments were due and all were repaid fully and in time (NABARD 2013).

The Bihar-Jharkhand project provided the targeted community an alternate source of energy for lighting purposes on a sustainable basis. The benefits out of the project implementation were better living conditions on account of sanitation, hygiene and health. Kerosene, a non-renewable source of energy is becoming costlier and is going out of the access of the poor and marginalized community. Further, the use of kerosene also has adverse health effects due to the smoke emissions. Solar lighting is providing an environment/health friendly and cost effective lighting to the poor villagers and uninterrupted study hours for school/

college going children. The project has also increased livelihood opportunities-increase in working hours for the household wage earners—the poor are mainly engaged in *bidi* (country cigarette) rolling, handicraft works, *tassar* (a type of silk) reeling, etc. In Deogarh district, it is reported that an additional earning of Rs. 15,000 was made by an SHG of weavers (12 members) within a span of 9 months through increased working hours of weaving activity. In Jamui district, the income of the women who are engaged in livelihood activities such as *bidi* rolling, handicraft works etc. has increased by 30 %. Drudgery reduction and health benefits were also reported (NABARD 2013).

The group based projects affirm to the reality that for large sections of the population at the lower end of economic strata installation and maintenance of solar home lighting systems is not a convenient option due to the time and efforts required for that. In group based projects that responsibility is pooled and handled by a trained and active member of the community.

Another measure of success of these projects is that they disprove the conventional theories that the poor require subsidized loans and subsidized interest rates. In all the projects though a possibility of subsidy availability from the government was there the beneficiaries went ahead with the projects without waiting for the subsidies from the government. In the West Bengal project the interest rates on the loan was at 12 % and in Bihar-Jharkhand project the rate charged was 13 %. Both these interest rates were market rates and were well above the average base rates charged by commercial banks. This implies that it is a potential bankable activity for bank credit on commercial terms and market related rates. If banks are not able to look at this opportunity and expand their business with rural clientele one has to doubt their business sense.

The major strength of the project which is derived from the role of NABARD is its ability to bring a variety of partners into the project and enlist their commitment and participation. NABARD is also able to leverage its soft loan and grant support for capacity building of all participants. At the same time NABARD's facilitation gives rise to expectations of subsidizing funding and a longer haul towards mainstream commercial financing of such projects. Balancing maximum private sector participation with minimal subsidies is a main challenge for new rural service delivery mechanism (Reiche et al. 2000).

However based upon the reasonable levels of success which these projects were able to achieve, NABARD should bring on board all banking institutions, especially regional rural banks and cooperative banks to introduce such group based financing schemes in their institutions to mainstream and upscale the financing of such variants of solar lighting systems. There is a vast untapped potential in rural India for these projects irrespective of grid connectivity as even in areas connected with grid, the power supply is highly erratic with regular shutdowns and outages.

The major recommendation which emanates from this paper is with regard to the combination of elements in designing a microfinance energy product in group mode. These elements are the facilitating institution, the financial institution-both for loan and grant, the end users and the environment-socio-political and economic. The manner in which each one of these elements has to play its role is clearly discernable from the case studies.

Further Research Demands

This paper studies the manner in which the projects were conceived, the innovative features of the projects and the early results of their implementation, based on the project reports, monthly progress reports and available MIS. As such it is a study based on primary and secondary data available at various institutions but not on the basis of information collected from actual users or clients. Therefore it requires a further rigorous analysis based on comprehensive data sets. The outcome of these projects has to be studied in a more comprehensive manner after a reasonable time frame of 2–3 years.

The mid-project product change in one of the projects could be an interesting issue for research on the intricacies of community mobilization. The objective reasons for success could be documented in detail.

Client satisfaction with the products, the benefits of the scheme and the cost benefit analysis, with a wide-ranging sample, should be studied at a later date. It would also be worthwhile for the technical aspects of the installed systems to be studied for their performance and utility after a period of, say, 3–5 years.

Notes

1. The currency used in the paper is Indian Rupee with the notation Rs. The exchange rate in the period 2010–2011 to 2012–2013 ranged from Rs. 52 to Rs. 60 per 1 US$.
2. The data and information used in the paper are based on the project proposals submitted by the NGOs, sanction letters issued to NGOs by NABARD, MIS from the NGOs, Annual reports of NGOs-Mlinda Foundation and ASSEFA and annual reports of the national bank for agriculture and rural development for the years 2010–2011, 2011–2012 and 2012–2013.
3. Inputs from my colleagues C.S.R. Murthy and Kaushal Kishore in preparation of this paper are gratefully acknowledged.
4. The scholarship provided by Appui au Developpement Autonome (ADA), Luxembourg enabled the author to participate and present this paper at the International Symposium 'Innovating Energy access for Remote Areas: Discovering Untapped Resources', organized by Berkeley Rural Energy Group and Berlin Technical University at University of California at Berkeley, USA. ADA's financial support is gratefully acknowledged.

References

Al-Saleh, Y. (2010). Systems of innovation as a conceptual framework for studying the emergence of national renewable energy industries. *World Journal of Science Technology and Sustainable Development, 7*(4), 309–334.

Bedson, J. (2007). *'Energy lending: Microfinance and access to modern, alternative energies', FDC briefing note*. Brisbane: The Foundation for Development Cooperation.

Kapadia, K., & Hande, H. (2004). *Issues and options: The potential role of microfinance to expand access to energy services*. Washington, DC: The World Bank.

Malerba, F. (2002). Sectoral systems of innovation and production. *Research Policy, 31*(2), 247–264.

Malerba, F. (2005). Sectoral systems of innovations. In J. Fagerberg, D. C. Mowery, & R. R. Nelson (Eds.), *The Oxford handbook of innovation*. Oxford: Oxford University Press.

Micro Energy International and PlaNet Finance Deutschland. (2010). *Fact sheet the potential of linking microfinance & energy supply*. Deutschland: PlaNet Finance.

Mohiuddin, S. (2006). Expanding the role of microfinance in promoting renewable energy, The Georgetown Public Policy Review. *Winter, II*(1), 2005–2006.

NABARD. (2013). 'Monthly Progress Reports' on RIF and UPNRM of West Bengal, Bihar and Jharkhand Regional Offices of National Bank for Agriculture and Rural Development, various months, Mumbai.

Patel, M. R. (2006). *Wind and solar power systems: Design, analysis and operation* (2nd ed.). Boca Raton, London: CRC Press, Taylor & Francis Group.

Rao, P. S. C., Jeffrey, B. M., Young, D. W., & John, B. B. (2009). Energy-Microfinance interventions in below poverty line households in India. *Energy Policy, 37*, 1694–1712.

Reiche, K., Covarbias, A., & Martinot, E. (2000). *Expanding electricity access to remote areas: Off-grid rural electrification in developing countries* (pp. 52–60). London: World Power.

Saghir, J. (2005) '*Energy and poverty: myths, links and policy issues' energy working notes no*. 4. Washington, DC: Energy and Mining Sector Board of World Bank Group.

SEEP. (2007) '*Using microfinance to expand access to energy services: The emerging experiences in Asia of SEWA, SEEDS, NUBL and AMRET*. Washington DC: The SEEP Network.

Sorensen, B. (2004). *Renewable energy: Its physics, engineering, use, environmental impacts, economy and planning aspects* (3rd ed.). Melbourne: Elsevier.

Terrado, E., Cabraal, A., & Mukherjee, I. (2008). *Designing sustainable off grid rural electrification projects: Principles and practices, technical report*. Washington DC: World Bank.

Wegstein, M. (2010). *Potential of solar home systems, biogas plants and micro-hydro in Nepal and opportunities for MFIs*. Kathmandu: Alternative Energy Promotion Centre.

Chapter 15
Innovative Energy Access for Remote Areas: "The LUAV-Light up a Village" Project

Izael P. Da Silva, Eliza Hogan, Benard Kalyango, Anne Kayiwa, Geoffrey Ronoh and Clint A. Ouma

Abstract The Light-up a village (LUAV) program is a rural development initiative designed to improve access to modern energy solutions in remote areas of developing countries. The initiative addresses the challenge of Pico PV market penetration by empowering rural communities to actively participate in lighting up their own villages using micro-solar systems. The LUAV business model was designed by an energy company, Barefoot Power (BFP), which began the LUAV field in 2012 in Uganda. The program incorporates local SACCOs and Community Based Organizations (CBO) as well as local governmental bodies in the identification and recruitment of participants. A LUAV program is designed to involve at least 100 households per community by providing each home with its own power generation solar system to run lighting and mobile device charging services. The participating households are given the option to either pay for the micro solar power system upfront or to pay for it in 3–12 monthly installments. For this pilot program, BFP sourced for funding from private investors to operate a revolving fund which is managed the SACCOs and CBOs who have the mandate to manage debt recovery

I.P. Da Silva (✉) · G. Ronoh · C.A. Ouma
Strathmore Energy Research Centre, Nairobi, Kenya
e-mail: idasilva@strathmore.edu

G. Ronoh
e-mail: gronoh@strathmore.edu

C.A. Ouma
e-mail: arthurclint@gmail.com

E. Hogan · B. Kalyango · A. Kayiwa
Barefoot Power Pty Ltd., Epping, Australia
e-mail: eliza@barefootpower.com

B. Kalyango
e-mail: benardk@barefootpower.com

A. Kayiwa
e-mail: annek@barefootpower.com

S. Groh et al. (eds.), *Decentralized Solutions for Developing Economies*,
Springer Proceedings in Energy, DOI 10.1007/978-3-319-15964-5_15

and keep the revolving fund active. Through this business model, 18 LUAV projects were implemented in Uganda during the 18 month trial period providing lighting and mobile charging services to 3,000 plus households. The program's success has a growing interest and plans are underway to replicate it in South Sudan, Rwanda and Kenya in 2014. According to the latest count more than 7,000 households have adopted the micro-system through LUAV.

Keywords Micro-Solar systems · Rural village electrification · Revolving fund

Introduction

In East Africa, the majority of the rural population (80–90 %) (Asamoah 2013) has no access to electricity. People light their homes with kerosene or candles, which produce inefficient light quality with poor illumination levels of about 1–10 % the recommended levels; in addition, these light sources pose several health risks to the user such as burns, respiratory ailments and blurry vision among others (Mills 2012).

Furthermore, it was found that the fuel based lighting systems are an expensive source of light (Miller et al. 2013) with some studies reporting savings of up to 400 % in households that completely replace their kerosene lamps with solar lanterns (GIZ 2010). Finally, in most rural communities, the people have no means to charge their mobile phones other than through central charging stations, which charge high prices and are often located far from their homes. Introduction of the solar lighting systems with phone charging options not only reduces the user's costs but provides an income generation avenue by charging the neighbors to use the service which was the case with about 70 % of the users in a study carried out in Uganda (GIZ 2010).

Solutions for these energy related problems have emerged in recent years. Solar powered LED home lighting and phone charging equipment is available throughout East Africa although rural penetration is about 4 % (Lighting Africa 2012). These products provide a safer, more cost effective alternative for rural communities over the typical energy sources mentioned above. However, there are three main challenges limiting the accessibility and uptake of these energy solutions by rural off-grid communities;

- Affordability: The modern energy products have high upfront costs.
- Consumer Awareness: Lack of trust and knowledge about the benefits of good quality solar products.
- Technical Expertise: Good quality installation and continued product service is not available.

Research Objectives

The LUAV program has been designed as response to three main challenges experienced when supplying lighting solutions to rural off grid populations as mentioned in the introduction of this document.

The main research question in this study would be *to find out whether installing community micro-solar lighting/charging systems is a viable way to increase electricity penetration in rural villages. Specifically the study aims at*:

1. Evaluating the willingness and ability to pay for solar lighting systems when presented as a community project as well as preferred payment period/method
2. Evaluate viability and scalability of the revolving fund in providing upfront capital for system installations.
3. Evaluating the receptiveness of the people to the technology and ease of awareness dissemination in the community setting.
4. To evaluate whether technical support is easier to deliver through community members or by technicians from BFP.

Additional Research

The Strathmore Energy Research Centre, SERC, is currently working with development partners to carry out additional research into the sustainability and scalability of the LUAV business model. In this, the long term socioeconomic impact of each LUAV as well as the opportunities and potential for replication in other developing countries is the main focus. As a result of the positive uptake of the LUAV in Uganda, SERC is working closely with BFP, the energy company, in its venture to replicate the program in South Sudan, Rwanda and Kenya in 2014. The main questions for research in addition to the current existing questions are listed below:

1. To evaluate the socioeconomic and lifestyle changes of residents within the successful LUAVs by comparing those who took up the system and those who did not participate in the program.
2. To evaluate the replication and scalability of the LUAV model by comparing the experience in Uganda with Kenya, Rwanda and South Sudan.
3. To measure the impact of capacity building as regards local technicians to repair and maintain the solar systems in the long run. This is an important aspect as in many cases before the trained technician would abandon the task because few systems would fail and thus for economic reasons the person decided to dedicate himself/herself to another kind of job (Fig. 15.1).

Fig. 15.1 Installation of LUAV micro-solar systems in a village by Barefoot Power Uganda

Specific Details of the Program

One of the main challenges of Pico PV market penetration in rural Africa is the lack of consumer awareness (Asamoah 2013). The first step in **awareness creation** for the LUAV program involves the identification of local NGOs, CBOs, SACCOs, community members, community associations and local government bodies or officials who will be willing and able to partner with the energy company in the implementation of the LUAV program. BFP evaluates each group and negotiates mutually beneficial partnerships towards the implementation of LUAV programs. Thus the awareness campaign is meant to address not only the targeted market of end users but also the government officials from the ministry of energy and the regulatory bodies which in the case of Kenya and Uganda are ERC—Electricity Regulatory Commission and the REA—Rural Electrification Agency and all the other above mentioned partners in the LUAV venture.

These partnerships are developed with the aim of raising support to recruit a minimum of 100 households per community who are ready to reduce or stop kerosene usage for lighting within the community by purchasing a micro solar system. It has been reported that local authenticity is one of the main success factors for new businesses in Africa (Accenture 2009). Local authenticity is developed by investing in local expertise and training the local people to run the activities on the ground. The partnerships with local CBOs and SACCOs increases local acceptance and makes the people more receptive to the product and the LUAV program.

Once a community has been educated and is ready to take up a LUAV project, the energy company revaluates the participating CBOs or SACCOs within the community to determine which one is most suitable to carry out the project implementation tasks and trains the members in the project procedures. Next, the energy company makes arrangements to finance the LUAV; in the early stages, BFP financed the venture by putting together a revolving fund from private investors, who initially believed in the concept. Today this process involves multiple stakeholders such as the SACCO member's deposits, NGOs donations, crowdfunding, and savings groups.

The identification of a credible and reliable CBO/ SACCO is crucial to the success of the project. This is because the CBO or SACCO plays the important roles such as managing community promotional campaigns, registering participating households, identifying community entrepreneurs, planning installations, and coordinating product repayment. This is done so that the energy company can focus on its main role of provision of reliable and relevant products, development of awareness campaigns and marketing strategies, continual evaluation of the CBO competence and training of users and local technicians to participate in product installation and maintenance.

The question of product affordability is addressed by the development of financial models that will allow the program participants to pay for the micro solar systems in installments. Although the financial models vary from case to case; the basic structure is that the systems are to be paid for upfront by the SACCO/CBO who thereafter collects the payment from LUAV participants.

In the case where a revolving fund is raised by an outside source, the community has to return it within an agreed time period (normally 12 months). Though, in some cases the source of the funding may allow for it to be utilized as a revolving fund to continue the initiative. The CBO is controlled by a clear MOU that ensures the funds are exclusively used for the LUAV. The CBO will manage the collections of the monthly installments of the households. After 12 months, all collections of installments will have regenerated the revolving fund at the CBO level, ready to finance the next LUAV. The products will also be priced to with 10–20 % margins required for sustainability. This margin can be used to pay incentives to the SACCO personnel in charge of debt collection.

To ensure that the CBO is successful in the initiative, they share in the profit margin to pay for the management of the initiative. This money can be shared among group members or utilized to buy assets for the group.

In May 2013, BFP was able to partner with, a crowdfunding organization to be a financing partner for the revolving fund. This loan is interest free and the borrower has 14 months to pay back. These funds will enable BFP to scale up the LUAV program to three countries.

Once the financial aspects and payment collection processes have been set in place for the LUAV implementation, the CBO will register the participating households, upon which BFP values the systems and deploys technicians to install the micro-solar systems on the houses of each participant. The installation is carried out in parallel with a technician training program to make technical assistance available to the LUAV participants. The local technicians will handle basic questions and trouble shooting of system challenges and the BFP technicians would visit the LUAV only to deal with major faults. One important detail here is the selection of the people to be trained as technicians for the micro system; they have to have already some basic skills such as repairing phones or TV set, etc. In the absence of this they may fail to raise a living from the support to the LUAV systems and thus give up the role of technicians.

Results

The LUAV program relies heavily on partnership building within the local eco-system as well as national and global partnerships. In the beginning, partners were difficult to engage without a proven model. On the other hand, communities were willing and enthusiastic to engage with Barefoot Power to take on the eradication of kerosene and supporting renewable energy as the main source of energy in their community. Over the last two years new partners have committed upon seeing the preliminary positive results. Currently LUAV is actively engaged with partners such as WWF, CARITAS and GIZ and national governments to expand to new countries such as Rwanda, Kenya and Ghana.

At the end of 2012/2013 trial period, BFP had completed 18 LUAV projects. These projects resulted in 3,000 plus households purchasing solar home systems.

These 18 completed projects counted on the support of partnerships with 11 NGO's, 3 SACCO's and one faith based organization. Furthermore, the initiative was supported by the Ugandan Rural Electrification Authority (REA) and the local governments in each village.

It was found that only one out of the 18 projects had not completed repayment of their micro solar systems within the stipulated 12 months and the default were due to the unreliable services of the CBO involved in that particular LUAV.

Although the projects were designed to cater for at least 100 households per community some projects had as low as 28 households while others exceeded the expectations and had up to 500 households signing up and successfully paying for the systems. Table 15.1 shows a summary of the results obtained from the first 14 LUAVs installed in Uganda, all payments were collected in duration of 12 months.

The product of choice in the LUAV program, shown in Fig. 15.2, is known as the Connect 600 from BFP. It consists of a 6 Wp polycrystalline panel with a 4 Ah AGM sealed battery and 4 LED lights which give light for a minimum 6 h once fully charged. Additionally two USB output allows for charging mobile devices such as phones and tablets. A 12 V output provides for radio or fan powering. Every unit comes with a standard two year warranty. The systems currently retail for about 130 USD.

The year of 2014 has seen the advent of more equipment in the market which can be powered by such micro system. One of such is a flat screen television which some smart feature which could play the role of a gateway to internet and thus completely change the level of awareness to the modern world in rural Africa.

Findings

The LUAV program has proven its success in Uganda due to the low delinquency rate and number of successful LUAVs. The findings of the research are currently being used to evaluate the expansion options into the neighboring countries of Kenya, South Sudan and Rwanda. Through the LUAV program, the research team reported the following findings;

Table 15.1 Summary of the results for the first 14 LUAVs installed in Uganda

	Name of project	Partner and/or CBO	No of HHs	Month started
1	Kiprotich village—Kapchorwa	BFPU/MESICS	100	July-12
2	Buswiriri LUAV—Bugiri	CARITAS-JINJA/MESICS	120	July-12
3	Kasese (LUAV)—Kasese	Karambi Sacco/WWF/MESICS	70	Dec-12
4	Kyabarungira Sacco—Kasese	Kyabarungira SACCO	90	Dec-12
5	Kalalu—Iganga	Mivule/solar links	162	Dec-12
6	Friends of nature—Kasese	Friends of nature/WWF	28	Dec-12
7	Okabi—Arua	GIZ/barefoot/community	130	Feb-13
8	Fofo—Nyo	GIZ/barefoot/community	132	Feb-13
9	Tororo LUAV1—Mbale	CARITAS-tororo/mesics	500	Jun-13
10	Tororo LUAV2—Mbale	CARITAS-tororo/mesics	500	Jun-13
11	Kiwani—Iganga	Mivule/solar links	140	Aug-13
12	Maddo LUAV-Masaka	CARITAS-masaka/MESICS	300	Sept-13
13	Kiyinda LUAV—Mityana	Kiyinda-Mityana diocese	200	Sept-13
14	sos children's village—fort portal	SOS children's villages	100	Nov-13
	Total		*2,572*	

Fig. 15.2 BFP Connect600 which is the micro system utilized in the LUAV project

- A 12 month payment period for micro solar systems is considered affordable in rural Uganda. The payments were between 10 and 20 USD per month and the delinquency rate was less than 10 %.
- Community Based Organizations (CBO's) play an important role in facilitating the projects. The sustainability of such a project has been found to depend

heavily on the CBO's ability to manage the local aspects of the project including addressing the community concerns and managing the finances involved at the community level (Da Silva et al. 2011).

- Offering installation services in addition to the technology reduces failure rate of systems and provides an opportunity to build local technical capacity.
- Local partnerships build brand trust and loyalty in African markets. The element of local authenticity helps to overcome the adverse effects of market spoilage and other challenges such as cultural relevance, product education and peer influence (Savannah Fund 2013).
- Financing partners, institutions were more willing to fund the project once it was clear that they would be dealing with groups of users, SACCOs and CBOs rather than individuals. This makes it easier for accountability, to follow up payments and provides better control mechanisms such as the tracking and documentation of the project's success rate.
- The project yielded low default rates as the members of the group encourage each other to complete payments through peer influence consequently reducing the risk of delinquency due to the individual human factors. As a result, only one out of the 18 LUAV projects launched had cases of delinquency.
- Strict criteria are necessary when selecting the CBO that will partner with the energy company to carry out the project implementation. Criteria of competence must be established by the energy company for each case. This is necessary because each community is different in terms of socioeconomic activities, cultural tendencies and social make-up.
- In the LUAV trial period the collection of payments by the CBO proved to be a major bottle neck in the project development. There is need to consider the implementation of a Pay As You Go technology which would support, manage and track the revolving funds without incurring extra man power expenditure especially with the rapid increase in number of CBOs, SACCOs and villages that stand to take up the LUAV program.
- The efficiency of the solar technicians trained under the LUAV program will not be fully measurable until the two year product warranty period expires. It is worth noting that there are currently no accreditation criteria or institutions for solar technicians in East Africa. However, once the criteria have been established, the energy regulatory bodies and certifying authorities will be able to extend training services in remote areas through the existing LUAVs.

Scaling up

The LUAV builds on the BFP Reverse Rural Electrification Model described by Da Silva and Sloet (2012) and promotes the notion of decentralized energy generation and distribution. This will have major long-term implications on the considerations of energy policy makers.

To ensure scalability of the LUAV program, BFP developed strategic partnerships with local governments which in turn led to the rapid adoption of the LUAVs in Uganda. These partnerships were set up to gain the authorization, endorsement and support of community leaders who would assist in the mobilization of the community to participate in the program. Led by enthusiastic government officials and popular community leaders, the recruitment of new members went beyond the initial expectations resulting in the recruitment of up to 500 participant households in some LUAVs such as Tororo LUAV 1 and LUAV 2.

The next initiative currently in its initial stages is to partner with local telecommunication companies in the marketing and promotion of the LUAVs as well as financial support. The telecommunication companies have an interest in the penetration of mobile phone charging technologies for rural areas in Africa as these technologies would increase the use and penetration of mobile phones. Leveraging the interest of telecommunication companies and mobile phone service providers is expected to improve the LUAV uptake rate further in remote areas.

References

Accenture. (2009). *Expansion into Africa: Challenges and success factors revealed*. Available from http://www.accenture.com/SiteCollectionDocuments/PDF/Accenture_Strategy_Expansion_into_Africa_POV.pdf. Accessed 10 Mar 2014.

Asamoah, N. (2013). *Catalyzing markets for modern lighting. GVEP ESME GRANTS*. Lighting Africa. www.lightingafrica.com

Mills, E. (2012). Health impacts of fuel-based lighting. In *3rd International Off-Grid Lighting Conference, Dakar, Senegal*.

Miller, C., et al. (2013). *Trust, demand and last mile distribution: The role of head teachers in building Africa's market for portable solar lights*. Available from http://www.solar-aid.org/assets/Uploads/Publications/Small-PV-Conference-Paper.pdf. Accessed 04 Dec 13.

Lighting Africa. (2012). Lighting Africa Market Trends Report 2012. Overview of the Off-Grid Lighting Market in Africa.

GIZ. (2010). *Solar lamps field test Uganda Final Report*. Available from https://energypedia.info/images/7/72/GIZ_Solar_Lamps_field_Report_Uganda_Webversion.pdf. Accessed 27 Jan 2014.

Savannah Fund. (2013). *The challenges and opportunities of start up marketing in Africa*. Available from http://www.savannah.vc/blog/2013/03/16/the-challenges-and-opportunities-of-startup-marketing-in-africa/. Accessed 10 Mar 2014.

Da Silva, I. P., & Sloet, B. (2012). *Barefoot power lights up rural Africa*. Available from http://www.mercatornet.com/articles/view/barefoot_power_lights_up_rural_africa. Accessed 17 Jan 2014.

Da Silva, I. P., Njuguna, P., & Njogu, M. (2011). Rural electrification using off-grid solar PV powered energy Kiosks. In *2nd Symposium Small PV Applications, Ulm, Germany* (pp. 275–283), 6–7 June.

Chapter 16
Exploring the Barriers to Impact Investing in the Sustainable Energy Area in West Africa

Dominique Diouf

Abstract Impact investing can be an important investment vehicle as part of the development of sustainable energy projects in West Africa, especially in rural areas. However, it is still facing major challenges in West Africa. Through an analysis of relevant documents on this subject, this research strives to identify and address barriers to impact investing as part of sustainable energy projects development in West Africa. It leads to actions that might be undertaken to make impact investing an effective instrument that can help boost sustainable energy in this region. Finally, this study identifies avenues for future research.

Keywords Sustainable energy · Impact investing

Introduction

According to the International Energy Agency, 1.3 billion people are without access to electricity and 2.6 billion people are without clean cooking facilities.[1] More than 95 % of these people are either in sub-Saharan Africa or in developing countries in Asia and 84 % are in rural areas. The level of access to energy for Economic Community of West African States (ECOWAS) countries, especially in the rural areas, is the lowest in the world (ECREEE, ITC and Casa Africa 2013). This is a surprising paradox insofar as energy resources are particularly abundant in West Africa. This not only deprives tens of millions of individuals of access to modern energy, but, in addition, it hinders health and environment of local communities and hence their socio-economic development.

[1] http://www.iea.org/topics/energypoverty/.

D. Diouf (✉)
Canada Research Chair in Sustainable Development Management Standards,
Faculty of Business Administration, Université Laval, Pavillon Palasis-Prince 2325,
rue de la Terrasse, Quebec City, QC G1V 0A6, Canada
e-mail: dominique.diouf.1@ulaval.ca; dominique.diouf@gmail.com

S. Groh et al. (eds.), *Decentralized Solutions for Developing Economies*,
Springer Proceedings in Energy, DOI 10.1007/978-3-319-15964-5_16

In recent years, efforts have been made in terms of improvement and extension of electricity systems and development of cleaner energy, such as renewable energy. As an example, between 2004 and 2009, investment in renewable energy in Africa grew from US$300 million to US$700 million and leapt to $3.6 billion in 2010 (Griffith-Jones et al. 2012). Important efforts have been also made to boost sustainable energy projects and policies in West Africa. The ECOWAS Centre for Renewable Energy and Energy Efficiency (ECREEE), based in Praia, Cabo Verde, was established in 2010 to create favorable framework conditions for renewable energy and energy efficiency markets in the 15 member states of the ECOWAS.

Despite these significant advances, the majority of African countries, especially those in West Africa, still face major challenges. Beyond the political, social, environmental and institutional challenges, access to financial services is often considered as one of the major barriers to the development of energy in West Africa.

In the context of renewable energy projects in Africa, the ability of end users to bear the costs related to energy services is certainly one of the major hurdles to the development of sustainable energy (ECREEE, ITC and Casa Africa 2013). To this end, inclusive finance, particularly microfinance, is often used as an alternative financing to access to modern energy. We argue that, beyond microfinance, it is urgent to design and test alternative and complementary financial systems better suited to the emergence of decentralized energy systems in rural areas. Impact investing may be an investment vehicle to boost sustainable energy development in West Africa.

Impact investing is "an investment designed with intent to generate positive social and/or environmental impact" (Morgan and The Rockfeller Foundation 2011) beyond financial returns. This type of investment is gaining popularity as individuals and institutional investors (pension funds, banks, insurance companies etc.) seek ways to obtain financial returns while making positive impact on society and environment (Daragh and Nurkholisoh 2012, p. 2). As impact investments typically target poor and disadvantaged population and the natural environment, they may contribute to stimulate economic growth, generate infrastructure projects and other initiatives (IISD 2013), including the many opportunities offered by the development of sustainable energy.

This research has several contributions. Firstly, it goes beyond inclusive finance by harnessing the power of impact investing for the sustainable energy sector, especially as part of decentralized and off-grid sustainable energy systems. Secondly, by investigating impact investment as an investment vehicle for sustainable energy in West Africa, this research is intended as a shift from traditional approaches in Africa based on philanthropy, development assistance or aid to a new track that consists of investing with impact in the renewable energy area. Thirdly, funders, micro-finance institutions, NGOs, governments and investors, especially impact investors, can refer to this study to better understand the challenges to impact investing in West Africa and consider solutions to address them.

The rest of this paper is laid out in the following manner: first of all, we will clarify our research question and objectives. This will be followed by the presentation of the methodological issues. Then, we will describe our findings. Finally, we will discuss the results, highlight both the scope and the limits and identify possible avenues of research.

Research Objectives

Impact investing could contribute to solve major social and environmental problems in the developing countries, particularly in Africa (Dalberg 2011) where governments lack sufficient resources to cope with social and economic challenges. More specifically, it would be an important vehicle for the governments who are committed to fighting against poverty, energy crisis, youth unemployment, agricultural crisis and food security etc. Nevertheless, actions taken by the government to boost impact investing are few and even scattered. To this, are added barriers to the growth of impact investing.

The following fundamental question structures this research: **What are the challenges associated with impact investing in the sustainable energy area in West Africa?**

The main objective of this research is to examine the barriers that might prevent from investing with impact in the sustainable energy sector in West Africa.

Impact investment funds vary with respect to the source of the capital, the location, expectations (Daragh and Nurkholisoh 2012). Freireich and Fulton (2009), in a report published by the Monitor Institute, identify two kinds of impact investors: impact first investors and financial first investors. Impact investors seek to optimize social or environmental impact with a floor for financial returns. They primarily aim to generate social or environmental good, and are often willing to give up some financial return if they have to (p. 31). Financial first investors seek to optimize financial returns with a floor for social or environmental impact. They are typically commercial investors who seek out subsectors that offer market-rate returns while achieving some social or environmental good (p. 31). Grabenwarter and Liechtenstein (2011) highlight five (05) key characteristics to impact investing: (1) Profit orientation; (2) Correlation between impact and financial return; (3) Intentional impact; (4) Measurable Impact; (5) Positive effect on society (p. 10). These authors state that there is no trade-off between profit and social impact because the two factors are positively correlated. Simon and Barmeier (2010) emphasize the fact that impact investing targets regions and sectors that traditional direct foreign investment does not. As part of this paper, we argue that, whatever the expectations, impact investing may be an important investment vehicle for the development of sustainable energy. In this respect, it is important to understand the barriers to this type of investment.

This research strives to not only examine the barriers to impact investing. Thus, it is intended to foresee solutions to barriers identified as part of this study and open up avenues of research in the field of impact investing applied to renewable energy in rural areas.

Methods

This research is mainly based on a review of existing papers, reports and other documents analyzing impact investing in the African (especially West African) context. Particular attention is given to the relationship between impact investing and renewable energy. Consulting secondary sources may be particularly useful in the early stages of research for generating sensible hypotheses or for other aspects of the research development (Cowton 1998, p. 429). In order to ground this research in a real context (Yin 2003), West Africa is chosen as a case study.

According to the data from the World Bank,[2] West Africa key facts are:

- 15 countries.
- A population of 245 million.
- About 65 % of them live in rural areas.
- Two sub-regional organizations: Economic Community of West African States (ECOWAS): Benin, Burkina Faso, Cape Verde, Cote d'Ivoire, Gambia, Ghana, Guinea, Guinea Bissau, Liberia, Mali, Niger, Senegal, Sierra Leone and Togo. ECOWAS works to promote co-operation in the region on a range of economic and political issues including conflict resolution. Eight countries in the region (Benin, Burkina Faso, Cote d'Ivoire, Guinea-Bissau, Mali, Niger, Senegal and Togo) are members of the West Africa Economy and Monetary Union (WAEMU) and share a common currency, a common central bank, a development bank, a regional stock exchange and a common banking regulator.
- Over 55 % of West Africans live with less than $1 a day. According to the Human Development Index of the UNDP, Ghana and Cape Verde were the only countries ranked in the medium human development category. The rest of the ECOWAS countries were classified at the low human development level.

West Africa is also a hub for several organizations:

- Multinational companies: extractive industries, telecom, banks etc.
- NGOs that contribute to eradicating poverty promoting education, preserving the environment.
- Foundations.
- UN organizations.
- Etc.

Despite political instability noted in some countries, West Africa continued to grow faster than other sub-regions of the continent (ACET 2013). However, it should be noted that West Africa is not a homogeneous sub-region. Some countries are considered as rich-resource countries, while others are less fortunate. The same applies to the adoption of policy and financing sustainable energy projects

[2] http://web.worldbank.org/WBSITE/EXTERNAL/NEWS/0,,contentMDK:20179737~pagePK:34370~piPK:42768~theSitePK:4607,00.html.

financing. As such, Ghana could be considered as a model for other countries in West Africa. Thus, barriers or challenges identified in this research are those most of West African countries face in terms of impact investing.

Results

The complexity of regulations, corruption, political instability and weak communication networks are, among others, the main challenges to consider when investing (even with impact) in emerging markets (Arosio 2011). This observation also applies to African markets. As for impact investing strictly speaking, the IISD (2013) has identified 8 factors as the most critical challenges to the growth of the impact investing industry in low-income and developing countries. These challenges are (1) lack of appropriate capital across the risk/return spectrum, (2) shortage of high-quality investment opportunities with track record, (3) difficulty exiting investments (exit options), (4) lack of common way to talk about impact investing, (5) lack of innovative deal/fund structures to accommodate portfolio companies' needs, (6) inadequate impact measurement practice, (7) lack of research and data on products and performance, (8) lack of investment professionals with relevant skills.

According to a report by Dalberg and Apix (2012), beyond the lack of awareness, the barriers to impact investing come from the demand challenges (lack of adequate financing sources, limited capacity building, no recognition of particular needs of impact enterprises etc.), supply challenges (lack of investment vehicles, limited deal flow, limited exit options) and directing capital challenges (lack of clarity under the fiscal framework). Even though West African Governments are making efforts to establish an institutional framework conducive to the development of business and investment flows, there are still major challenges, particularly in the field of impact investing. One of these challenges is how to get institutional investors (commercial banks, pension funds, insurance companies etc.) on board with impact investing. Despite the important role they play in the West African economy, investors are still very conservative "and stick to investing in treasury bills and real estate" (Dalberg 2011, p. 39). This conservatism is even greater when it comes to investing in the sustainable energy sector, especially the rural areas. This sector is still considered as unprofitable and high risk and investors, particularly financial-first-investors, are often wary even suspicious.

As previously mentioned by the IISD (2013), lack of reliable data is one the most critical challenge to the development of impact investing. Investors with a long-term time horizon, such as impact investors, are, indeed, more and more interested in environmental, social and governance issues and ask, beyond the financial reports, more detailed information (CICA 2010). The lack of information on financial and non-financial performance as well as the absence of credible reports on the impact of their activities are some of the great challenges most of impact investors are facing in West Africa (Dalberg 2011). This can have a negative impact on investment and capital flows in Africa, especially in the context of sustainable energy development projects. In West Africa, the non-availability of reliable and updated energy information creates a major constraint for investors and project developers in the sustainable energy sector in the ECOWAS region.[3]

Another barrier to impact investing in West Africa is related to the electrification system. According to ECREEE, ITC and Casa Africa (2013), rural electrification is highly dependent on the national electricity plan. This leaves too little room to autonomous, decentralized and cheaper rural mini-grid systems fueled by renewable energy. This also inhibits entrepreneurship and projects development in sustainable energy sector in West Africa. Therefore, investments, including impact investing, in renewable energy in the rural areas are still low.

Discussion

The following question structured this research: what are the challenges associated with impact investing in the sustainable energy area in West Africa? The main objective was to examine the barriers that might prevent from investing with impact in the sustainable energy sector in West Africa. To do that, we relied on both academic and professional literature to identify some of the major constraints to impact investing in West Africa. We found that barriers to impact investing in West Africa come from regulatory, financial, fiscal, political etc. challenges and the lack of reliable data. Moreover, the national electricity systems in place tend to limit the establishment of autonomous and cheaper rural power systems and, thus, to inhibit entrepreneurship and projects development in this sector.

The scope of this research lies in its exploratory nature. By identifying and analyzing barriers to investing impact, this study recognizes the potential of such an investment to help fight energy poverty in West African rural areas. However, such challenges can only be met if concrete measures are taken by policy makers. Because it does offer benefits, impact investing has valid arguments for garnering the support of governments (IPCV and IRI 2011). We believe that the current context of fuel poverty requires open and bold actions. For this purpose, it is important to undertake tree major actions.

Firstly, the countries of West Africa should promote autonomous and decentralized energy systems using renewable energy. As they better adapt to different socio-economic contexts and be rooted in local cultures, such energy systems may contribute to an integrated community development. Secondly, we think it is also important that institutional investors embark on impact investing by exploring and investing in areas such as renewable energy. To date, as mentioned by ECREEE, ITC and Casa Africa (2013), few benefits are granted to private capital to invest in renewable energy in West Africa. In addition, investments in this area are often funded by official development assistance (ODA). As private investments are

[3] http://www.ecreee.org/page/knowledge-management-and-awareness.

largely dependent on the policies in place (IPCV and IRI 2011), institutional investors should, however, benefit from a favorable environment to help boost energy development in rural areas. Future studies might explore the ways governments can tailor investment incentive programs to target the different types of impact investors and involve them in the promotion of sustainable energy in West Africa. While our recommendations suggest that solutions should come from policy makers, an ecosystem approach could also be considered in the search for solutions to the challenges of access to energy. This approach would better highlight the complexity surrounding the sustainable energy issues.

Despite its scope, however, this research has limitations. One of the key limitations is that it is based solely on documentary research. Interviews conducted in situ with major stakeholders could give a more practical character. Future research might also rely on renewable energy projects in rural areas as case studies. Future research might also use a critical approach by exploring, for example, the extent to which impact investing might crowd local investments.

References

African Center for Economic Transformation (ACET). (2013). *Good governance aspects of sustainable development challenges and opportunities in the mineral, oil, and gas sector in West Africa.* Sustainable Resources Series.

Arosio, M. (2011). *Impact investing in emerging markets, responsible research.*

Canadian Institute of Chartered Accountants (CICA). (2010). *Environmental, Social and Governance (ESG) Issues in Institutional Investor Decision Making.*

Cowton, C. J. (1998). The use of secondary data in business ethics research. *Journal of Business Ethics, 17*(4), 423–434.

Dalberg. (2011). *Impact investing in West Africa, Dalberg Global Development Advisors.*

Dalberg, & Apix. (2012). *Assessment of impact investing in Senegal.*

Daragh, L., & Nurkholisoh, A. (2012). *Impact investing in emerging countries, insights from the due diligence process.*

ECOWAS Centre for Renewable Energy and Energy Efficiency (ECREEE), Casa Africa, & Instituto Tecnológico de Canarias (ITC). (2013). *Renewable energy in West Africa: Status, experiences and trends.*

Freireich, J., & Fulton, K. (2009). *Investing for social and environmental impact: A design for catalyzing an emerging industry.* Monitor Institute Available online http://www.monitorinstitute.com/impactinvesting/documents/InvestingforSocialandEnvImpact_FullReport_004.pdf.

Grabenwarter, U., & Liechtenstein, H. (2011). *In search of gamma-an unconventional perspective on impact investing.*

Griffith-Jones, S., Ocampo, J. A., & Spratt, S. (2012). *Financing renewable energy in developing countries: Mechanisms and responsibilities.*

International Institute for Sustainable Development (ISSD). (2013). *Overcoming Barriers to Scale: Institutional impact investments in low-income and developing countries.*

IPCV & IRI. (2011). *Impact investing, a framework for policy design and analysis.*

Morgan, J. P., & The Rockfeller Foundation. (2011). *Impact investments—An emerging asset class.*

Simon, J., & Barmeier, J. (2010). *Impact investing for development.* Chicago.

Yin, R. K. (2003). *Case study research: Design and methods* (3rd ed.). Thousand Oaks, CA: Sage.

Chapter 17
The Synergies Between Mobile Phone Access and Off Grid Energy Solutions

Michael Nique and Helene Smertnik

Abstract Mobile connectivity has grown beyond the electricity grid in most emerging markets: the slow growth of grid access over the last 10 years compared to the rapid expansion of mobile networks has widened the existing gap between access to mobile and access to electricity. The GSMA estimates that this gap is equivalent to more than 643 million people covered by mobile networks without access to electricity, representing up to 53 % of the global off grid population. According to five mobile channels, namely the off grid telecom tower infrastructure, mobile operators distribution networks, machine to machine connectivity, mobile payments and mobile services, the mobile industry offers innovative pathways to achieve reliable energy access for currently underserved communities.

Keywords Mobile phone · Mobile payment · Off grid telecom tower

Introduction

1.2 billion people don't have access to electricity and 2.8 billion have to rely on wood or other biomass fuels to cook and heat their homes (Banerjee et al. 2013). In order to reach universal, affordable and sustainable access to energy by 2030 (as stated by the United Nations), there is a need to accelerate the current pace of expansion of electricity access to keep up with the demographic growth but also develop innovative ways to provide decentralized and sustainable energy to off grid populations.

In a context of poor access to basic services such as electricity and water, access to mobile services and mobile phone ownership have grown in the last 10 years.

M. Nique (✉) · H. Smertnik
GSMA, Mobile Enabled Community Services (MECS), London, UK
e-mail: mnique@gsma.com

H. Smertnik
e-mail: hsmertnik@gsma.com

S. Groh et al. (eds.), *Decentralized Solutions for Developing Economies*,
Springer Proceedings in Energy, DOI 10.1007/978-3-319-15964-5_17

The growth and increased maturity of mobile markets, based on ubiquitous population mobile coverage, extensive distribution networks and affordability of mobile services and mobile devices, is offering new market-based opportunities for energy practitioners.

Research Questions

- What is the current status of mobile availability and uptake in emerging markets?
- What is the size of the opportunity for energy practitioners to leverage mobile?
- What are the different mobile channels available to support energy access?
- How innovative business models such as mobile-based Pay As You Go solutions can disrupt the off grid solution market?

Methodology

This article was implemented based on the work the GSMA Mobile Enabled Community Services program carried out in 2013. The sections on the five mobile channels to support access to energy and the Pay As You Go example were developed from interviews with Energy Service Companies leveraging mobile infrastructure, technology and services part of their business model.

Related to the market sizing section, the energy addressable market is an estimate of the number of people who live within range of GSM networks and have no access to electricity, but could be hence impacted by the deployment of mobile-enabled services. For a total of 114 developing countries, the energy addressable markets were calculated by overlaying the following data in urban and rural locations on a country basis:

- The percentage of the population with access to electricity (2013 from IEA "electrification rate" and World Bank "access to electricity");
- The percentage of the population being covered by GSM networks (the most recent data available from mobile operators and GSMA GIS based estimates).

The Ubiquity of Mobile in Emerging Markets

As of 2013, more than 4 out of 5 people living in emerging markets were covered by mobile networks and 2.3 billion people had a mobile phone subscription, with an estimated growth of up to 3 billion in 2017.[1] The Fig. 17.1 illustrates how the

[1] GSMAi 2013 on the number of Unique Subscribers.

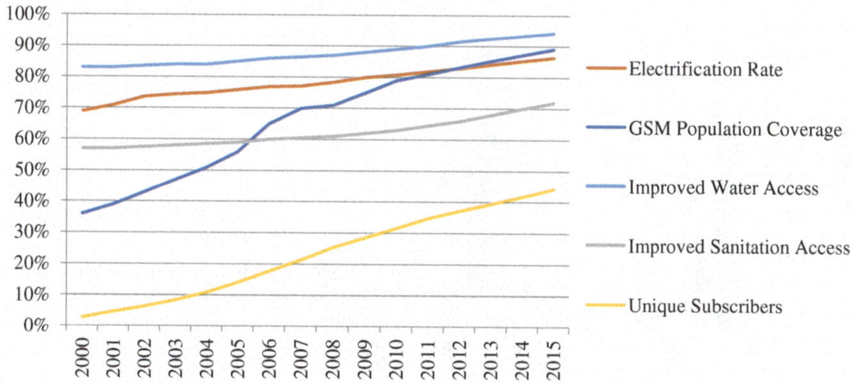

Fig. 17.1 Mobile growth versus utility access growth in emerging markets

global growth of mobile networks has outpaced the expansion of the electricity grid, clean water and sanitation systems in emerging markets.

Leveraging this mobile ubiquity and increasing maturity of the mobile industry, and in conjunction with the cost reduction of renewable energy sources in recent years (PV module prices have fallen by 80 % since 2008 and by 20 % in 2012 alone (Liebreich 2013)], there are strong opportunities for public and private players to leverage the mobile infrastructure, technologies and services to provide sustainable access to energy to current off grid communities, or communities with poor access to the grid.

A Majority of the Off Grid Population is Covered by Mobile Networks

As of mid-2013, the GSMA Mobile Enabled Community Services program estimated the global energy addressable markets, e.g. the number of people covered by mobile networks without access to electricity, at more than 643 million people, representing up to 53 % of the global off grid population (Nique 2013). Out of this total, more than 476 million people live in rural areas.

As presented in Fig. 17.2, the largest addressable market is Sub Saharan Africa (359 million people) where the reach of electricity networks remains limited (~32 % of the population[2]) but where mobile networks cover more than 74 % of the population. In East Africa, Kenya, Tanzania and Uganda accounts for more than 82 million people who could benefit from the access to mobile-enabled energy services.

[2] According to the electrification rate in 2012 from the International Energy Agency.

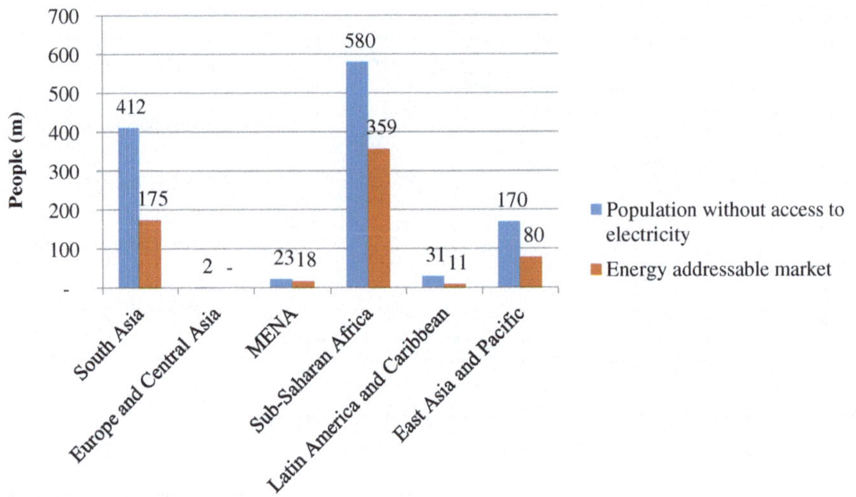

Fig. 17.2 The largest addressable market is Sub Saharan Africa

Five Mobile Channels to Enhance Access to Energy

Based on the current footprint and maturity of the mobile industry, five mobile channels can support access to energy solutions (Table 17.1 summarizes impact of these channels):

- **Mobile Tower Infrastructure**: The mobile tower infrastructure has grown beyond the electricity grid by relying on its own power solutions, mainly diesel generators. At the end of 2011, more than 346,000 towers[3] were operating in off-grid environments. In a context where diesel costs are rising, leading to high operational charges, increasing synergies exist between operators and Energy Service Companies willing to provide stable energy to telecom towers and communities under an energy hub or micro-grid model.
- **Mobile Operator Distribution Network**: The mobile operator's airtime distribution channels and mobile money agent networks are becoming national distribution networks providing last mile access to mobile communication solutions. For example in Kenya, there were more than 96,000 mobile money agents across the country in April 2013 (DiCastri and Gidvani 2013). These kiosks and shops carry a trusted brand, reach remote communities and can scale-up responsively to the needs of potential customers.
- **Machine to Machine Connectivity**: Machine to machine (M2M) connectivity enables point to point communication thanks to embedded wireless chipsets within the utility system. With GSM networks serving as the communication

[3] GSMA Green Power for Mobile.

Table 17.1 Five mobile channels impact summary

Channels	Impact
Mobile tower infrastructure	Increase sustainability of decentralized micro-grids by providing power for consumptive and productive use
Mobile operator distribution network	Support last mile delivery services for off grid products (e.g. home solar systems)
	Improve customer awareness and trust in emerging energy solutions by co-branding products
Machine to machine connectivity	Improve maintenance through remote monitoring
	Enable Pay As You Go functionality, improving energy solutions affordability
	Improved user centric design thanks to consumer usage patterns collection
Mobile financial services	Increasing system affordability through Pay As You Go solutions
	Improving payment efficiency
	Enabling private energy connection finance
	Proposing smart tariffs based on customers energy usage or time of usage
Mobile services (SMS, USSD, Applications)	Improve utility agents business capability through mobile tools usage (e.g. customer relationship management)
	Enable the collection of crowd-sourced information directly from customers
	Improve supply chain management through mobile platforms

backhaul, decentralized energy solutions can be remotely monitored, allowing real time information to be channeled directly from energy solutions to service providers and enabling an on/off switch function, critical for the Pay As You Go (PAYG) business model.

- **Mobile Financial Services**: More than a money transfer platform, mobile financial services are now enabling subscribers to buy products, pay their bills and save money through their mobile phone. As of the end of 2012, there were more than 30 million active mobile money subscribers worldwide with more than 150 live deployments in 72 countries (Penicaud 2013). The ability to leverage mobile phones for energy payments is now enabling the development of PAYG solutions tailored to low-income customers.
- **Mobile Services (SMS, USSD, Apps)**: Mobile handsets are increasingly available and affordable in the developing world through formal and informal distribution channels. In Kenya, the monthly total cost of ownership[4] of a low

[4] The Total Cost of Ownership of a mobile handset refers to the amount a mobile user has to spend per month to own and use a mobile handset. This include the price of the mobile handset and an average use of mobile services (voice, SMS, data).

cost mobile handset decreased from EUR10.11 in 2009 to EUR2.83 in 2011.[5] This pervasiveness of mobile phones in rural and underserved locations can be leveraged in different ways to enhance sustainable access to utility services and products.

Emerging Mobile-Based Energy Business Model—The Pay as You Go Example

While underserved populations spend an important proportion of their income on hazardous energy solutions (up to 30 % of their yearly income mainly on fossil fuels, such as kerosene (Hammond et al. 2007)], part of the same population cannot afford to buy clean energy solutions due to their high upfront costs. There are however evidence supporting the hypothesis that offering payment plans to customers increases the rate of adoption of solar products as they become more affordable.[6]

The Pay As You Go model, developed on the brink of mobile financial services deployment, are now allowing entrepreneurs to offer home solar systems under a micro-financed or a "solar as a service" scheme, enabling low income customers to pay for energy directly via their mobile phones:

- **Lease to own (micro-financed)**: An Energy Service Company (ESCo) offers a micro-loan solution to their customers to afford a home solar system; customers first have to make a down-payment to have the home solar system installed at their house and then repay for the full price of the unit part of their energy consumption via daily, weekly or monthly instalments. Once they repay for the full cost of the product, they own the home solar system and can use it freely (if the unit is GSM enabled, the unit internal switch will be permanently unlocked without any agent intervention). Products usually come with a warranty of 1–3 years according to contract terms.
- **Solar as a Service**: an Energy Service Company provides a service to its customers while the home solar system remains the property of the service provider. An installation fee is charged to new customers and the service is then provided on a prepaid basis (amount according to the solution capacity). Energy prices are usually lower than in the "lease to own" model as the solar system and other products provided by the ESCo (lights, TV, fridge, …) remain its property. Full maintenance is also ensured under the service agreement with the end user.

[5] Source Nokia from GSMA MDI website: https://mobiledevelopmentintelligence.com/statistics/70-monthly-total-cost-of-ownership-usd#.

[6] Information collected from several Energy Service Companies providing Energy Pay As You Go solutions.

Adding a GSM component to an energy system is the most seamless solution for PAYG, as remote monitoring and credit update on the unit meter can be done over the air, without an agent or a user intervention. Service providers receive real time information under a SMS format about the unit operations (power consumption, battery charge/discharge), customers' payments (frequency of payments, credit) and any maintenance/theft issues. ESCos are also building extensive databases on unit operations, which are then analyzed to offer better products and services from a user centric perspective. M-KOPA, providing home solar systems under a Pay As You Go model in Kenya, has been to date the most successful Pay As You Go provider, reaching 50,000 units sold as of January 2014. The deep integration of mobile technology in their business model (as seen in Fig. 17.3), M2M connectivity coupled to mobile payments, added to their distribution partnership with leading mobile operator Safaricom, has enabled this emerging ESCo to provide reliable products and stable services.

Most of the PAYG solutions are currently available in East Africa as shown in the Table 17.2; with poor access to electricity, good mobile penetration and

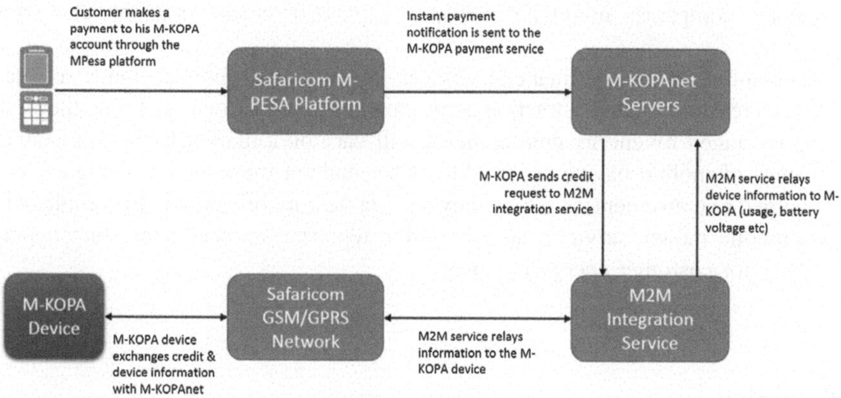

Fig. 17.3 M-KOPA Pay As You Go model

Table 17.2 Example of energy Pay As You Go providers (as of the end of 2013)

Companies	Country operations	Service model	Payment type
M-KOPA	Kenya, Uganda	Lease to own	Mobile money
Mobisol	Tanzania, Rwanda	Lease to own	Mobile money
Off grid electric (OGE)	Tanzania, Ghana	Solar as a service	Mobile money
Angaza design	Kenya, Tanzania	Lease to own	Mobile money
Fenix international	Uganda	Lease to own	Mobile money
Econet solar	Zimbabwe, Lesotho, Burundi	Solar as a service	Airtime billing
Simpa networks	India	Lease to own	Scratch cards or mobile payments

increasing mobile money services traction, this region has become the cradle for mobile-based energy PAYG solution deployments. However, the opportunity to deploy and scale PAYG is real and important in most of the global off grid regions, provided the right ingredients are present: quality energy products, mobile payments capability (using mobile money but also mobile airtime), working capital available to energy entrepreneurs financing the risk of customers default and an efficient distribution network.

Challenges of Such Mobile Enabled Solutions

- Financing—while pay as you go solutions improve energy affordability, the financing burden falls on Energy Service Companies providing such products or services. Entrepreneurs have to wait for the duration of the payback periods to recoup their sales revenue, a period that can be as long as 36 months for some providers. This puts an important pressure on cash flow availability, especially for customers with high default risks. Without debt financing tools and working capital, companies might be unable to expand or ensure efficient after sales services.
- Availability of mobile financial services—even though mobile money services are increasingly gaining traction across markets, their growth and how they can be leveraged by energy entrepreneurs will vary on a market basis. The convenience of mobile payments should not be made at the expense of higher fees charged to consumers each time they pay via their mobile phone. In complement to mobile money services, airtime billing represent another interesting opportunity for customer energy payments.

Discussion

The success of mobile telecommunications in emerging markets is enhancing the opportunity for energy practitioners to develop innovative access models tailored to off grid and underserved communities' ability to pay. According to the GSMA, five mobile channels, based on the reach and impact of the mobile infrastructure, technologies and services, appears key to support such innovation. The development of Pay As You Go (PAYG) solutions under a micro-loan or solar as a service model, where units can be remotely monitored through machine to machine connectivity over mobile networks and where customers can make payments directly via their mobile phones, can act as a paradigm in displacing hazardous fossil fuels for the off grid households with access to mobile. As mobile markets mature and mobile financial services get more traction, most PAYG pilots could move in a commercial phase in 2014, with new entrants in different parts of Africa, Asia and

Latin America. Still at a nascent stage of development, technology and business models innovation should lead to the emergence of more synergetic models, coupling mobile, energy but also water, some of the key pillars to socio-economic empowerment.

References

Banerjee, S. G., Bhatia, M., Azuela, G. E., Jaques, I., Sarkar, A., Portale, E., et al. (2013). *Washington overview. Global tracking framework. Sustainable energy for All D.C.* Washington, D.C: World Bank.

DiCastri, S., & Gidvani, L. (2013). *The Kenyan journey to digital financial inclusion. GSMA mobile money for the unbanked.* Bankable Frontier Associates.

Hammond, A., Kramer, W. J., Tran, J., Walker, C., & Katz, R. (2007). *The next 4 billion—Market size and business strategy at the base of the pyramid.* World Resources Institute.

Liebreich, M. (2013). *Scaling up financing to expand the renewables portfolio.* Bloomberg New Energy Finance.

Nique, M. (2013). *Sizing the opportunity of mobile to support energy and water access.* GSMA Mobile Enabled Community Services.

Penicaud, C. (2013). *State of the industry—Results from the 2012 global mobile money adoption survey.* GSMA Mobile Money for the Unbanked.

Chapter 18
Optimizing Device Operation with a Local Electricity Price

Bruce Nordman and Mattia Bugossi

Abstract Making optimal use of available electric power is important for efficiency, functionality, and to reduce capital costs, particularly in developing countries. This paper shows the results from simulating the behavior of refrigerators and freezers that vary their operation according to a local price and price forecast. The price is set to the availability of local photovoltaic (PV) power and is used to adjust the temperature setpoints of the devices. For off-grid systems, this can be used to concentrate consumption during times of PV availability, to increase efficiency and reduce battery size. Our simulations show a reduction of up to 26 % of the energy used by the devices at night.

Keywords Off-grid · Nanogrid · Local price

Introduction

In electricity systems of many scales, matching supply and demand is a critical need. This is a basic function of any grid—utility grid, microgrid, or nanogrid (Marnay et al. 2011). For energy access deployments, there are often significant limits on system capacities for energy and power, and these may vary from day to day. This paper demonstrates that a local price of electricity can be used to shape demand to use it more optimally. The context addressed is a standalone system

B. Nordman (✉) · M. Bugossi
Lawrence Berkeley National Laboratory, 1 Cyclotron Road, 90-2000, Berkeley, CA 94720-8130, USA
e-mail: BNordman@lbl.gov

M. Bugossi
UNIVPM—Università Politecnica delle Marche, P.zza Roma 22, 60121 Ancona, Italy
e-mail: MBugossi@lbl.gov

S. Groh et al. (eds.), *Decentralized Solutions for Developing Economies*, Springer Proceedings in Energy, DOI 10.1007/978-3-319-15964-5_18

powered with local photovoltaic (PV) power backed up with a battery. Refrigerators and freezers are modeled to show how price can be used to better match their consumption to electricity availability.

Research Objectives

The purpose of this study was to explore the general principle of using local prices as a way to control devices—or rather, so devices can control themselves. This is a core principle of the technology of Local Power Distribution (LPD) (Nordman 2012; Nordman et al. 2012 and Nordman and Christensen 2013). LPD is based on a network model of power, rather than the conventional single unitary grid. A network of local grids can be attached to a large scale utility grid, a small mini-grid, or a local micro-grid. The basic unit of power distribution in LPD is the nanogrid, and each nanogrid has its own local price and price-forecast. This paper considers only a single nanogrid, including integral storage, plus a connection to local photovoltaic power.

Data Sources

With the purpose of this study being exploratory, it was not necessary to be detailed in our modeling. For example, the efficiency of storing energy into and out of a battery is assumed to be constant, and we do not consider the interaction between charge level and capacity to store or withdraw power. The details of compressor operation are not modeled; each minute of operation is assumed to have an identical effect on reducing temperature in our refrigerators and freezers.

For PV data, we used sample data from a convenient source.[1] The loadshape of produced power was simply scaled to match our system needs and the hourly output linearized. Figure 18.1 shows one day of PV system output, unscaled, along with a price signal trivially derived from the PV output. For batteries, we assume that 10 % of electricity is lost in the charge/discharge roundtrip.

Our base case of device operation was modeled on typical behavior of a recent models of refrigerators and freezers.[2] Our units are manual defrost so there is no automatic defrost cycle to schedule (though these could be readily set to be at times of high electricity availability). We did not model door openings.

[1] Adapted from http://www.seia.org/research-resources/potential-impact-solar-pv-electricity-markets-texas.

[2] Personal communication from Lloyd Harrington, 2013.

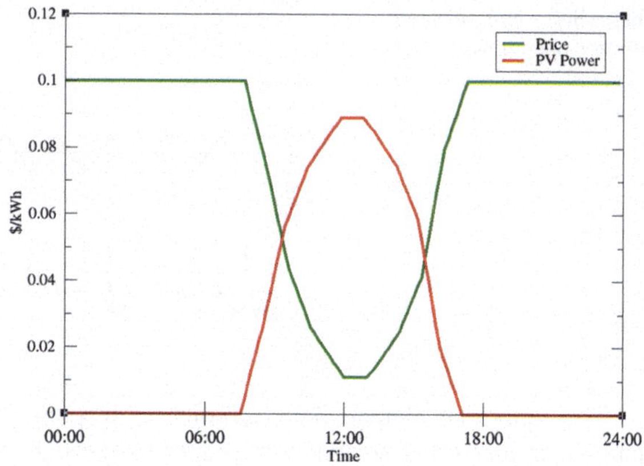

Fig. 18.1 PV daily electricity loadshape and local price

Prior Work

Matching supply and demand is not only of concern for off-grid systems, but arises in most electricity systems. For improved operation of conventional utility grids, a project at U.C. Berkeley (Taneja et al. 2011, 2013) demonstrated a phase change material used in the freezer compartment of a typical refrigerator (with integral freezer) to shift demand. Energy is required to "charge" the material as it freezes; it later absorbs heat as it melts. They considered how such a system would behave in the context of a fixed time-of-use price regime. They also operated it to follow renewable generation in the local utility grid and select among a few different temperature setpoints to shift energy demand to the desired times.

Our project builds on this concept but with different context and goals. We take the appliances as-is, without physical modification; a local price forecast is the only information considered; and the setpoint is directly manipulated in a continuous manner.

Individual Device Operation

Figure 18.2 shows a six hour period of our basecase freezer operation. There is a constant setpoint (−10 °C) so that the compressor cycles do not vary, with compressor on-times and off-times of about 20 min each. The unit turns on the compressor when the internal temperature reaches 0.5 °C above the setpoint, and turns off the compressor when 0.5 °C below the setpoint is reached. Our basecase of refrigerator operation is similar. We assume that the freezer setpoint can range from −3 to −18 °C, and the refrigerator setpoint from 1 to 6 °C. These are to give the system maximum flexibility without compromising the integrity of the units function.

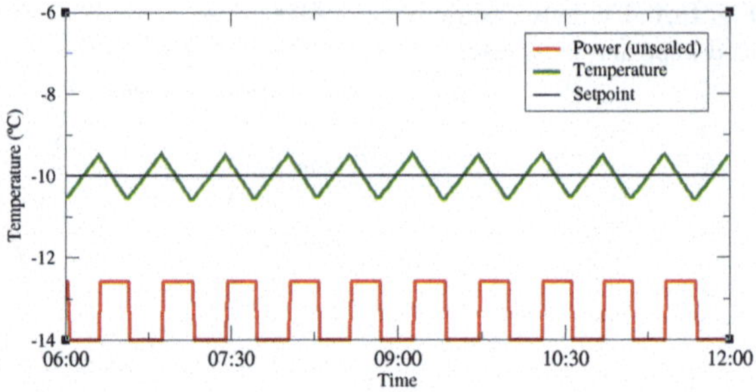

Fig. 18.2 Constant price freezer operation

The units can change their setpoint in response to the local price. This can happen at any time, so that a setpoint change may lead to immediate change in compressor state, or may cause the expected duration of compressor operation (or period of non-operation) to be longer or shorter than would have occurred without the setpoint changing. Compressors have a 10 min minimum cycle time as short cycles are energy inefficient.

Optimizing Demand Patterns

When the sun is shining, power is more available than when it is not. Consuming at night requires storing power in a battery which entails capital cost for the battery and associated hardware, as well as energy losses from the round-trip through the battery. Thus, it is advantageous to concentrate electricity demand during the day, and to match as well as possible the loadshape of PV output. Doing this can reduce the size of the battery needed as well as the energy losses.

The context of our analysis is a system with a nanogrid controller which assesses available electricity supply (from solar and battery), current demand, and past history, to set a current price and forecast of future prices. The system reduces the price of electricity in accordance with actual and forecast solar output, with the lowest price corresponding to the highest output, as shown in Fig. 18.1. Devices can use the price to concentrate their consumption in the middle of the day.

In general, the price and forecast can change at any time. Periods of constant price can be of any length and do not need to be the same as other periods. When the price or forecast changes, then devices may adjust their operation.

We used periods of 10 min for the price forecast, so that the setpoint for each unit also changed every 10 min. In our analysis, we did not change the actual price trajectory from that in the forecast; however, a nanogrid controller can change the forecast at any time. The shape of the price should match the supply/demand condition, so in this case it is simply the inverse of the supply.

Compressor operation is based only on the setpoint value (and minimum cycle time) so is not directly a function the electricity price. Similarly, the setpoint is based only on the price so it not directly a function of the PV output (or supply/demand

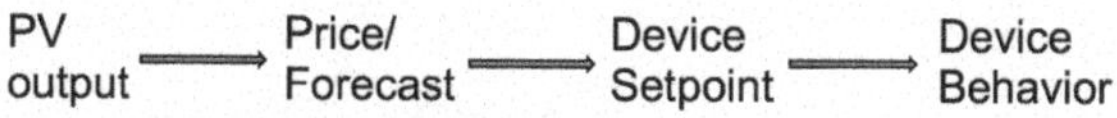

Fig. 18.3 Basic sequence of control operations

balance generally). Each step of the process is as simple as possible, to enable more complex systems while keeping the complexity contained.

Figure 18.3 shows these distinct steps of operation in this system. In this case, the effect of changing the initial signal has a clear and direct effect on the device behavior, but this separation of functionality enables much more complex systems as described later in the paper.

Dynamic refrigerator operation is similar to that of a freezer, except that with a smaller temperature range available, the degree that consumption can be concentrated during the day is reduced.

Single Device Results

Figures 18.4 and 18.5 show freezer operation with a dynamic price. Figure 18.4 presents the entire day. Figure 18.5 is a close-up to show incremental setpoint changes and the how the unit changes operation from night (compressor mostly off) to day (compressor mostly on). Figure 18.6 shows the power consumption of the freezer operation from Fig. 18.4; it is plotted as average power over each cycle of compressor on-time and off-time. For the variable price case, the freezer understands that there is a ‘day mode’ and ‘night mode’ of operation, where prices are lower during the day. It detects which mode it is in and for how long it will remain in that mode by looking at the trajectory of future prices. In night mode, it sets the

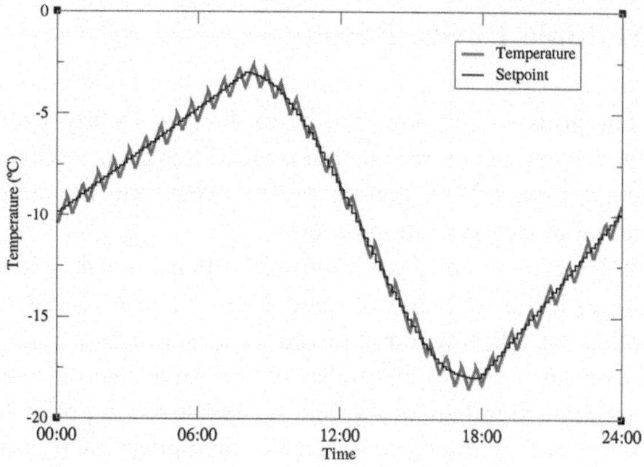

Fig. 18.4 Variable price freezer operation—one day

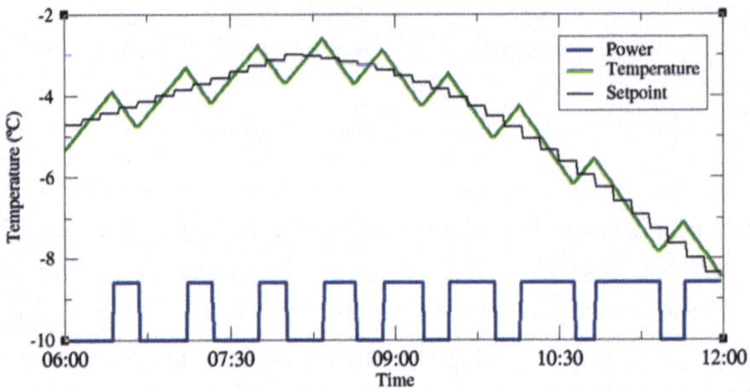

Fig. 18.5 Variable price freezer operation—6 h

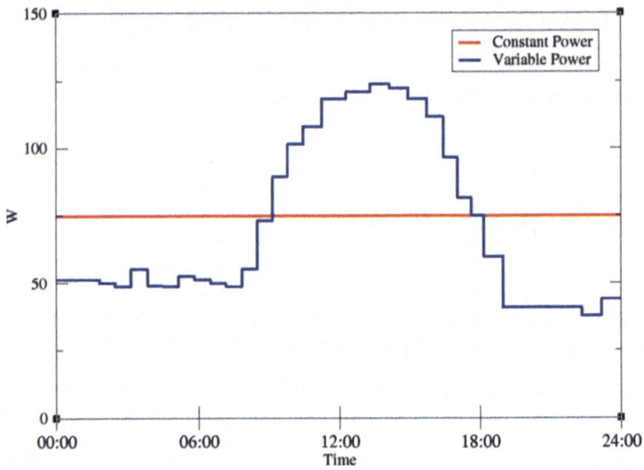

Fig. 18.6 Variable price freezer operation—power

setpoint to linearly move from its current point to its maximum point over the course of the night. In day mode, it adjusts the setpoint non-linearly, to have the greatest change in setpoint when the price is lowest, at the middle of the day.

Multiple Device Results

Nanogrids enable power to be exchanged locally within a building or community. Exchanges occur when prices are different in adjacent nanogrids, indicating different availability. Sharing power evens out the spikes of consumption that individual device operations produce.

To explore this, we simulated fifteen freezers (and fifteen refrigerators), with power levels and rates of temperature increase and decrease similar to our basecase units, but slightly randomized. The starting condition (internal temperature and compressor status) were also randomized. We did not want many units to change operation simultaneously, and so updates to the price/forecast was delayed by up to 9 min to that at most two receive the update each minute. The compressor cycling of ten freezers is shown in Fig. 18.7; the overall pattern is consistent—more on-time during the day—but their cycling details are random.

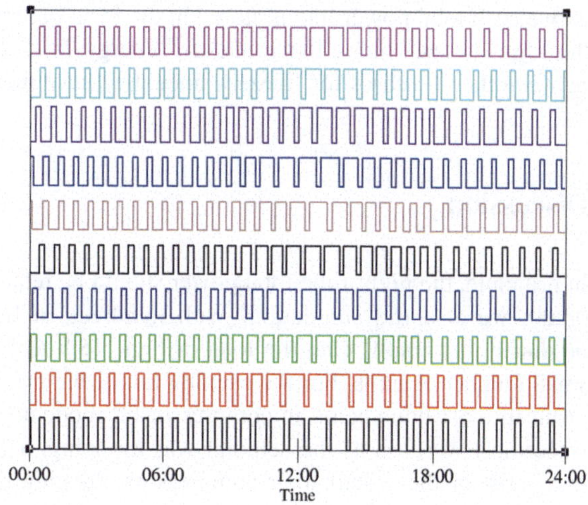

Fig. 18.7 Compressor cycling for ten (of 15) freezers

The total energy use for the fifteen units of each type are shown in Figs. 18.8 and 18.9. These are plotted as averages for each hour to eliminate short-term variations which obscure the overall pattern. We simulate both a constant price a (and so constant setpoint) and a variable price. Also on these figures we show the PV output required for the constant price scenario. The power under the PV output curve but

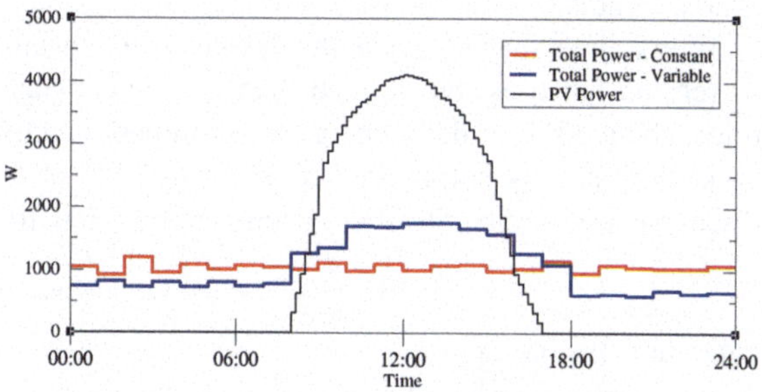

Fig. 18.8 Constant and variable 15-freezer operation

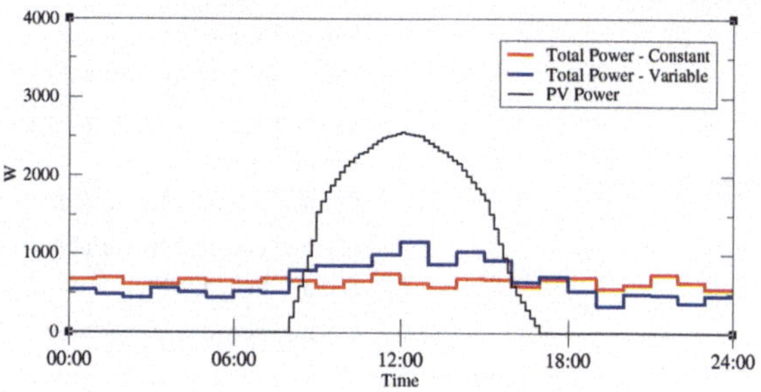

Fig. 18.9 Constant and variable 15-refrigerator operation

above each total power line is stored in the local battery, to be withdrawn for use in the night hours, minus the round-trip energy loss. There is a greater difference between the two cases for freezers since they have more ability to shift energy use.

Discussion

In principle, the night-time total power should be relatively flat from midnight until 08:00; we also did simulations with hundreds of freezers and refrigerators and produced that result. That the total in Figs. 18.8 and 18.9 is not flat is due to the small number of units. For the evening hours, the cycling of refrigerator power is a result of most units being in compressor off mode after the setpoint has stabilized, since most were on as the setpoint was dropping. Then, as the setpoint begins to rise, most of these start their compressors, creating the pattern. Further randomization could avoid this alignment.

Table 18.1 shows the primary quantitative results, which compare each constant price case with the percent change that occurs when variable prices are introduced. The table shows absolute energy values for the constant price case, and the percent change when moving to the variable case. Direct PV energy is the amount used directly from the PV panels by the units. Generation in excess of what the units consume is stored in the battery. Battery energy is the amount required from the battery. Excess PV consumption must be 10 % greater than the energy drawn from the battery. The total unit consumption between the constant and variable cases is nearly identical.

With a time-varying price, the devices still consume power during the night and so rely on batteries, but less so than for the constant price case. With a constant price, about 35 % of daily energy is consumed directly from the PV system. 16 and 10 kWh of battery output are needed, with 10 % of these values lost in the charge/discharge cycle. So, the PV system would need to be 1.4–1.6 % larger in the

Table 18.1 Results for one day for 15 units of each type all values except % are energy in kWh/day

	Freezer		Refrigerator	
	Const.	Var.	Const.	Var.
Device energy	25.1		15.5	
Direct PV energy	8.6	12.8	5.4	7.5
Direct TV fraction of total (%)	34	51	35	48
Change constant to variable (%)	49		37	
Battery energy	16.4	12.2	10.1	8.0
Change constant to variable	4.2		2.1	
Change constant to variable (%)	−26		−21	
Battery loss (kWh)	1.64	1.22	1.0	−0.8
Change constant to variable	0.42		0.21	

constant price case to deal with the additional battery roundtrip losses. With a variable price, about 50 % of energy is consumed directly from the PV system output, and the battery size (and corresponding energy loss) is reduced 21–16 %.

These energy and battery capacity savings are the most obvious and valuable benefits of this change in device behavior, but not the only ones. The rate at which PV power must be absorbed by the battery is also reduced. For these cases, the maximum hour of excess PV power has is reduced by 24 and 21 %. Battery chemistries have different limits on energy densities, charge rates, and discharge rates, so that being able to reduce this is an advantage.

There are two ways to interpret these benefits. One is that less hardware is needed to accomplish the desired tasks—less generation, conversion, and storage. The other is that for the same infrastructure, more benefit can be obtained—more energy available, and more storage available.

This analysis is for simple and fixed variations in generation and hence supply/demand balance, but the ability of devices to respond to changing prices can be used in complex grid topologies with many sources and end-use devices. All that is required is for each local grid to set a local price to reflect its own conditions, and use that in exchanges with other grids to help shape the local prices of adjacent grids.

Figure 18.10 shows a more general flow of information for a local grid context. The grid controller may have multiple potential sources of power, and may be able to transfer power to other grids. Each of these has a price/forecast associated with it to take into account. The battery state-of-charge will also affect price determination, depending on whether the controller is actively seeking to charge the battery, or let it discharge. Some of the information flows are actually bidirectional, such as with other sources. Power can also flow along many of these links, and in some cases, such as with the battery and other local grids, can reverse direction.

Additional Issues

Varying the temperature in a refrigerator or freezer is not the normal mode of operation, but should not impact general food storage. There are limits that should not be passed certainly: temperatures that are too cold can freeze vegetables; temperatures too high can lead to premature spoilage, particularly in sensitive items such as milk. Users of a refrigerator and freezer can use knowledge of what they are putting in it to select appropriate limits, and to change them over time as their usage changes.

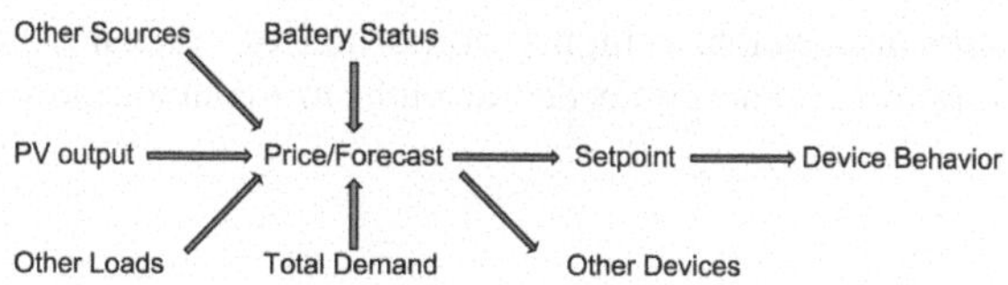

Fig. 18.10 Information flows in local grids

Given the high cost of electricity in many energy access contexts, attention has been given to making devices that are more efficient than might be economically justified for a unit that faces low electricity prices. These would most likely be different from more conventional units in having much better insulation. This would reduce the total electricity requirement, but would also mean that temperatures would rise more slowly, which would enhance the ability of a price-modulated control to concentrate energy use during the day. Using DC power also has efficiency advantages when coupled with local DC generation and/or storage (Garbesi et al. 2011).

Further Work

Any local grid will have a variety of devices, with electronics and lighting being among the most common. All of these can take a price and forecast into account in their operation. Lights might dim as prices get high, and turn off entirely when a threshold is reached. Electronics can modulate consumption, such as dimming displays or powering down to sleep more quickly. Water pumps can similarly serve as storage, operating when electricity is inexpensive and storing the water in tanks for use at a later time. The local price and forecast will drive decisions about when to charge and discharge a battery, and battery capacity in turn can help shape the local price.

While this project had price a simple function of local PV power, the forecast can be more complex and variable. With more complicated price trajectories, a more sophisticated algorithm would be needed to shape demand across multiple periods of high and low price during one day. In addition, for units with automatic defrost, defrost cycles need to be scheduled about once a day and can be set to occur at a time of relatively low cost. To avoid many units defrosting at the same time, they should randomly select a defrost time across a range of times where the cost is reasonably low.

If power suddenly becomes scarce during the day, the local price and forecast will be changed to quickly rise in response. This could then cause refrigerators to stop dropping their setpoint so that units that have their compressor off will delay turning them on, and units that are on will turn off more quickly. When the situation reverses, devices will automatically change operation.

The refrigerator characteristics used were from commercial units in industrialized countries. In the energy access context with more expensive power, it could be beneficial to insulate the units to a greater degree; this would reduce energy use overall, and would also enable a higher proportion of concentrating energy use during the day due to less energy needed at night to maintain the setpoints.

Conclusion

A local electricity price is a simple and universal mechanism to reflect the local supply/demand condition. Our analysis showed that such a price can be used to change freezer and refrigerator operation to make better use of local generation, and reduce hardware needed for battery power and losses associated with using battery storage.

References

Garbesi, K., Vossos, V., Sanstad, A., & Burch, G. (2011). Optimizing Energy Savings from Direct-DC in US Residential Buildings. LBNL-5193E.

Marnay, C., Nordman, B., & Lai, J. (2011). *Future Roles of Milli-, Micro-, and Nano-Grids, presented at the CIGRE International Symposium: The Electric Power System of the Future—Integrating Supergrids and Microgrids*. Bologna, Italy. LBNL-4927E.

Nordman, B. (2012). Dynamic Nanogrid Concepts. U.S. Army Research Laboratory, Advanced Microgrid Concepts and Technologies Workshop, June 7–8.

Nordman, B., & Christensen, K. (2013). Local power distribution with nanogrids. In *Proceedings of the International Green Computing Conference*.

Nordman, B., Christensen, K., & Meier, A. (2012). Think globally, distribute power locally: The promise of nanogrids. *IEEE Computer*, 89–91. ISSN: 0018-9162.

Taneja, J., Lutz, K., & Culler, D. (2011). Demand response for everyone: Using freezer energy storage to balance the grid. Paper presentation to LoCal Summer Retreat.

Taneja, J., Lutz, K., & Culler, D. (2013). The impact of flexible loads in increasingly renewable grids. In *Fourth IEEE International Conference on Smart Grid Communications (SmartGridComm'13)*.

Part III
Implementation and Decision-Making

Chapter 19
Financing Energy Efficiency and Climate Adaptation Measures on Household Level in Kyrgyzstan—Market Based Approaches in a Post-soviet Country?

Anastasia Bakteeva and Jonas van der Straeten

Abstract This paper explores the potential of innovative financing tools to foster the implementation of energy efficiency measures such as thermal insulation, using the example of two regions in Kyrgyzstan. The country is challenged by cold winters, national energy crises and poorly insulated housing inherited from Soviet times. Based on a household survey the paper investigates the problem of climate vulnerability and high energy expenditures on household level, identifies financing gaps for investments into efficiency measures and discusses financing tools to address them, such as green microfinance. The results of the study show that there is a considerable potential for market-based approaches if they target a more efficient use of thermal energy in households, particularly for space heating.

Keywords Energy efficiency · Kyrgyzstan · Green finance · Thermal insulation

Problem Statement

The former Soviet states in Central Asia are profoundly challenged by climate change in two respects. Firstly, the region's diverse climatic zones are highly vulnerable to the ecological impacts of climate change, including droughts, heat waves, crop losses,

A. Bakteeva (✉)
MicroEnergy International, Potsdamer Str. 143, 10783 Berlin, Germany
e-mail: bakteeva.a@gmail.com

J. van der Straeten
Research Group Microenergy Systems, Center for Technology and Society,
Technische Universität Berlin, Secr. HBS 1, Hardenbergstr. 16–18, 10623 Berlin, Germany
e-mail: straeten@ztg.tu-berlin.de

S. Groh et al. (eds.), *Decentralized Solutions for Developing Economies*,
Springer Proceedings in Energy, DOI 10.1007/978-3-319-15964-5_19

decrease in precipitation and melting of mountain glaciers and snowfields (Cruz et al. 2007; Sorg et al. 2012). Secondly, the region has inherited an inefficient energy system shaped by the Soviet tradition of extensive, state-subsidized energy services (Wiedemann et al. 2012). The case of Kyrgyzstan shows how climate change impacts and power supply crises are interlinked: with its high dependency on hydropower which accounts for 90 % of the total electricity generation, water shortages dramatically reduce the total electricity generation capacity. In addition, subsidized electricity prices (in 2009 0.01 USD/kWh for residential sector, Gassmann 2012), not only discourage modernization of obsolete equipment but also stimulate growth in energy demand that cannot be met in the future. This leads to the paradox situation, that Kyrgyzstan, rich in hydropower and currently a net-exporter of electricity, is expected to face winter power supply deficits from 2014 onwards, which could reach an annual 2,750 GWh by 2020 (World Bank 2013).

With winter temperatures as low as −20 to −30 °C in some regions of the country, the aggravating problems in the energy sector affect many households on an existential level. As little emphasis was put on thermal insulation during Soviet times, the country was left with poor quality housing, with unsatisfactory quality outdoor insulation, high air permeability of windows and inadequate air exchange mechanisms (Energy Efficiency Protocol 2011; Kraudzun et al. 2014). Recent studies show that households in mountainous areas of Kyrgyzstan spend up to 22 % of their household budget on energy, mostly for coal, firewood and electricity for heating (Liu and Pistorius 2012). In Kyrgyzstan the share of expenses for heating is up to twice as high for poor households than it is for wealthier households (Wu et al. 2004).

In total, the residential sector accounts for 63 % of the country's total electricity consumption (Energy Efficiency Protocol 2011), most of which is used on space heating (Wu et al. 2004). Therefore, increasing energy efficiency at the household level is not only advocated as one of the priority climate mitigation and adaptation measures (UNFCCC 2007), but can also contribute to relieving the tense situation in the Kyrgyz energy sector and improving the livelihood of households, particularly in rural areas (Braubach et al. 2011; Sarkar and Singh 2010). From an economic point of view, investments into energy efficiency measures such as retrofitting of houses to improve thermal insulation, often promises high returns by saving fuel expenses. Nonetheless, few households in Kyrgyzstan and other poor countries actually invest into thermal insulation. This fact suggests that there is a financing gap or at least a awareness gap for energy efficiency investments.

Effective means to fill this gap remain scarce. Multilateral financial agencies have had little success in financing energy efficiency projects in developing countries so far, as a review of recent literature shows. Approaches ranged from projects with utility companies and Energy Service Companies (ESCOs) to subsidies, grants or specific credit lines. These models proved to be effective in developed countries (Rezessy and Bertoldi 2010) but they are often not adapted to countries with emerging economies. Building up ESCOs takes decades and credit lines often miss their target because of undeveloped financial sectors, legislative frameworks or high transaction costs (Sarkar and Singh 2010). There is still a need for sustainable financing mechanisms, particularly on the level of households as well as micro, small and medium enterprises (MSMEs) (Painuly et al. 2003).

In recent years international development agencies and banks have increasingly paid attention to "green finance"; financial tools which are addressed at companies and projects that promote sustainable development, including resource efficiency, pollution reduction and mitigation of environmental damage (International Finance Corporation 2011). An application of "green" financial instruments in microfinance offers a great potential to address energy and climate change adaptation needs of rural households using a market-based approach (Glemarec 2012; Gujba et al. 2012). Kebir et al. also see it as important to provide access to green finance for MSMEs, as they are often crucial stakeholders in local value chains for energy efficiency measures (Kebir et al. 2013).

Research Objectives

Based on a household survey in Kyrgyzstan, this paper examines the problems of climate vulnerability and high energy expenses on household level and explores the potential of innovative financing mechanisms to mitigate them. The underlying research question of the study was: Can green finance foster the implementation of climate adaptation measures such as housing thermal insulation in Kyrgyzstan? The questions of the survey were aimed at identifying and quantifying the demand for housing retrofitting and thermal insulation. Based on this data, it was asked if there is a financing gap, which prevents this demand from being fulfilled. The household survey was accompanied by a comprehensive analysis of the country's microfinance sector and of supply chains and quality of products and services for thermal insulation.

Methods

The study is based on a quantitative survey of 384 households in the regions of Naryn and Issyk-kul. Households were randomly selected based on the official Kyrgyz census data. The survey was conducted on an individual basis and questions were addressed to the household head or the person responsible for housekeeping.

The results of the survey were complemented by qualitative interviews with ten households in the two regions that took place prior to the survey. Both the interviews and survey covered a range of aspects related to the demand for micro loans for thermal insulation: housing conditions, household energy consumption, satisfaction and awareness about energy efficiency and thermal insulation, experience with microfinance and household cash flows.

Results

The survey showed that 75 % of the respondents' houses were built more than 20 years ago during Soviet times. Nearly half of these houses were built of *saman*, a locally available construction material made of haulm and clay with much higher heat conductivity then for example bricks. Households spend on average at least 20 % of their monthly income on space heating. Despite the low prices for electricity, thermal fuels are also widely used for heating (coal by 93 % of respondents, firewood by 74 %, and manure cakes by 47 %). In order to cut down the heating cost, families often live in just one or two rooms for the winter period and "seal off" the rest of the house. Still, 22 % of respondents indicated that they suffer from a lack of heat in winter.

The capability and/or willingness to change that situation differed considerably between the two regions and between rural and urban respondents. While in the economic more advanced region of Issyk-Kul, two thirds of the rural and half of the urban population indicated that they were "planning to make any changes to make [their] house warmer within the next three years",[1] this was only the case for less then one third of rural and urban residents in the region of Naryn (see Fig. 19.1).

The replacement of heating systems, windows and doors were by far the most preferred retrofitting measures to make the house warmer (see Fig. 19.2).

However, an assessment of the underlying motivation for implementing retrofitting measures using a 5-point scale showed that comfort was ranked highest in both regions. However, opinions on other factors differ. In poorer and colder Naryn region "Saving money for heating fuel" was ranked as the second most important reason but only fourth in Issyk-Kul. This might be explained by the lower total income and higher share of energy expenses in Naryn as well as the higher potential savings, due to the colder climate in this region.

Despite the general willingness to invest in housing insulation measures, the price of such retrofitting often exceeds household disposable income: the majority of respondents have a household income of less than 10,000 KGS (156 EUR) a month and 20–30 % have an income of less than 5,000 KGS (78 EUR). Thus, the purchase and installation of, for example, five PVC-windows which costs between 23,000 and 35,000 KGS (539 EUR) is way beyond the budget of typical household.

Respondents who already conducted housing retrofitting in the past resorted to different sources of financing, with savings (64 %) as the most popular answer, followed by sales of livestock (12 %) and loans from financial institutions (9 %) (see Fig. 19.3).

Despite the low usage of loans for house retrofitting until now, the framework conditions for the disbursement of thermal insulation loans seem fairly favourable.

---

[1] The question were asked this way as the qualitative interviews had revealed that retroffiting measures were are not perceived as "thermal insulation" but as means to "make It warmer" ("*Uteplenie*").

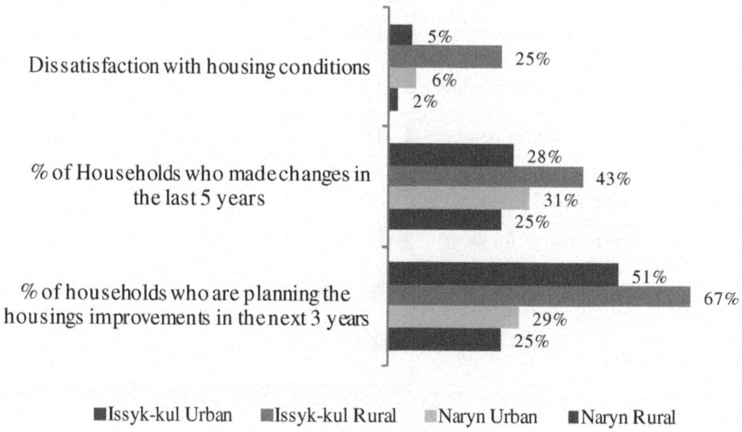

Fig. 19.1 Comparison of demand patterns for thermal insulation measures across districts (n = 384)

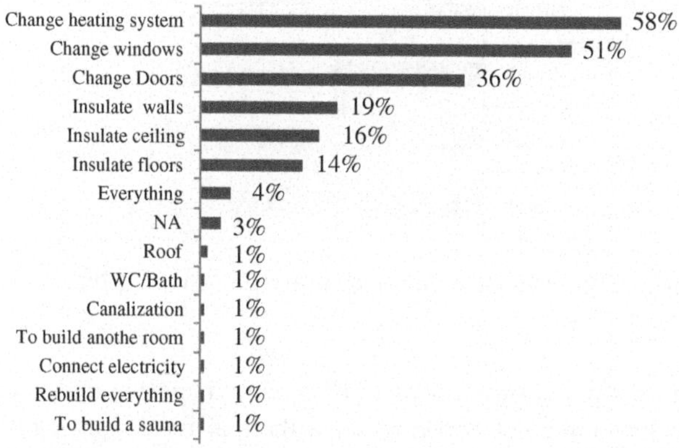

Fig. 19.2 Types of planned improvements within the next 3 years in % (n = 172, multiple answers possible)

The outreach of the financial sector in Kyrgyzstan is generally very high; in the survey sample 45 % of the respondents have experience with taking a loan from a credit institution, 12 % of which were dedicated to housing. There are however striking differences between the two regions. In Naryn, where respondents generally rely much more on informal sources of financing (e.g. relatives friends), the percentage of loans dedicated to housing was significantly lower as in Issyk-Kul (only 4 % in rural Naryn). On the other hand, nearly half of the *total* number of respondents in rural Naryn (47.9 %) expressed a general willingness to apply for an insulation loan (urban Naryn 21.1 %), compared to 25.5 % in rural and 9.6 % in urban Issyk Kul (see Fig. 19.4).

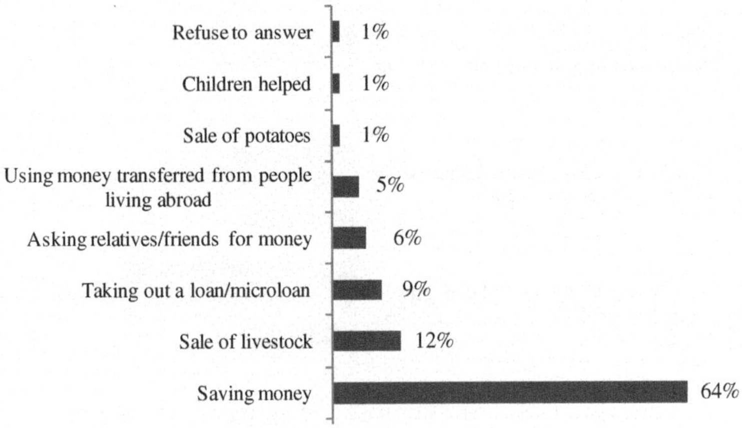

Fig. 19.3 Financing sources for housing retrofitting (n = 130, multiple answers)

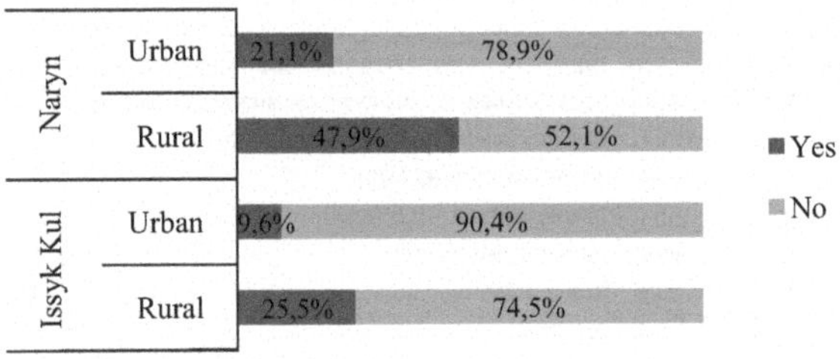

Fig. 19.4 Willingness to apply for an insulation loan (N = 384, n = 379)

Besides an aversion to owe money (74 %), high interested rates were mentioned as prohibitive factor for not applying for a thermal insulation loan (30 %).[2] For more than a fourth of existing loans, interest rated were higher than 25 %, being as high as 35–40 % in some cases. High interest rates of microfinance intuitions have been subject to controversy debates and political exploitation in Kyrgyzstan lately.

Discussion

The results of the study show, that even in the challenging environment of Kyrgyzstan with its Soviet heritage of state-subsidized energy (especially electricity) services, there is a considerable potential for market-based approaches if they target

[2] Choosing multiple answers was allowed in this question.

a more efficient use of thermal energy in households, particularly for space heating. There is evidence of poor housing conditions, for example in regard to indoor winter temperatures, which leads to large heating fuel expenses among the population. It was shown in the study that this need does indeed translate into a demand for retrofitting measures: about one third of the population have already undertaken housing improvement measures and approximately 40 % are planning to do so in the next three years. This demand is however rather driven by the desire for more comfort, then by economic calculations of households, and it is more the wealthier urban households than those in poorer, rural areas, who have the financial means to fulfill this demand.

The study reveals the paradox situation, that the households in colder and more remote regions, with poorer housing conditions, a higher savings potential and a higher motivation to realize it, express the highest willingness to apply for thermal insulation loans but find it hardest to access these loans. If market-based approaches are supposed to address the energy and climate change adaptation needs of the most vulnerable households, they have bridge this financing gap with innovative and adapted financing tools, such as green microfinance.

The design of these loans has to take the following aspects into account: Interest rates need to reflect the long lifetime but relatively low yearly return on investment of energy efficiency measures. They furthermore should take into account the seasonality of income of many households. Loan programs rely on the local availability of products and services in the sufficient quality to ensure efficiency gains. In the case of an existing program in Tajikistan, the "Warm comfort" program for example, this implied a high donor investment in building up and enforcing supply chains for high quality thermal insulation products, for example windows and doors. For Kyrgyzstan, however, the study showed that supply chains for some products are better developed. Finally, loan programs need to be accompanied by technical assistance and capacity building within financial institutions, to enable them to control and monitor the quality of energy efficiency products and services.

Research on the potentials and impacts of green finance instruments is still in its infancy. More evaluation of the few existing projects needs to be done e.g. on households factual energy savings. More attention should be paid to potential synergies with housing programs or to the role of utilities.

References

Braubach, M., Jacobs, D., Ormandy, D., World Health Organization, & Regional Office for Europe. (2011). *Environmental burden of disease associated with inadequate housing: A method guide to the quantification of health effects of selected housing risks in the WHO European region : Summary report*. Copenhagen: World Health Organization, Regional Office for Europe.

Cruz, R. V., Harasawa, H., Lal, M., Wu, S., Anokhin, Y., Punsalmaa, B., Li, C. (2007). *Asia. Climate change 2007: Impacts, adaptation and vulnerability. Contribution of working group II to the fourth assessment report of the intergovernmental panel on climate change* (pp. 469–506). Cambridge: Cambridge University Press.

Energy Charter Protocol on Energy Efficiency and Related Environmental Aspects. (2011). *Regular review of energy efficiency policies*. Kyrgyzstan.

Gassmann, F. (2012). *Switching the lights off: The impact of energy tariff increases on households in the Kyrgyz Republic*. MERIT working papers 066, United Nations University - Maastricht Economic and Social Research Institute on Innovation and Technology (MERIT).

Glemarec, Y. (2012). Financing off-grid sustainable energy access for the poor. *Energy Policy, 47*, 87–93. doi:10.1016/j.enpol.2012.03.032.

Gujba, H., Thorne, S., Mulugetta, Y., Rai, K., & Sokona, Y. (2012). Financing low carbon energy access in Africa. *Energy Policy, 47*(Suppl 1), 71–78. doi:10.1016/j.enpol.2012.03.071.

International Finance Corporation. (2011). *Climate finance: Engaging the private sector*. Retrieved from http://www.ifc.org/wps/wcm/connect/5d659a804b28afee9978f908d0338960/ClimateFinance_G20Report.pdf?MOD=AJPERES.

Kebir, N., Spiegel, N., Schrecker, T., Groh, S., Scott, C., & Aliaga Ferrufino, G. (2013). *Exploring energy SME financing in emerging and developing countries*. California: Sustainable Business Institute.

Kraudzun, T., Vanselow, K. A., & Samimi, C. (2014). Realities and myths of the Teresken syndrome—an evaluation of the exploitation of dwarf shrub resources in the Eastern Pamirs of Tajikistan. *Journal of Environmental Management, 132*, 49–59. doi:10.1016/j.jenvman.2013.10.019.

Liu, M. F. M., & Pistorius, T. (2012). Coping with the energy crisis: Impact assessment and potentials of non-traditional renewable energy in rural Kyrgyzstan. *Energy Policy, 44*, 130–139. doi:10.1016/j.enpol.2012.01.025.

Painuly, J. P., Park, H., Lee, M.-K., & Noh, J. (2003). Promoting energy efficiency financing and ESCOs in developing countries: Mechanisms and barriers. *Journal of Cleaner Production, 11*(6), 659–665. doi:10.1016/S0959-6526(02)00111-7.

Rezessy, S., & Bertoldi, P. (2010). *Financing energy efficiency: Forging the link between financing and project implementation*. Ispra: Joint Research Centre of the European Commission. Retrieved from http://ec.europa.eu/energy/efficiency/doc/financing_energy_efficiency.pdf.

Sarkar, A., & Singh, J. (2010). Financing energy efficiency in developing countries—lessons learned and remaining challenges. *Energy Policy, 38*(10), 5560–5571. doi:10.1016/j.enpol.2010.05.001.

Sorg, A., Bolch, T., Stoffel, M., Solomina, O., & Beniston, M. (2012). Climate change impacts on glaciers and runoff in Tien Shan (Central Asia). *Nature Climate Change, 2*(10), 725–731. doi:10.1038/nclimate1592.

UNFCCC. (2007). *Climate change: Impacts, vulnerability, and adaptations in developing countries*. Retrieved from http://unfccc.int/resource/docs/publications/impacts.pdf.

Wiedemann, C., Salzmann, S., Mirshakarov, I., & Volkmer, H. (2012). Thermal insulation in high mountainous regions: A case study of ecological and socioeconomic impacts in the Eastern Pamirs, Tajikistan. *Mountain Research and Development, 32*(3), 294–303. doi:10.1659/MRD-JOURNAL-D-11-00093.1.

World Bank. (2013, June 26). *A more efficient Kyrgyz electricity sector could reduce the power deficit by half*. Retrieved January 28, 2014, from http://www.worldbank.org/en/news/press-release/2013/06/26/more-efficient-kyrgyz-electricity-sector-could-reduce-power-deficit-by-half.

Wu, X., Lampietti, J., & Meyer, A. S. (2004). Coping with the cold: Space heating and the urban poor in developing countries. *Energy Economics, 26*(3), 345–357. doi:10.1016/j.eneco.2004.04.011.

Chapter 20
Demand Assessment of Solar Electrification in Off-Grid Rural Communities of Pakistan Through Microfinancing of Solar Home Systems

Warda Ajaz and Hadley Taylor

Abstract The following paper documents the methodologies used and findings of a demand assessment for microfinanced Solar Home Systems in rural households in Pakistan. The purpose of the study was to design appropriate technology packages and a financing scheme tailored to the energy needs and affordability of rural households to facilitate a scale-up of the Pakistan Poverty Alleviation Fund's off-grid solar electrification program. A gender disaggregated, qualitative as well as quantitative energy demand assessment was conducted in three districts i.e. Chakwal, Khushab and Thatta, through household surveys and focus group discussions. Methodologies used to determine the energy needs, ability and willingness to pay for a SHS is outlined. The results of the assessment allowed for a targeted design of system sizing and packages as well as a tailored financing option for the rural households. The paper moreover concludes that that introducing a microfinanced SHS program in rural off-grid communities is highly demanded by the rural, poor, off-grid residents, and has the opportunity to make a significant environmental and social impact in the country.

Keywords Microfinance · Solar home system · Energy access

All data was collected through household surveys in the districts of Chakwal, Khushab and Thatta in December 2013 and January 2014. The report and data collection was commissioned by the GIZ on behalf of the PPAF, who have approved the presentation of the data for academic purposes.

W. Ajaz (✉)
Partners in Sustainable Development (PSD), Islamabad, Pakistan
e-mail: warda_ajaz@hotmail.com

H. Taylor
MicroEnergy International (MEI), Berlin, Germany
e-mail: hadleydltaylor@gmail.com

S. Groh et al. (eds.), *Decentralized Solutions for Developing Economies*,
Springer Proceedings in Energy, DOI 10.1007/978-3-319-15964-5_20

Introduction

One of the biggest hurdles in sustainable development throughout the world is addressing energy poverty. It is therefore, a well-established fact that any country which wants to enhance the pace of sustainable development must first address the constraints related to mass access to clean and affordable energy by the poor, especially rural poor (Srivastava and Rahman 2006; Bhattacharyya 2006).

The premier and most important step in addressing the energy poverty in rural areas is rural electrification. There are various reasons as to why rural electrification is considered indispensable for addressing the rural energy needs; firstly, rural electrification saves the need of using polluting lighting sources like kerosene oil, hence reducing the indoor air pollution and improving health. Secondly, the use of electrical appliances such as mobile phones or televisions enhance access to information and knowledge (Buragohain 2012).

According to the International Energy Agency Statistics, around 1.44 billion people across the world still live without access to electricity (IEA 2013; Palit and Singh 2011). While providing grid electricity to such a large number of people seems like an overwhelming feat, off-grid renewable electricity technologies like solar home systems have shown a great potential for rural electrification in the developing world (Urmee and Harries 2009; Mala et al. 2009; Cherni and Hill 2009; Chakrabarti and Chakrabarti 2002). Various studies have proven that the use of SHS in rural areas not only improves the socio-economic condition of the communities but also reduces the environmental pollution (Wijayatunga and Attalage 2005; Mondal and Klein 2011). Nevertheless, large scale Solar Home System rural electrification programs require considerable financial resources which are generally hard to mobilize in most developing countries (Sovacool 2013). Therefore, utilizing private finance options seems like the only feasible option in the current scenario.

Pakistan is currently experiencing an ever-worsening electricity crisis. IEA statistics reveal that currently, only 67 % population in Pakistan is connected to the national grid, of which 93 % are located in urban areas (IEA 2013). Furthermore, almost all the off-grid communities depend on mustard and kerosene oil for lighting and more than 64 % rely on biomass for cooking (IEA 2013). This alarming situation is contributing a great deal in receding the country's economic growth and it has therefore become need of the hour to address the issue of rural electrification on a priority basis. Looking at the successful example of the SHS program initiated by the Infrastructure Development Company Limited, in Bangladesh, the following study sought to assess the feasibility of initiating a rural electrification scheme through microfinanced SHS in Pakistan (Grameen Shakti 2014).

In this context, the Pakistan Poverty Alleviation Fund (PPAF)[1] is implementing a grant based village electrification program in off-grid rural communities. While

[1] The Pakistan Poverty Alleviation Fund (PPAF) is an autonomous organization that aims to serve the poor, marginalized and disadvantaged households across Pakistan by facilitating their access to resources and opportunities.

several thousand of such systems are currently in operation, PPAF is exploring the feasibility of an exponential expansion of this program, using a microfinancing mechanism for solar home systems (SHS).

For this purpose, the following demand assessment was conducted to assess the feasibility and furthermore design SHS to be microfinanced through PPAF and its Partner Organizations (POs).

This paper outlines the methodologies used in this study, discusses the results and findings and presents the conclusions of this study regarding the possibility of solar electrification in the off-grid rural areas of Pakistan through micro financing.

Research Objectives

The study was conducted with the following objectives:

1. To conduct a demand assessment for a rapid expansion of the on-going solar electrification program by introducing SHS using microfinance, in off-grid rural communities from three districts namely Chakwal, Khushab and Thatta, as selected by PPAF.
2. Designing an appropriate SHS for the target communities, based on the results of the demand assessment.
3. Designing an accompanying microloan based on the energy needs, ability and willingness to pay of the target communities, as revealed by the demand assessment.
4. Determine the demand for an SHS with the appropriately sized SHS packages and end-user financing scheme.

Methods

In order to fulfill study research objectives, the methodology of sample survey explained in Fig. 20.1 was used:

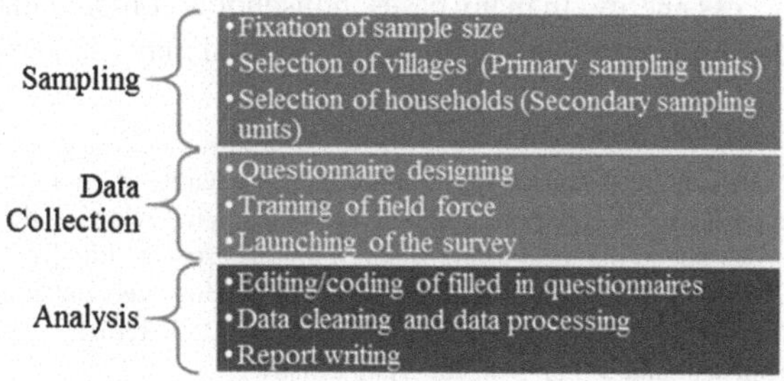

Fig. 20.1 The survey methodology

As a first step, random sampling technique was used to prepare an appropriate sampling plan. The target Union Councils (UCs) in each district were selected after consultation with PPAF's POs working in the respective districts. In the Khushab and Chakwal districts, the sampling locations were narrowed down to 4 UCs located in these regions and then to 10 villages where 200 questionnaires were conducted. In Thatta region, the sample size of 211 households was proportionally distributed amongst the off-grid villages of target UCs.[2]

Subsequently, a through and detailed questionnaire was prepared using secondary data and inputs from sector experts including Energy, Microfinance, Gender and Statistics. After finalization, the questionnaires was translated in Urdu language for smooth administration on the rural population with little or no education. The questionnaire was pre-tested before launching the actual survey. The pretesting helped to make various changes in the questionnaire in the light of problems faced during pre-testing.

The interviewers were then sent in the field for conducting household surveys according to the sampling plan. Each of the interviewers was trained beforehand on how to how to approach the respondents for each question through a specially prepared Manual of Instructions.

Finally, Software Package for Social Sciences (SPSS) was used for data analysis which was then formed a basis for SHS and accompanying Microloan designing.

Results

In order to appropriately design the SHS, and accompanying loan terms, it was critical to determine the market demand for SHS in off-grid regions. To assess the demand, MEI and PSD developed a household questionnaire and focus group discussion (FGD) guideline to determine respondents' energy needs, ability and willingness to pay for a SHS through a microfinancing scheme.

Energy Needs

Without a grid connection, households are forced to find alternative solutions to access energy. In many cases, households combine many sources to meet their basic energy needs. For example, a household may use kerosene for lighting, charge their

[2] Female enumerators were included in the teams of enumerators particularly to address female respondents. Keeping in mind the difficulty in interviewing the female respondents which usually arises due to the local customs and religious aspects, initially it was decided to cover minimum of 20 % female respondents in the target area. However, the competitiveness and hard work of the female staff made it possible to cover 38.92 % female respondents in Chakwal, 42.65 % in Khushab and 39.58 % in the Thatta district.

mobile phone at a local shop, and listen to a radio run dry cell batteries. These traditional fuels are not only damaging to the environment (and in many cases the health of the users) but also are very expensive. It was found that almost all of the surveyed population's household energy needs could be replaced by a SHS with the exception of cooking. Therefore fuels and appliances used for cooking were out of scope of this study.

Lighting

The respondents of the survey stated that they mainly use lanterns (kerosene lamp), or Chinese made torches to light their homes. Kerosene/mustard oil and dry cell batteries are the prevalent energy sources for lanterns and torches respectively. The percentage of households with respect to the lighting source being used is given in Fig. 20.2 by district.

Appliances

The results reflected that the mobile phone is the most popular appliance in the surveyed areas. Analysis showed that on average, 1.26 mobile phones, 0.14 fans, 0.07 TV and 0.13 radios per household are available. Figure 20.3 shows the percentage of households owning radio, TV, fan and mobile phone.

Furthermore, the respondents were asked which appliances they plan to buy in the next 3 years, if given access to a reliable energy source. The responses, in order of preference, were fans, TV, bulb/tube lights and Radio.

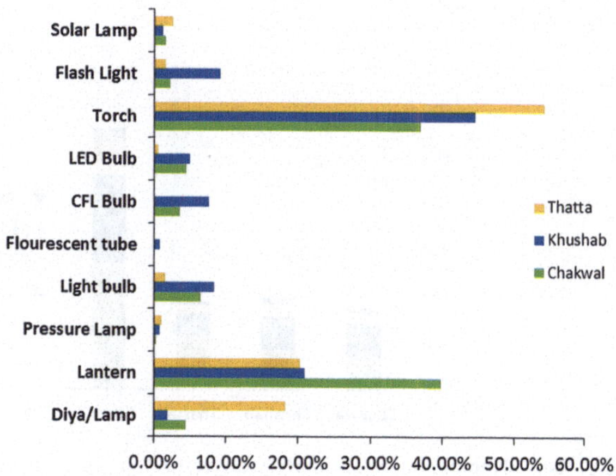

Fig. 20.2 Percentage of households with respect to the source of lighting used, (multiple answers possible), (Chakwal = 309, Khushab = 263, Thatta = 197)

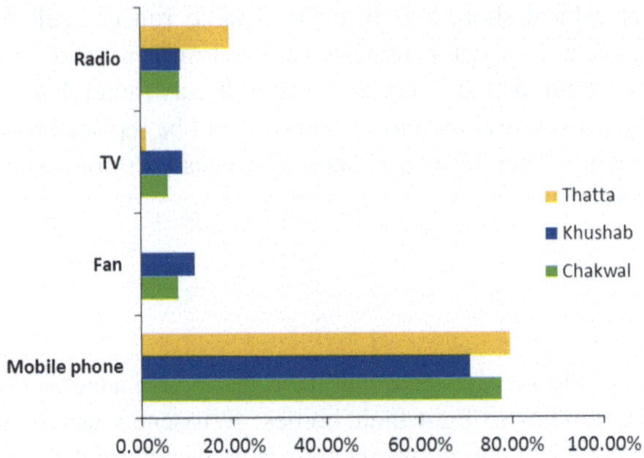

Fig. 20.3 Percentage of households with respect to availability of appliances, (multiple answers possible), (Chakwal = 201, Khushab = 267, Thatta = 167)

Ability to Pay

The ability to pay for a SHS is determined by income characteristics, monthly savings, decision-making power in the household and the current energy expenditures and appliances that can be replaced by SHS.

Average Monthly Household Income

The results revealed that most of the surveyed households have a monthly income of less than PKR 10,000. The detailed income dispersion in each district is given in Fig. 20.4.

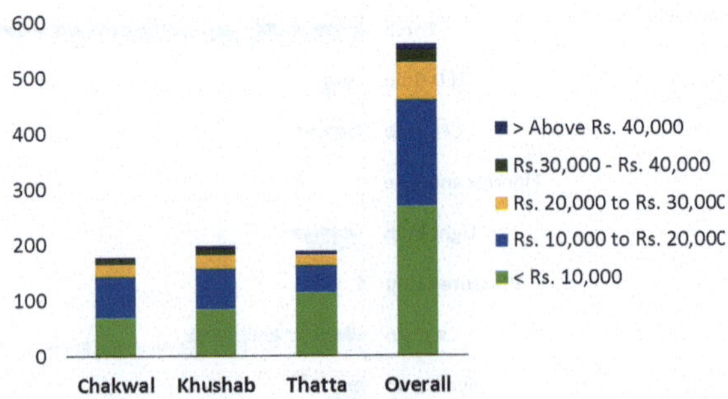

Fig. 20.4 Number of households with respect to their average monthly income (N = 581, Chakwal = 185, Khushab = 204, Thatta = 192)

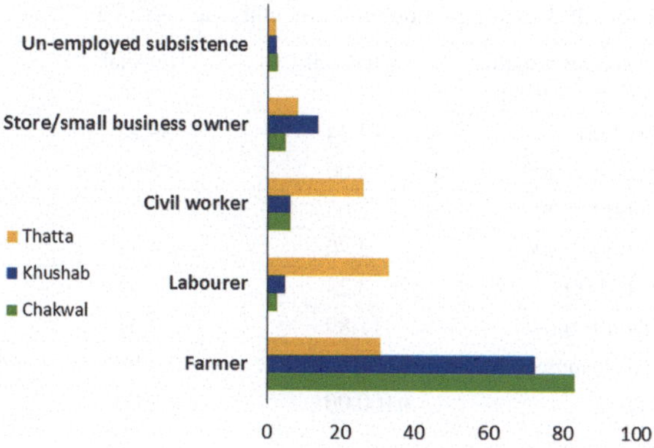

Fig. 20.5 Percentage of households with respect to their primary source of income, (N = 581)

Sources of Income

A clear majority of respondents in Chakwal and Khushab stated that the primary source of income in their household is agriculture. In Thatta however, labor was the dominant source of income. The percentage of households with respect to their primary source of income is given in Fig. 20.5.

Moreover, out of total 581 sampled households, 166 households reported that there are other persons in the households working to earn money in addition to the respondents/head of household. The source of this secondary income is described in Fig. 20.6.

Due to the nature of income generation in the studied regions, the majority of respondents either receive their income monthly (24 %), or seasonally (45 %).

Average Monthly Savings

Out of total surveyed households in Thatta district, 57 % reported that they are able to save less than PKR 1,000/- per month. This is followed by 43 and 39 % in Chakwal and Khushab respectively. Only 6 % in Khushab, 4 % in Thatta and 3 % in Chakwal reported that their households are able to save above PKR 10,000 per month (Table 20.1).

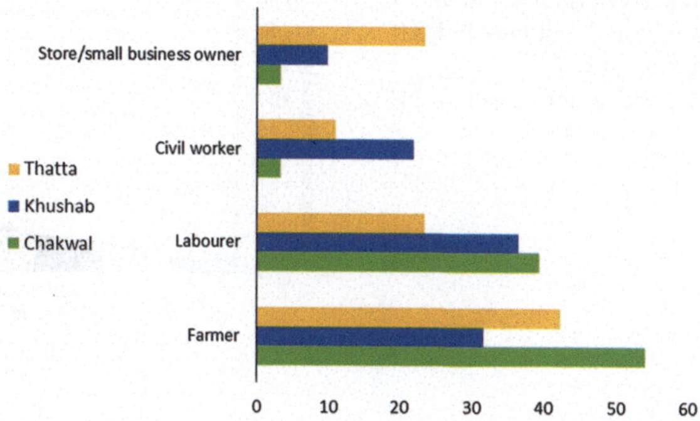

Fig. 20.6 Percentage of households with respect to their secondary source of income, (N = 166)

Table 20.1 Average household monthly savings (%)

Average monthly savings (PKR)	Chakwal (%)	Khushab (%)	Thatta (%)	Total (%)
<1,000	43.24	39.22	57.29	46.47
1,000–4,000	20.00	23.53	15.10	19.62
4,000–8,000	7.03	12.25	4.17	7.92
8,000–10,000	2.70	4.41	1.56	2.93
>10,000	3.24	6.37	3.65	4.48
Don't know	11.89	9.31	9.90	10.33
No response	11.89	4.90	8.33	8.26
Total	100.00	100.00	100.00	100.00

Current Energy Expenditures

Households spend significant proportions of their monthly income to meet their most basic of energy needs through traditional fuels, and paying neighbors and nearby shops to recharge their devices. The average household energy expenditure was found to be PKR 589/- per month. On an overall basis 46 % of respondents are spending PKR 827 per month on kerosene/mustard oil and 14 % are paying PKR 380 per month to recharge their appliances at a neighbor's house or local shop. For distribution of energy expenditures that can be replaced by a SHS, please see Fig. 20.7.

Decision Making Power in the Household

Identification of the most appropriate target group is critical to develop marketing and consumer awareness campaigns to support the scale-up of a microfinanced SHS program. Respondents were asked who in the household decides on purchases larger than PKR 5000. 94 % of respondents in Khushab, 91 % in Chakwal and 75 % in Thatta all responded that the "male head of household" is the decision maker in this regard. As a general practice in Pakistan, authority on spending

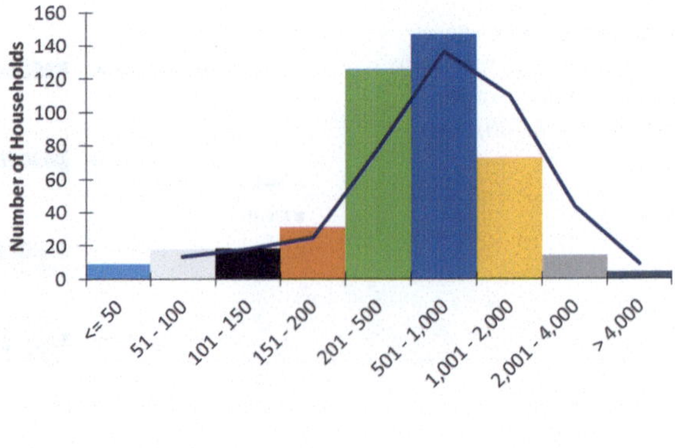

Fig. 20.7 Average monthly energy expenditures in PKR (Chakwal = 162, Khushab = 160, Thatta = 118)

belongs on the income generator, however in many cases, as was revealed during the FGDs, the female head of household expresses her needs to the male head of household, and he is then responsible for fulfilling this need.

Willingness to Pay

The questionnaire included the questions on respondents' willingness to pay for a SHS. Willingness to pay was assessed by evaluating perceptions about SHS, expectations for national grid connection, as well as their desire to take out a microloan to finance a SHS, if the opportunity were available.

Relationship to the National Grid

78 % households in Khushab, 73 % in Chakwal and around 29 % in Thatta reported that they are not expecting to get an on-grid electricity connection in the near future. Whereas 18 % households in Thatta district, 3 % in Khushab and only 1 % in Chakwal district stated that they are expecting to get an electricity connection from the national grid in the coming year.

Perceptions of SHS

Overall, 71 % respondents showed their willingness to buy a SHS. Results dissagregated by district reflected a figure of around 93 % for Chakwal, 91 % for Khushab and 76 % for Thatta. Moreover, the gender dissagregated analysis shows that the percentage of men who were willing to purchase SHS is greater in all districts as compared to women (Table 20.2).

Of those who did not show their willingness to buy a SHS, around 42 % of the respondents stated the reason that it is too expensive. See Fig. 20.8 for the variations in answers.

Willingness to Microfinance a SHS

Furthermore, the respondents were asked if they would be willing to take a microloan to finance a SHS. The percentage of respondents who showed their willingness is shown in Fig. 20.9.

Table 20.2 % of respondents who would like to purchase a SHS, (%), (N = 412)

	Men (%)	Women (%)	Total (%)
Chakwal	59.15	33.54	92.68
Khushab	53.76	37.10	90.86
Thatta	47.50	28.33	75.83

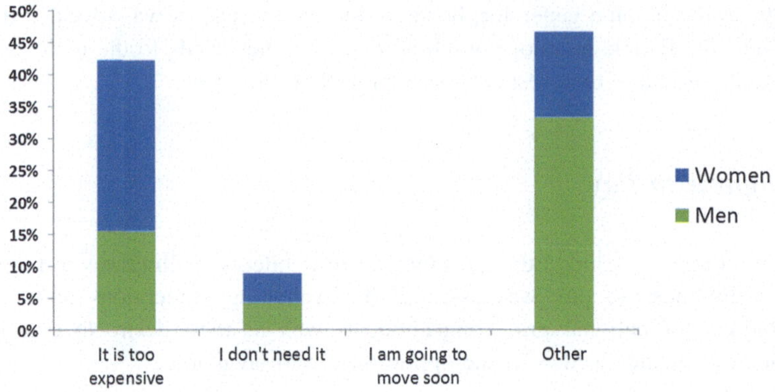

Fig. 20.8 Percentage of respondents reporting different reasons for unwillingness to purchase a SHS, disaggregated by gender, (N = 45)

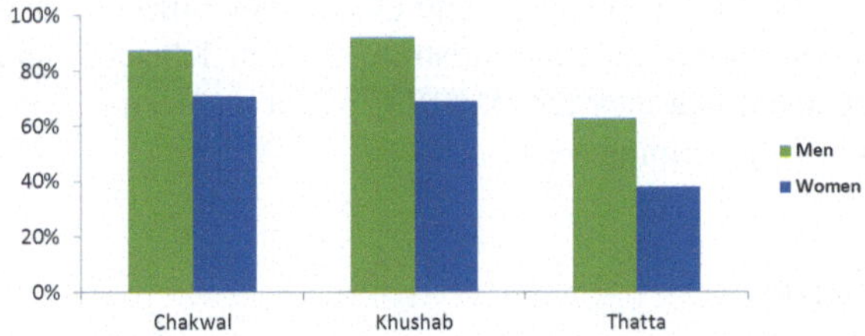

Fig. 20.9 Percentage of respondents willing to take a microloan to finance a SHS, disaggregated by gender, (N = 421)

As can be seen, the ratio of respondents willing to take loan in Khushab is higher than in the other two districts. The reason is that the PPAF PO responsible for the region is already implementing a microfinance program for the purchase of SHS and therefore the communities are fully sensitized. Chakwal, although slightly below Khushab, also shows a high percentage of respondents willing to take microloan for purchase a SHS as the local PO is working very closely with the communities. Although they are not providing microfinancing for SHS, the communities are fully aware of the potential of solar energy. In Thatta, the communities did not report any loan under any microfinance arrangement from their respective PO or any other source. This is because in Thatta, local PO provides community mini-grids for household clusters, and therefore the public perception about solar energy is understood as a communal asset rather than that of the individual household.

Overall, more than 72 % responded 'yes' to this question. Of those who stated 'no', the reasons are detailed in Fig. 20.10.

Analysis of gender disaggregated data depicts that out of total respondents in the Chakwal district, 54 % men and 28 % women are interested in buying a SHS

Fig. 20.10 The reasons for unwillingness to take microloan for SHS

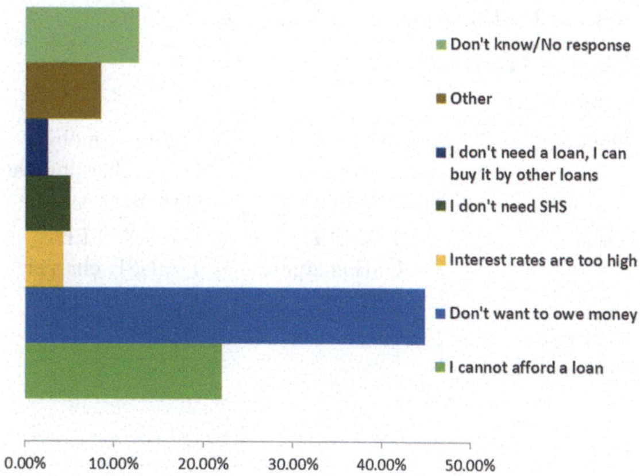

through a microloan against 8 % in each case of men and women who are not interested in this arrangement. A similar picture is observed in the Khushab district, while in the Thatta district, 40 % men and 14 % women are interested in buying a SHS through microloan against 19 % men and 16 % women who are not interested in taking microloan.

Discussion

Considering the results of the surveys and inputs from the communities during the FGDs, it can be concluded that in these regions, SHS is highly demanded by off-grid rural communities. However, due to low income levels in the region, the technology is largely unaffordable to the population without a microfinancing mechanism. In the following section, preliminary recommendations for implementation of a microfinanced SHS program in Pakistan are presented.

SHS Technical Design

The SHS should be designed for three distinctive categories of the clients to address the heterogeneity in ability to pay and energy needs, as well as solar radiation and housing conditions across rural Pakistan. Graduated system sizing will allow the program to incorporate more segments of the off-grid population. For example, of the respondents who already owned a SHS, limited generation capacity was the most frequently cited reason for dissatisfaction with their system. While at the same time, the majority of respondents had an income of less than PKR 10,000 per month. Therefore, designing a range of SHS tailored to the energy needs of the different income groups, from very small and affordable to the very poor market

Table 20.3 SHS sizing and technical specifications

Solar panel (W)	40	65	100
Battery (Ah)	40	75	120
Functions	Lighting, mobile phone charging and basic cooling	Lighting, mobile phone charging and cooling	Lighting, mobile phone charging, cooling and entertainment
Loads	3 × 3 W LED, 1 × USB charger, 1 × 6 W fan	4 × 3 W LED, 1 × USB charger, 1 × 15 W fan	5 × 3 W LED, 1 × USB charger, 1 × 15 W fan, 1 × 24 W television
Total price estimate (PKR)	25,170	39,495	57,340

segments, to a larger system to satisfy the energy demands of the less vulnerable poor is recommended. By employing MEI's SHS technical sizing methodology, as well as a sensitivity analysis to price, and critical inputs from the household surveys, the following SHS packages are recommended to be integrated into a microfinanced SHS program in rural off-grid regions in Pakistan (see Table 20.3).

In addition, to ensure that the SHS technical design incorporates women's needs in rural Pakistan, one detachable light should be included in all system packages for two reasons:

1. The majority of respondents did not have an attached bathroom on their home. As this can be a major safety concern for women after dark, a mobile light would help mitigate this danger.
2. Most households surveyed had an outdoor kitchen, which keeps harmful smoke inhalation to a minimum. A detachable light will help ensure that families do not move their kitchens inside the home to make use of the light while cooking.

During the FGDs, the respondents expressed apprehensions with regard to the quality of the equipment. They were worried that if the equipment stopped working, they would be trapped into a repayment scheme for non-functioning hardware. Therefore, quality products, proper warranty for replacement of all components, adequate after-sales services and end-user training must be provided to maintain the integrity of the program, and a sustained demand in the region.

SHS Loan Design

The end-user microfinancing scheme to accompany the SHS packages was tailored to meet the needs of the respective income groups for which they were designed. Sensitivity analyses were conducted when designing loans in a reiterative process with SHS design, to achieve harmony between quality and affordability of SHS and accompanying loan installments. Loan terms were designed in accordance to the respondents' ability to pay for energy access, in addition to utilizing the loan framework currently employed by the PPAF and its POs, to allow an ease of incorporation of the SHS packages into the financial portfolio of the POs.

Loans for the SHS should be distributed to the entire household through family loans, as this is the loan type that the POs are most familiar with, utilizing social collateral as the main form of guarantee. Also, since a SHS is used by all members of the household, this formulation is most appropriate.

Loan period should be limited to 2 years, as it should not exceed the minimum life of the battery included in the SHS. It is recommended to set the interest rate at 28 %, allowing the PO the maximum profit margin for financing the SHS. As incorporating a renewable energy technology into an otherwise financial portfolio can stress processes within the institution, and requires significant capacity building for the loan officers, the maximum margin allowed under the PPAF financing framework is recommended to incentivize the program at the PO level. Additionally, a down payment of 20 % should be required upfront, to finance installation costs of the SHS.

It is recommended that repayment schedules should be somewhat flexible to incorporate the income characteristics of the many different income groups and regional economies to incorporate more segments of the population. Therefore, MEI designed repayment schemes with monthly, biannual and annual installments to be negotiated between the client and the loan officer based on their income source (Table 20.4). Biannual and annual installments were designed particularly to incorporate the produce farmers and fisherman's income frequencies as stated during the demand assessment. A grace period is not needed for the SHS loan, as energy savings are incurred immediately, by the eradication of traditional fuel expenditures.

Overall, the methodology developed and used to determine the demand for microfinanced SHS, while also collecting the data inputs needed to design the technical specifications and SHS packages, in addition to a financial product proved to be successful. In future however, the following lessons learned from the project will be useful for further projects seeking to implement a microfinanced SHS program for rural households:

1. Determine market segmentations prior to collecting data to ensure that all segments are represent a statistically significant sample in relation to population size

Table 20.4 Installments for SHS microloan displayed by SHS package and repayment scheme

	40 W SHS package	65 W SHS package	100 W SHS package
Monthly installments (PKR)	604	947	1375
Biannual installments (PKR)	7764	12,397	18,263
Annual installments (PKR)	16,334	26,081	38,421

2. Data concerning energy and household spending as well as income is most useful as absolute numbers rather than categorical data
3. Female enumerators and discussion groups are imperative to understanding a women's perspective, particularly in contexts wherein women's decision making power is low in the household

Overall the demand assessment is imperative to the design and success of an electrification program, however, it must be noted that it is only one side to the equation. For implementation a significant institutional assessment of the implementing institutions and POs as well as an assessment of the supply environment and need for capacity building is necessary to design successful frameworks and processes that ensure financial viability and quality of a program. Only with all these aspects considered, can a program be designed to facilitate rural electrification at scale through microfinanced SHS.

Conclusions

The results of the demand assessment showed a significant awareness of the benefits of a SHS, and demand for the systems. However, the main barrier to acquiring such a system for the households was ability to pay upfront. Therefore, a microfinancing scheme for SHS would achieve PPAF's goal of a rapid scale-up of solar electrification in Pakistan, ultimately alleviating energy poverty in the country, and significantly improving the livelihoods of rural households. Based solely on the results from the demand assessment and utilizing simple extrapolation from the sample size, it is estimated that if the program is implemented at a national scale, approximately 5.1 million rural households in rural Pakistan could electrify themselves through such a scheme.

References

Bhattacharyya, S. C. (2006). Energy access problem of the poor in India: Is rural electrification a remedy? *Energy Policy, 34*, 3387–3397.

Buragohain, T. (2012). Impact of solar energy in rural development in India. *International Journal of Environmental Science and Development, 3*(4), 334–338.

Chakrabarti, S., & Chakrabarti, S. (2002). Rural electrification programme with solar energy in remote region—a case study in an Island. *Energy Policy, 30*, 33–42.

Cherni, J. A., & Hill, Y. (2009). Energy and policy providing for sustainable rural livelihoods in remote locations—the case of Cuba. *Geoforum, 40*, 645–654.

Grameen Shakti. (2014). Available at http://www.gshakti.org/.

IEA. (2013). World energy outlook. International Energy Agency. Available at http://www.worldenergyoutlook.org.

Mala, K., Schlapfer, A., & Pryor, T. (2009). Batter or worse? The role of solar photovoltaic (PV) systems in sustainable development: Case studies of remote atoll communities in Kiribati. *Renewable Energy, 34*, 358–361.

Mondal, A. H., & Klein, D. (2011). Impacts of solar home systems on social development in rural Bangladesh. *Energy for Sustainable Development, 15*, 17–20.

Palit, D., & Singh, J. (2011). Lighting a billion lives—empowering the rural poor. *Boiling Point, 59*, 42–45.

Sovacool, B. K. (2013). Expanding renewable energy access with pro-poor public private partnerships in the developing world. Energy Strategy Reviews, *1*, 181e192.

Srivastava, L., & Rahman, I. H. (2006). Energy for sustainable development in India: Linkages and strategic direction. *Energy Policy, 34*, 643–654.

Urmee, T., & Harries, D. (2009). A survey of solar PV program implementers in Asia and the Pacific regions. *Energy for Sustainable Development, 13*, 24–32.

Wijayatunga, P. D. C., & Attalage, R. A. (2005). Socio-economic impact of solar home systems in rural Sri Lanka: A case-study. *Energy for Sustainable Development, 9*(2), 5–9.

Chapter 21
Scale Versus Substance? Lessons from a Context-Responsive Approach to Market-Based Stove Development in Western Kenya

Temilade Sesan

Abstract Improved stoves targeted at the 2.6 billion people worldwide that use solid biomass for cooking have not been taken up in the numbers expected by donors and practitioners. Following widespread critique of the subsidy-based dissemination models popular in the 1970s and 1980s, donors have begun to emphasise the potential of market-based models to increase stove adoption rates. In analysing the USEPA project implemented by Practical Action in western Kenya, this paper examines how a market-based approach has translated in the kind of informal economy operated by many biomass-reliant communities. The paper concludes that a context-responsive approach is likely to facilitate the dissemination of locally appropriate interventions, but it may not always be compatible with mainstream visions of large-scale stove deployment.

Keywords Improved stoves · Market-based development

Introduction

According to the most recent estimates given by the International Energy Agency, 2.6 billion people worldwide lack access to modern cooking fuels such as liquefied petroleum gas and electricity, relying instead on the traditional use of solid biomass fuels (fuelwood, charcoal, crop residue, animal dung) to meet their cooking and heating needs (OECD/IEA 2013). 'Traditional' here refers not to the quality of biomass fuels in themselves, but to the practices by which they are burnt in 'inefficient' open fires or 'primitive' stoves (World Bank 2011) by energy-poor populations concentrated in developing economies in Africa, Asia and Latin

T. Sesan (✉)
Centre for Petroleum, Energy Economics and Law, University of Ibadan, Ibadan, Nigeria
e-mail: temi@gbengasesan.com

S. Groh et al. (eds.), *Decentralized Solutions for Developing Economies*,
Springer Proceedings in Energy, DOI 10.1007/978-3-319-15964-5_21

America (OECD/IEA 2010). The widespread reliance on traditional biomass by these populations—the majority of whom live on a subsistence basis or on very low incomes in rural areas and are only partially engaged in the market economy (Friedmann 1992; Kanagawa and Nakata 2007)—has been associated with health and environmental hazards ranging from deforestation in the 1970s, to indoor air pollution in the 1990s, to global warming and climate change more recently (Sesan 2014).

For over four decades now, improved stoves have been promoted, mainly by national governments and international organisations but also by local non-governmental organisations, to address the various problems attributed to traditional biomass reliance over the years (Foell et al. 2011; Shastri et al. 2002). Designed to burn biomass fuels more cleanly and/or efficiently than traditional cooking devices (Larson and Rosen 2002; Mahiri and Howorth 2001), improved stoves have been widely promoted on the basis of a combination of health, environmental, resource conservation, female empowerment and economic benefits they offer to individual households and communities at large (Burki 2011; GACC 2011; Lee et al. 2013; Lewis and Pattanayak 2012). However, notwithstanding the increasing sophistication and variety of the technologies promoted on improved stove programmes (Kshirsagar and Kalamkar 2014; Quadir et al. 1995), the demonstrated potential of those technologies to mitigate the negative impacts of traditional biomass use (Smith-Siversten et al. 2009; Wilkinson et al. 2009), and the emergence of innovative dissemination models especially aimed at 'bottom of the pyramid' populations (Chaurey et al. 2012), the rate of improved stove uptake by energy-poor households has been perennially slow (Bailis et al. 2009; GACC 2011).

While there is little debate over the need to scale up the dissemination of improved stoves s among poor households, there is a considerable degree of uncertainty in the stove development field as to what would constitute an effective approach to scaling up. Beginning in the 1980s, stove development organisations sought to facilitate local acceptance and use of improved stoves by moving toward more context-responsive forms of engagement with the populations at whom those interventions were targeted. These notions of context-responsiveness were premised upon theories of participatory development, which is conventionally represented as emerging out of a recognition of the shortcomings of top-down development approaches, and is credited with having the potential to give rise to more socially and technologically appropriate solutions with greater probability of widespread adoption (Guijt and Shah 1998; Schumacher 1993). Almost concurrently, generous subsidy regimes were introduced to assist poor households with the costs of purchasing new improved stoves, with the result that many households ended up acquiring the stoves at little no cost to themselves (Barnes et al. 1994). However, households often proved unwilling to purchase new stoves at full market value when their subsidised stoves fell into disrepair, leading experts in the field to retrospectively denounce subsidy-based dissemination models as having hindered the potential of many stove projects of the period to continue unaided (Hanbar and Karve 2002).

Following the general criticisms levelled at subsidy-based approaches, major funders of improved stove interventions have begun to emphasise market-based dissemination models (see for example Hoffman et al. 2005). Increasingly, stove development organisations are subscribing to the ideology that the only way improved cooking technologies can reach the millions of poor households that need reaching is to adopt the practices associated with a fully functioning market system of the kind found in rich countries. Indeed, very few voices of caution or dissent can be heard today amidst the growing enthusiasm to establish market routes to scaling up cooking interventions—voices, like Bailis et al. (2009) and O'Neal (2005), which crucially reflect an understanding of the context of poverty in which most stove interventions are implemented, rather than an unquestioning belief in the 'power' of the 'market' to provide appropriate solutions in all contexts.

The market movement in stove development was consolidated with the advent of the United Nations Foundation-led Global Alliance for Clean Cookstoves (GACC) in 2010 (Smith 2010). The GACC promotes the activities of institutionalised market actors on the national and international scenes, but it also acknowledges the role that local women in particular have to play in taking market-led solutions to the last-mile customer (Hart and Smith n.d.). This latter function is especially important in the context of the mostly informal economies in which solid biomass is widely used. This paper discusses the potential of informal market models to deliver targeted outcomes through the work of Practical Action, a UK-based international non-governmental organisation that has implemented improved stove programmes in Kenya since the mid-1980s. Practical Action's stove programme, despite being specifically targeted at poor and marginalised populations, has historically favoured market-based dissemination approaches over the giving of subsidies and handouts. Practical Action's subscription to a market approach however seeks to take into account the realities of the socio-economic contexts into which improved stoves are introduced, in effect privileging the needs of the poor over the inflexibility of market operations. The organisation employs a participatory approach in working with local women's groups to establish small-scale supply chains for improved stoves in their communities. Using the case of the USEPA smoke alleviation project implemented by the organisation in Kadibo, western Kenya from 2009 to 2010, the paper describes some of the market development elements of the context-responsive approach taken by the organisation to establishing stove enterprise groups and discusses the outcomes for the poor households targeted by the programme. First, though, the methods used in gathering the data presented will be briefly outlined.

Methods

Fieldwork for this study was carried out between November and December 2009 using qualitative methods, specifically semi-structured interviewing and non-participant observation. Of the eight 'locations' within Kadibo division that were involved in the USEPA project, I was only able to gain access to one—West Kochieng—due to time and resource constraints which Schatzman and Strauss (1973) identify as having the potential to restrict the researcher's access in the field.

The sample size was restricted to twenty-one—not sufficiently large to elicit generalisable observations, but small enough to facilitate in-depth analysis of the interview data generated.

Interviews were conducted across three actor categories, as follows: three Practical Action staff (4 interviews); two project community authorities (2 interviews); and thirteen local energy users (15 interviews, four of which were with members of the only stove enterprise group in West Kochieng involved in the project at the time of fieldwork). Access to the local energy users was negotiated by the aforementioned Practical Action staff, with whom I had established email contact prior to going into the field.

In addition to membership (or otherwise) of the stove enterprise group, the local energy users in the sample were selected on the basis of their adoption (or otherwise) of the improved cooking interventions introduced by the USEPA project. As such, four different sub-categories of local energy user emerged: stove enterprise group members that had adopted at least one of the six interventions promoted on the project; group members that had not adopted any of the interventions; non-group members that had adopted at least one of the interventions; and non-group members that had not adopted any of the interventions. This multi-faceted sample structure afforded insight into both the demand and supply-side dynamics of the project in West Kochieng, though the lessons drawn out in this paper relate mostly to the latter. The Practical Action staff interviewed were selected on the basis of organisational hierarchy (for instance, the sample included a senior member of staff with a good overview of the organisation's stove programme in the region) and degree of involvement in field implementation of the USEPA project. The questions asked varied across the different respondent categories: the interviews with Practical Action staff focused on project implementation processes and outcomes relative to specified targets; those with community authorities discussed their role in facilitating links between local stove enterprise groups and external organisations such as Practical Action; while the interviews with local energy users explored issues relating to stove adoption decisions, group membership and identity, project participation, and household priorities.

All but one of the interviews with local energy users were with women, as they were the main focus of Practical Action's intervention in the region. A typical interview took place in a woman's household and lasted about an hour when it was not planned ahead to coincide with food preparation times. Five interviews were scheduled to take place around the time that each of the women planned to cook either breakfast or lunch on prearranged days. This strategy was employed to enable firsthand observation of the way that the women organised and performed their everyday cooking tasks using various stove technologies, both traditional and improved. When making appointments for these 'fireside interviews', I arranged to arrive at the households about an hour prior to the commencement of food preparation. The aim was to allow some time to build a level of rapport with each interviewee, so as to establish common ground for conversation and make the women more comfortable with opening up their private domain to an outsider. These informal opening conversations would continue in the kitchen area throughout the duration of food preparation, which ranged from about 25 min to an hour. When 'hanging around' in this way, I usually offered to help the women with tasks I could manage in the hope of mitigating the observer effect on the situation. This degree of immersion in the women's day-to-day affairs facilitated observation of naturally occurring clues that shed further light on the investigation and provided context for interpreting interview data (q.v. Kvale 1996). The detailed field notes taken during and after each interview/observation session later provided a starting point for coding the data gathered as well as an ongoing reference point during analysis of those data.

My conduct of interviews and observations in the privacy of individual households required sensitivity to the kind of behaviour that was expected of a visitor in different circumstances. This not only enhanced my level of acceptance in the project community but also ensured that local energy users were able to participate in the research on their own terms. Nonetheless, my status as educated, urban, and middle-class firmly located me as an outsider relative to the informants. In these situations, the existence of a power differential was evident between the researcher and the researched (Wolf 1996) which posed a challenge to field interactions. It is possible that the interviewees felt a need to impress with their responses, as a way of compensating for the differences in socio-economic standing between them and the researcher; however, many of the women took the opportunity to voice their aspirations to a fellow female; aspirations which may have been more difficult to unearth by a researcher of the opposite gender perceived as belonging on the 'other' side with the men in the community. The wide gap between these women's ambitions—such as an oft-expressed desire to complete formal education—and their lived realities, aspects of which are discussed in the following section, provided a basis for understanding the significance of the USEPA project in the context.

The USEPA Smoke Alleviation Project

In January 2009, Practical Action commissioned a two-year project to develop market systems for the dissemination of six improved cooking interventions—the Upesi stove, the fireless cooker, the solar cookit, the smoke hood, the liquefied petroleum gas (LPG) stove, and eaves spaces—in western Kenya (Interview Practical Action Staff 1). The project was christened the USEPA project after the donor organisation, the United States Environmental Protection Agency, which funded its implementation. Under the USEPA project, stove enterprise groups mostly acted as retailers and installers of the improved cooking interventions; production, mainly of fireless cookers out of locally available materials, was limited to a few women within the groups. This section goes on to describe the form that the project took in West Kochieng where fieldwork was conducted, but first outlines some of the peculiarities of the informal economy in the location to provide context for the discussion that follows.

A Different Kind of Marketplace

West Kochieng is situated in Nyanza province, which is one of 8 administrative provinces in Kenya and home to the Luo people, who constitute the third largest ethnic group in the country. The Luo are a close-knit people who live communally: the unit of spatial demarcation is not the household, but the 'homestead' which comprises several individual homes—occupied by extended family members—arranged around an open courtyard. The Luo, particularly those who reside in rural areas, attach great significance to the observance of tradition and custom, eschewal of which would cause an individual/household to be regarded by society as an outcast. The influence of culture is all-pervasive, touching on every area of individual and communal life, from living and cooking arrangements to hospitality codes to the attribution of gender roles.

In particular, women are culturally assigned a subordinate position to men in the region. This is a reality that is reflected even in the routine of everyday life: in the absence of her husband, for instance, a woman is expected to simply tell visitors who come knocking that 'no one' is at home, a response which tacitly discounts her own existence as an individual. In the absence of broader societal affirmation, women typically band together in groups of fifteen to twenty 'to uplift one another as members' (Interview West Kochieng Resident 1) both socially and economically, oftentimes via the platform of cooperative savings or 'merry-go-round' schemes. In adopting these women's groups as the focal point for its stove intervention in West Kochieng, Practical Action is in effect operating within the boundaries prescribed for women by culture while seeking to empower them socio-economically so they gain a measure of influence as providers and users of improved cooking technologies.

Similarly, the effect of tradition on market exchanges in West Kochieng is very significant. I observed in the course of fieldwork that some of the traditional and time-honoured practices valued by people in the community would be considered as violating the modern economic norms of commoditisation and profit maximisation. A good example is the way that land is appropriated for building and farming purposes. Empty structures belonging to dead people are retained as they are, rather than being sold off or turned over to more 'lucrative' purposes. In fertile areas, individual *shambas*, or farm plots, grow progressively smaller as land is divided and re-divided among however many sons are born into the household. Smaller farm plots definitely mean a decrease in individual farm yield, yet family land is divided as many times as is necessary because that is the way prescribed by tradition. As such, the widely proclaimed 'efficiencies' of a modern market system do not come into play in this context. For instance, according to the women who run stove enterprises in West Kochieng, conventional marketing and advertising tactics such as the use of 'memorable' radio jingles as suggested by Brewis (2005) are not very effective in reaching prospective customers. The women understand that their peers in the community respond better to more personalised forms of advertising such as one-on-one marketing and public demonstrations, and they respond accordingly:

> 'Advertising on the radio would help, but the more effective one is, bring it to the market and to public *barazas*… Because some people who have never heard about it don't believe… So when they demonstrate, the people are actually ready to wait and see. And when they see that, then they actually buy and some will say, 'okay, I'll give you the deposit'… So the direct marketing has really helped.' (Interview West Kochieng Resident 2)

Besides the market and community *barazas*, other popular demonstration outlets for stoves include schools and churches—places where community members gather for social purposes not normally associated with buying and selling. Though sales and marketing of interventions are done individually, the burden of advertising is sometimes shared among members of a stove enterprise group.

Credit management is another aspect of this marketplace that has been modified to fit the requirements of local enterprise. According to the stove enterprise group members interviewed, the credit models which have been proven to work best are those that harness the power of the group. Such schemes rest on the principle that members who take out individual loans will hesitate to default on repayments because they are accountable to their fellow group members, which is often the case. However, the peculiar challenges of living on low incomes in rural areas can sometimes undermine that premise: according to Practical Action Staff 1, a woman may take a loan for the purpose of expanding her small business, but the moment an emergency shows up in the form of a sick or hungry child, she promptly diverts the funds to healthcare or food as the case may be. The relatively flexible credit provisions of this marketplace, though not conducive to a conventional profit-maximising enterprise model, are essential to the viability of market-based interventions seeking to improve aspects of residents' livelihoods.

Context-Responsive Stove Market Development in West Kochieng

It is against this background that the USEPA project set out to provide local women with start-up support to enable the stove enterprises they initiated to eventually take off unaided. By November 2009 when fieldwork for this study was conducted, the project was only assisting stove retailers with making financial and logistical arrangements for bulk stove purchase. From the point of delivery of the stoves, each woman was expected to find her own buyers, sell the stoves without assistance from the project, take out her profit, and give the capital back to the project towards the purchase of another batch of stoves. The women were not required to pay at the point of collection of the stoves, but only after they had sold the stoves and realised a profit. This system was adopted to circumvent the women's inability to gain access to adequate start-up capital. At the time of fieldwork, the possibility of facilitating access to capital through the channel of village savings and loans schemes was already being explored:

> 'But now the locations that we have started working in, we have now created 6 groups, and these groups have currently started what is called… village savings and loans. And so when they raise money here, we are trying to talk to them, that when their loans get to the level that they can get 100 stoves by themselves, we will leave that. They get 100 stoves, and they come and sell. So it becomes like revolving for them.' (Interview Practical Action Staff 1)

Such community savings and loan schemes—succinctly referred to as 'COSALO' in the localities where they operate—are usually initiated by development agencies working to improve different aspects of livelihoods, and they appear to be gaining widespread acceptance in Kadibo division as a whole. The loan amounts that can be taken out by individuals are usually proportional to the value of their contribution to the fund. A small interest rate is usually applied which constitutes the main source of income for the group and goes towards building up the group capital. Practical Action expects that the COSALO scheme will contribute to resolving the challenge of limited access to capital present at all levels of the local stove distribution chain.

According to one scenario laid out by Practical Action staff, if 10 group members take loans of 2,000 Kenyan shillings (Kshs) each and pool their individual sums together, with the bulk sum they can arrange for purchase and delivery of a batch of 100 stoves. When the stoves are delivered, each woman collects her quota and, as usual, conducts the marketing and sales by herself. The potential challenge with this model, however, is that the COSALO platform is purely transactional and members are free to use their loans to pursue any commercial activity as long as they can repay. As such, there is no guarantee of getting up to 10 women who will be willing to invest their capital in stoves at any given time. The observations I made in the course of fieldwork, as well as conversations I had with women running stove enterprises, indicate that access to credit/capital is highly valued in West Kochieng. However, these facilities are often sought for the purpose of initiating or expanding a range of micro-businesses that may not be related to stove enterprise. Thus in attempting to persuade COSALO group members to invest their resources in stove enterprise, Practical Action seeks to influence priorities on the supply side.

This may be particularly difficult to achieve as only a few group members, some of whom are seen by their peers as the stove 'experts', rely on stove enterprise as their main or only source of income. Although stove businesses offer higher profit margins than most local micro-businesses do, many women in West Kochieng stated that they found it difficult to establish the market links required to derive a steady income out of the enterprise. This indicates that, even where formal or informal credit platforms have been introduced, the entry barriers to stove enterprise need to be further lowered for many local women by offering them training in 'soft' skills such as networking and marketing. In addition, such women will likely require longer-term donor support—both financial and logistical—for their fledgling businesses than is normally allowed for in programme design, an argument made more generally by Bailis et al. (2009) in their critical assessment of the current market orthodoxy in stove development. The next section summarises the opportunities and limitations of the context-responsive approach to stove market development exemplified by Practical Action's intervention in West Kochieng.

Discussion and Conclusions

> 'There are gross inequalities which continue to grow, and that sometimes we leave in healthy tension. Using the case of improved stoves, I don't know if you've heard of any improved stoves programme that was 100 percent successful. I would say that for two households willing to take up improved stoves, there are six households for whom survival is more key than environmental issues… you'd wonder, India has been the home to a great number of stoves programmes. How come we still have more initiatives going on?' (Interview Practical Action Staff 2)

This paper began by discussing how, in spite of a proliferation of improved stove programmes in developing countries and the promise that increasingly sophisticated interventions hold to significantly improve cooking and living conditions in energy-poor households, those interventions have achieved much lower dissemination rates among target populations than originally envisaged by enthusiastic stove promoters. It is against this background that stove dissemination models have been continuously reviewed and refined over the years, with a current global emphasis on harnessing market platforms toward the goal of achieving scale, largely defined in quantitative terms.

In particular, the paper highlighted Practical Action's attempts to develop local improved stove supply chains in informal economies within western Kenya through its USEPA project. Practical Action's approach to stove development in western Kenya was shown to demonstrate an appreciable degree of sensitivity to the significance of non-technical networks in technology dissemination as well as an awareness of the contingency of the local context in which it was operating. The organisation's strategy, while premised on the use of market tools in the empowerment of marginalised groups of energy users, privileged a bottom-up approach which did not give primacy to those tools but started from the realities of local people. Nonetheless, in narrowly promoting access to credit for an enterprise that many women in West Kochieng could not prioritise without them addressing other barriers to substantive participation embedded in their everyday experiences, it would appear that the organisation did not fully recognise the limitations of its own strategy in the context. Indeed, the responses of the women to the community savings and loan scheme as structured by Practical Action suggest that the notion of a transfer of priorities from stove development organisations to local energy users may be a problematic one, as poor people's lived realities often dictate a different set of priorities than those prescribed by well-meaning outsiders.

Still, Practical Action's attempt at a context-responsive approach enabled a degree of engagement with the economic realities of the context and facilitated the uptake of some improved cooking technologies by poor biomass-reliant households with peculiar marketing needs. The merits of using local women as the main actors in stove enterprise are evident in this case: apart from empowering otherwise marginalised women socio-economically, the final costs of the improved technologies to users turned out to be highly sensitive to local incomes. The Upesi stove liners, for instance, produced, marketed and sold by members of various women's groups in western Kenya, are the least expensive of the technologies introduced by the USEPA project at 350 Kshs. Fireless cookers, also made and sold locally by the women, are available on the market at an average cost of 900 Kshs. The smoke hood on the other hand, which is manufactured and installed by city-based artisans and meant to be used in conjunction with the Upesi stove, comes at a final cost of 5,500 Kshs—fifteen times as expensive as the Upesi and the equivalent of 2 months' wages in West Kochieng. The result is that the Upesi stove was the most widely adopted technology in the sample—taken up by 7 of 13 households—while the smoke hood, present in only one relatively affluent household, was one of the least accessible of the technologies promoted. As such, Practical Action's approach resulted in a greater likelihood of its most participatorily developed interventions being directed toward local energy users at or near the base of the income pyramid.

The lessons offered by this case are especially pertinent in the current phase of stove development when, as noted earlier, a near-consensus has been achieved on the desirability of adopting market strategies over dissemination approaches which incorporate subsidy elements, on the basis that the former route is more financially sustainable over the long term and is potentially more value-adding than the latter. As has been shown, Practical Action's context-responsive approach to market development required the organisation to operate within the provisions of the 'economy of affection' (Hyden 1980) in West Kochieng. The provisions within the community for certain 'market' functions such as advertising and credit arrangements were seen to deviate from the rational, profit-maximising norms of formal markets. Those informal provisions have, however, been critical to sustaining the local market for improved stoves. Where the approach needs further strengthening is in the provision of complementary skills and sustained donor support that will better equip local women to take advantage of the market platforms that have been adapted to fit their context.

Notwithstanding the potential for impact demonstrated by Practical Action's context-responsive approach to market dissemination, such approaches may not always be compatible with mainstream visions of 'scaling up' which emphasise universal reach and access of, sometimes, 'efficient' stove technologies which are developed out of context and which therefore may not be economically or culturally appropriate. Though Practical Action's context-responsive approach resulted in low-cost and culturally appropriate interventions such as the Upesi stove being directed toward local users on the lowest rungs of the energy ladder, progress has been achieved slowly and incrementally (Sesan 2012). With respect to alleviating energy poverty among biomass-reliant households, therefore, the outcomes of the context-responsive approach to market-based stove dissemination taken by Practical Action suggest the possibility of a trade-off between scale and impact on stove programmes specifically directed at the poorest energy users: such an approach may yield small scale but is likely to have high impact, which means that it is better able to reach the neediest households with technologies they can afford, even if the benefits derived from those technologies are only incremental. Indeed, several examples are now emerging in the literature of market-led stove programmes that are on their way to achieving significant scale while leaving out the poorest households (q.v. Sesan et al. 2013; Shrimali et al. 2011).

On the basis of the findings presented here alone, it would seem that Westhoff's (1995) assertion that the context-responsive approaches espoused by participatory development proponents from the 1980s onward have facilitated the identification of more appropriate technologies and dissemination models has been borne out to an extent. An important caveat to note, however, is that the ultimate success of a stove programme cannot be measured in terms of the appropriateness of the technology or rates of dissemination, but—in keeping with alternative views of development which support subjective interpretations of the ideal—in terms of the degree to which it reflects and enhances the priorities of local energy users. In other words, context-responsive approaches to stove development need to be employed more for the freedom they afford energy-poor households to input their priorities into decision-making processes on the global development scene than for the potential they hold to facilitate the realisation of pre-set project priorities. Perhaps the greatest challenge for stove development experts in this phase will be to balance the goal of getting the poorest to participate substantively in the burgeoning global stove market with ensuring that such people are able do so on their own, context-specific terms.

References

Bailis, R., Cowan, A., Berrueta, V., & Masera, O. (2009). Arresting the killer in the kitchen: The promises and pitfalls of commercialising improved cookstoves. *World Development, 37*(10), 1694–1705.

Barnes, D., Openshaw, K., Smith, K., & van der Plas, R. (1994). *What makes people cook with improved biomass stoves? A comparative international review of stove programmes.* Washington, DC: The World Bank.

Brewis, A. (2005). Ten top tips for successful scaling up. *Boiling Point, 50*, 5–6.

Burki, T. (2011). Burning issues: Tackling indoor air pollution. *The Lancet, 377*(9777), 1559–1560.

Chaurey, A., Krithikaa, P. R., Palita, D., Rakesha, S., & Sovacool, B. K. (2012). New partnerships and business models for facilitating energy access. *Energy Policy, 47*(Supplement 1), 48–55.

Foell, W., Pachauri, S., Spreng, D., & Zerriffi, H. (2011). Household cooking fuels and technologies in developing economies. *Energy Policy, 39*(12), 7487–7496.

Friedmann, J. (1992). *Empowerment: The politics of alternative development.* Cambridge, MA and Oxford: Blackwell.

GACC. (2011). *Igniting change: A strategy for universal adoption of clean cookstoves and fuels.* Washington, DC: Global Alliance for Clean Cookstoves.

Guijt, I., & Shah, M. (1998). Waking up to power, conflict and process. In I. Guijt & M. Shah (Eds.), *The myth of community: Gender issues in participatory development* (pp. 1–24). London: Intermediate Technology Publications.

Hanbar, R., & Karve, P. (2002). National programme on improved Chulha of the government of India: An overview. *Energy and Sustainable Development, 6*(2), 49–56.

Hart, C., & Smith, G. (n.d.). *Scaling adoption of clean cooking solutions through women's empowerment: A resource guide*. Washington, DC: Global Alliance for Clean Cookstoves.

Hoffman, K., West, C., Westley, K., & Jarvis, S. (2005). *Enterprise solutions to poverty: Opportunities and challenges for the international development community and big business*. London: Shell Foundation.

Hyden, G. (1980). *Beyond ujamaa in Tanzania: Underdevelopment and an uncaptured peasantry*. London: Heinemann.

Kanagawa, M., & Nakata, T. (2007). Analysis of the energy access improvement and its socio-economic impacts in rural areas of developing countries. *Ecological Economics, 62*(2), 319–329.

Kshirsagar, M. P., & Kalamkar, V. R. (2014). A comprehensive review on biomass cookstoves and a systematic approach for modern cookstove design. *Renewable and Sustainable Energy Reviews, 30*, 580–603.

Kvale, S. (1996). *Interviews: An introduction to qualitative research interviewing*. London: Sage.

Larson, B. A., & Rosen, S. (2002). Understanding household demand for indoor air pollution control in developing countries. *Social Science and Medicine, 55*(4), 571–584.

Lee, C. M., Chandler, C., Lazarus, M., & Johnson, F. X. (2013). *Assessing the climate impacts of cookstove projects: Issues in emissions accounting*. Stockholm: Stockholm Environment Institute.

Lewis, J. J., & Pattanayak, S. K. (2012). Who adopts improved fuels and cookstoves? A systematic review. *Environmental Health Perspectives, 120*(5), 637–645.

Mahiri, I., & Howorth, C. (2001). Twenty years of resolving the irresolvable: Approaches to the fuelwood problem in Kenya. *Land Degradation and Development, 12*(3), 205–215.

O'Neal, D. (2005). Designing stoves for mass production. *Boiling Point, 50*, 8–11.

OECD/IEA. (2010). *Energy poverty: How to make modern energy access universal? Special early excerpt of the world energy outlook for the UN general assembly on the millennium development goals*. Paris: International Energy Agency.

OECD/IEA. (2013). *World Energy Outlook*. Paris: International Energy Agency.

Quadir, S. A., Mathur, S. S., & Kandpal, T. C. (1995). Barriers to dissemination of renewable energy technologies for cooking. *Energy Conversion and Management, 36*, 1129–1132.

Schatzman, L., & Strauss, A. (1973). *Field research: Strategies for a natural sociology*. Englewood Cliffs, NJ: Prentice-Hall.

Schumacher, E. (1993). *Small is beautiful: A study of economics as if people mattered*. London: Vintage.

Sesan, T. (2012). Navigating the limitations of energy poverty: Lessons from the promotion of improved cooking technologies in Kenya. *Energy Policy, 47*, 202–210.

Sesan, T. (2014). Global imperatives, local contingencies: An analysis of divergent priorities and dominant perspectives in stove development from the 1970s to date. *Progress in Development Studies, 14*(1), 3–20.

Sesan, T., Raman, S., Clifford, M., & Forbes, I. (2013). Corporate-led sustainable development and energy poverty alleviation at the bottom of the pyramid: The case of the CleanCook in Nigeria. *World Development, 45*, 137–146.

Shastri, C. M., Sangeetha, G., & Ravindranath, N. H. (2002). Dissemination of efficient ASTRA stove: Case study of a successful entrepreneur in Sirsi, India. *Energy for Sustainable Development, 6*(2), 63–67.

Shrimali, G., Slaski, X., Thurber, M. C., & Zerriffi, H. (2011). Improved stoves in India: A study of sustainable business models. *Energy Policy, 39*(12), 7543–7556.

Smith, K. (2010). What's cooking? A brief update. *Energy for Sustainable Development, 14*, 251–252.

Smith-Siversten, T., Diaz, E., Pope, D., Lie, R. T., Diaz, A., McCracken, J., et al. (2009). Effect of reducing indoor air pollution on women's respiratory symptoms and lung function: the RESPIRE randomized trial, Guatemala. *American Journal of Epidemiology, 170*(2), 211–220.

Westhoff, B. (1995). One of the oldest technologies in the world—from the open hearth to the microwave. In B. Westhoff & D. Germann (Eds.), *Stove images: A documentation of improved and traditional stoves in Africa, Asia and Latin America* (pp. 18–19). Frankfurt: Brandes and Apsel.

Wilkinson, P., Smith, K. R., Davies, M., Adair, H., Armstrong, B. G., Barrett, M., et al. (2009). Public health benefits of strategies to reduce greenhouse gas emissions: Household energy. *The Lancet, 374*(9705), 1917–1929.

Wolf, D. L. (1996). Situating feminist dilemmas in fieldwork. In D. L. Wolf (Ed.), *Feminist dilemmas in fieldwork* (pp. 1–55). Colorado: Westview Press.

World Bank. (2011). *Household cookstoves, environment, health, and climate change: A new look at an old problem*. Washington, DC: The World Bank.

Chapter 22
How Big is Small? Enough to not Breathe Oil!

The Peruvian Case of Diesel-Fuelled Wick Lamps for Lighting

Angel Verástegui Gubler and Verónica Pilco Mamani

Abstract Health risks due to indoor air pollution (IAP) from inefficient domestic burning processes for cooking or lighting are not breaking news. But the presence of high levels of sulfur dioxide in burnt wood emissions from traditional cookstoves; its remaining high levels in the air after two hours from turning off the source; and the fact that this gets even worse with an oil-fuelled wick lamp that pollutes almost the same as a second traditional cookstove in the same room for at least one hour each day for 20 % of the world's population, perhaps are. This paper shows first evidence from Peru's rural context in the simultaneous lack of modern energy devices for lighting and cooking.

Keywords Sulfur dioxide · Indoor air pollution · Diesel wick lamp

Introduction

Worldwide there are about 1400 million people without access to electricity (OECD 2010). Of these, it is estimated that 500 million people still use fossil fuels, among them mainly kerosene, to produce light (Lam et al. 2012a, b).

In Peru, about three million people lack access to electricity (MEM 2011). Unlike in other countries, in Peru no one is using kerosene-fuelled wick lamps because kerosene has been banned by law since 2010, since it is used in the production of illegal drugs (narcotics). However, there are many families in rural

A.V. Gubler (✉) · V.P. Mamani
Deutsche Gesellschaft für Internationale Zusammenarbeit—GIZ,
Energising Development Peru, Pje. Bernardo Alcedo 150, San Isidro, Lima 27, Perú
e-mail: angel.verastegui@giz.de; angel.verastegui@gmail.com

V.P. Mamani
e-mail: veronica.pilco@giz.de; veronica_pilcom@hotmail.com

S. Groh et al. (eds.), *Decentralized Solutions for Developing Economies*,
Springer Proceedings in Energy, DOI 10.1007/978-3-319-15964-5_22

areas of the rainforest which have replaced kerosene with diesel, using it as fuel for wick lamps.

In addition, almost all families using wick lamps cook in open fires (traditional stoves). The negative impact of traditional stoves in open fires has long been researched (Fullerton et al. 2008; Smith et al. 2004); however, there is no evidence about the exposure to both indoor air pollutants at the same time.

The smoke from diesel combustion contains many chemical components in the form of gas or ultrafine particle emissions (particulate matter, black carbon, etc.).

Resulting products from these emissions are carbon dioxide (CO_2), carbon monoxide (CO), sulfur dioxide (SO_2) and mono nitrogen oxides (NOx) (Morawska et al. 2004). The International Agency for Research on Cancer (WHO—Organización Panamericana de la Salud 2013) classified engine diesel exhaust as carcinogenic to humans.

The risk is much higher with fuel combustion in wick lamps, since one tenth of the fuel burned is converted to black carbon, compared to a diesel engine where this relation represents only one-thousandth (Jacobson et al. 2013). In addition there is some evidence that indoor pollutants from fuel-based lamps may have a correlation with cataracts and tuberculosis, but this requires further study (Mills 2012).

Research Objectives

This paper aims to study exposure levels to the polluting gases produced by diesel-fuelled wick lamps (DFWL), and to the levels from its simultaneous use with wood burning traditional cooking stoves. The first research question was to discover if the use of DFWL results in dangerous exposure levels of the same dangerous gases that traditional cookstoves produce, mainly particular matter 2.5 ($PM_{2.5}$) and carbon monoxide (CO).

In addition, it was tested if the concentration levels of sulfur dioxide (SO_2) are higher than those recommended by the World Health Organization (WHO) as the highest exposure limits.[1] The focus on sulfur dioxide responds to the concern about the high level of sulfur that Peruvian diesel has, one of the highest in Latin America as Fig. 22.1 shows.

According to the Peruvian Environment Ministry (MINAM 2013), the high content of sulfur in diesel is responsible for an increase in the last three years of SO_2 levels in the air by nearly 500 % in the Peruvian capital Lima. The last occurred by stable levels of nitrogen dioxide NO_2 and ozone O_3 in the same time period.

The last research question was to measure the emission levels of these three gases from DFWL in simultaneous use with a traditional cookstove (3-stone-fire), and to evaluate its remaining levels over time after turning off the sources.

[1] SO_2 is one of the most common air pollutants according to the WHO beneath $PM_{2.5}$, nitrogen dioxide (NO_2) and ozone (O_3).

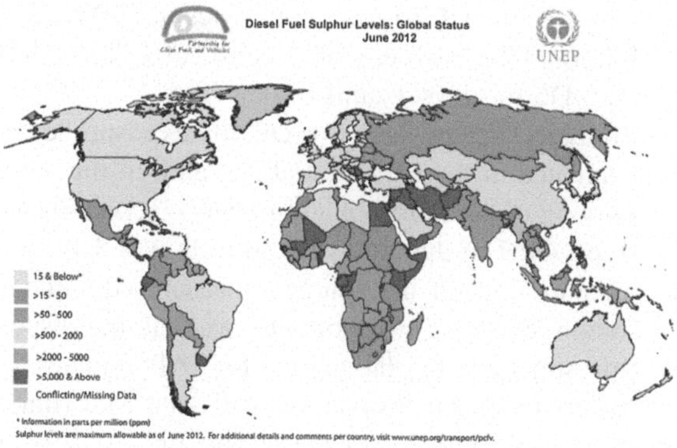

Fig. 22.1 Diesel fuel sulfur levels: global status (UNEP 2012)

The measure of carbon dioxide (CO_2) was only for discovering if there are any health implications regarding its emissions levels in the discussion around IAP, since it wasn′t recorded in Peruvian traditional stove tests.

Methods

Two DFWL with different types of wick (Type A: cotton and Type B: old cloth) were collected from households in two different towns in the Amazon area (the provinces San Martín and Amazonas, respectively) and used as polluting sources. Tests were conducted on indoor air pollutant concentration levels of dangerous gases ($PM_{2.5}$, CO, SO_2 and carbon dioxide CO_2), resulting from the burning of these two types of DFWL with diesel fuel.

The equipment used for measuring $PM_{2.5}$ and CO was the Aprovecho IAP-Meter with resolutions of 0–60,000 $\mu g/m^3$ (red laser scattering photometer) and 0–1000 ppm (electrochemical cell) respectively. For SO_2 and CO_2, Aeroqual was used with resolutions of 0–15 ppm (Gas Sensitive Electrochemical—GSE Sensor) and 0–5000 ppm (Non dispersive Infra Red—NDIR Sensor) respectively. Both devices were calibrated in December 2011 for the last time.

The equipment belongs to Housing Ministry's improved cookstove certification laboratory in Lima, where these tests were conducted. The equipment was located in the room in a simulated position of a regular nose of a typical user, following conventional protocols used for testing cookstoves (Aprovecho Research Center 2014).

The environment chosen had a ventilation rate of 4.29 h-1, which was determined with the window and door closed, as recommended by the new protocol for IWA (International Workshop Agreement GACC 2012) on improved stoves.

During the trials for each type of test, it was intended to homogenize some variables such as:

- initial background measure of all tested gases (30 min) for setting a baseline of concentrations in the room;

- length of the DFWL test; both types of DFWL A and B were evaluated for 3.5 h burning time each day on 3 consecutive days during similar hours respectively (D1–D6 for days 1 until 6);
- Similarly, during days D7–D9, the exposure levels of the traditional stove were evaluated in the mornings alone, and, in the afternoons, simultaneous with the burning of the most polluting DFWL according to the results of tests on D1–D6 (type A). The duration of both tests was 1 h;
- after turning off the sources, measure equipment remained recording concentration levels for two hours to evaluate the dispersion speed of the gases;
- infrastructure; the laboratory for IAP imitates a rural house made of typical material and dimensions for walls and roof (mud bricks and corrugated iron);
- same characteristics of diesel and firewood for all tests; the laboratory's firewood is standardized for stove tests regarding humidity, wood type and origin;
- technical evaluator; same person for all tests with prior experience in stove evaluations; and
- the approximate level of light emitted by the DFWL; the lighting level should remain constant Durant the tests, so the evaluator pulled the wick during the tests for keeping its light constant).

To control the environmental variables that could influence the results of the tests, the Davis Vantage Pro Weather Station was used. This equipment took values for inner temperature of the room (°C), relative humidity (%), wind speed (m/s) and solar radiation (W/m^2). Average values of these units during the evaluation days can be seen on Table 22.1.

The results on concentrations levels of these gases should be compared to the recommended highest exposure levels made by WHO ($PM_{2.5}$, SO_2 and CO) and by the Occupational Safety and Health Administration (OSHA, CO_2) respectively. These levels can be seen on Table 22.2.

Results

The DFWL type A showed in all tests the highest concentration of gases. This DFWL showed the highest fuel consumption with an average of 101 g of diesel versus 55 g for 3.5 h burning with DFWL type B respectively. All the results listed below are taken from the tests conducted with DFWL type A.

The average $PM_{2.5}$ levels of the most polluting DFWL was around 10,435 $\mu g/m^3$ with peaks nearly 30,000 $\mu g/m^3$. Figure 22.2 shows $PM_{2.5}$ concentration levels for the three days of measurement and its average for DFWL A.

This concentration of $PM_{2.5}$ particles reaches approximately 60 % of the emission levels of a traditional cookstove as the only pollution source, which showed average levels of 14,841 $\mu g/m^3$ and peaks near 35,000 $\mu g/m^3$ in this research. The levels of $PM_{2.5}$ for a traditional cookstove as the only pollutant on D7–D9 and its average can be seen in Fig. 22.3.

Table 22.1 Average values of environmental variables during evaluation days

Average values	D1–D3	D4–D6	D7–D9
Inner temperature (°C)	32.9 ± 0.4	32.5 ± 1	41.9 ± 1.8
Relative humidity (%)	56 ± 0.6	54.7 ± 2.1	42.7 ± 2.0
Wind speed (m/s)	1.2 ± 0.3	1.3 ± 0.1	1.2 ± 0.2
Solar radiation (W/m$^{2)}$	578.9 ± 53.9	593.3 ± 56.5	476.4 ± 49.5

Table 22.2 Recommended exposure levels

Air pollutant	Exposure time	Recommended level (ppm)	Institution
SO_2	10 min	0.17 ppm	WHO
	24 h	0.007 ppm	WHO
$PM_{2.5}$	24 h	25 µg/m^3	WHO
	365 days	10 µg/m^3	WHO
CO	30 min	50 ppm	WHO
	1 h	25 ppm	WHO
CO_2	15 min	30,000 ppm	OSHA
	8 h	1000 ppm	OSHA

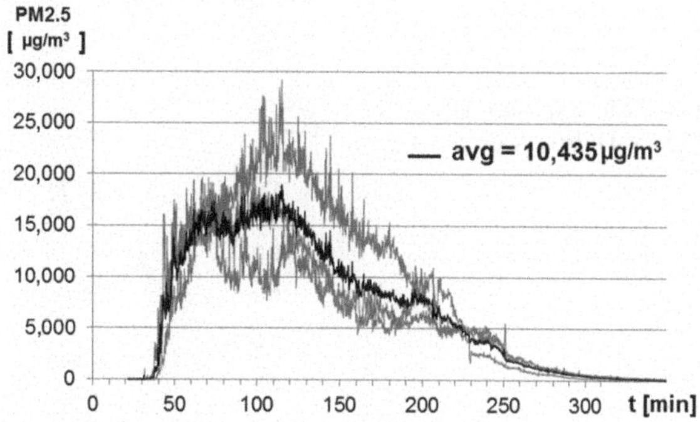

Fig. 22.2 Exposure levels of SO_2 for D1–D3 for DFWL A

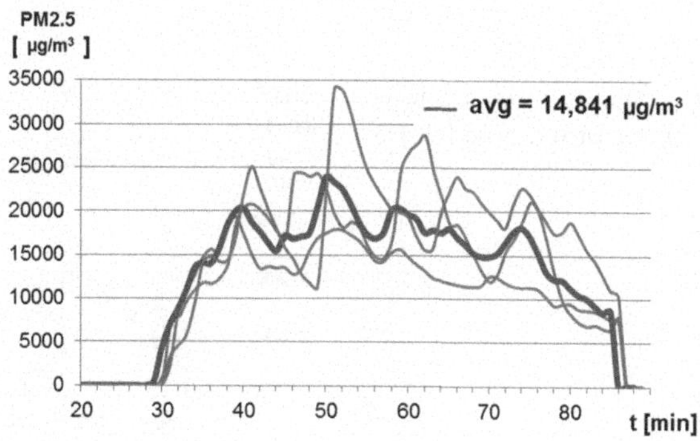

Fig. 22.3 Exposure levels of $PM_{2.5}$ for traditional stove for D7–D9 for DFWL A

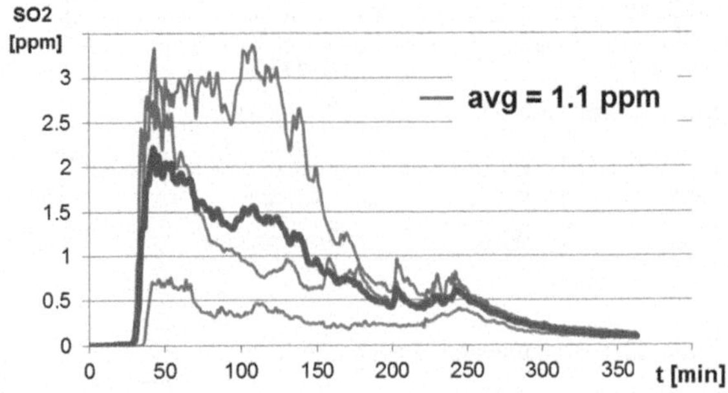

Fig. 22.4 Exposure levels of SO_2 for DFWL A on D1–D3

The tests with the same DFWL showed that the average concentration of SO_2 emitted after the first 10 min of burning was 1.1 ppm, exceeding almost up to seven times the limit allowed by the WHO for 10 min exposure of 0.17 ppm. There were peaks over 3 ppm. Figure 22.4 shows the exposures levels for SO_2 taken on D1–D3 and its average.

The concentrations of CO and CO_2 from both DFWL didn't show risky levels with 4.8 and 136.4 ppm in average for the worse results with DFWL A respectively. This can be seen on the curves for both gases with DFWL A for D1–D3 in Figs. 22.5 and 22.6.

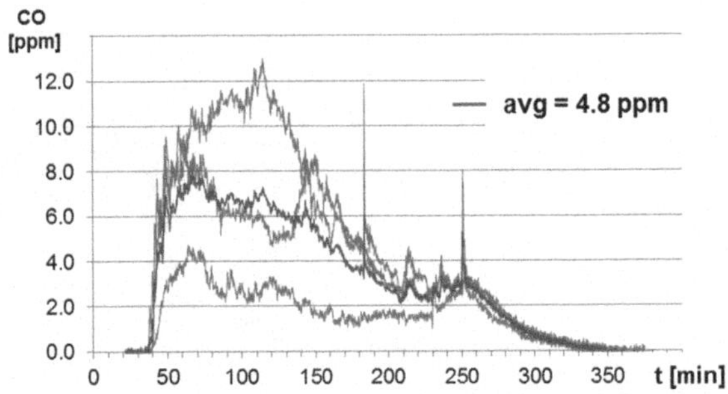

Fig. 22.5 Exposure levels of CO for DFWL A on D1–D3

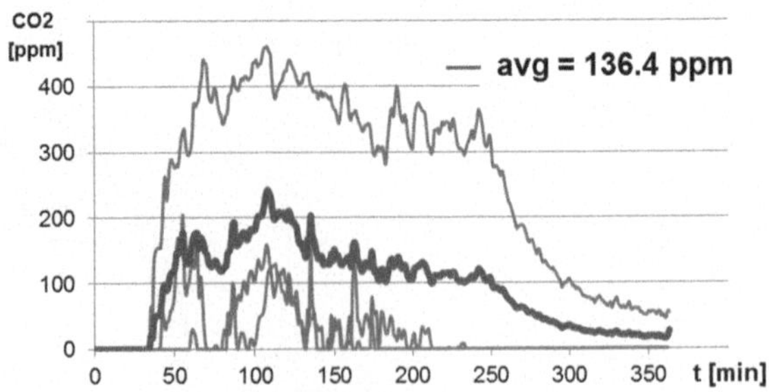

Fig. 22.6 Exposure levels of CO_2 for DFWL A on D1–D3

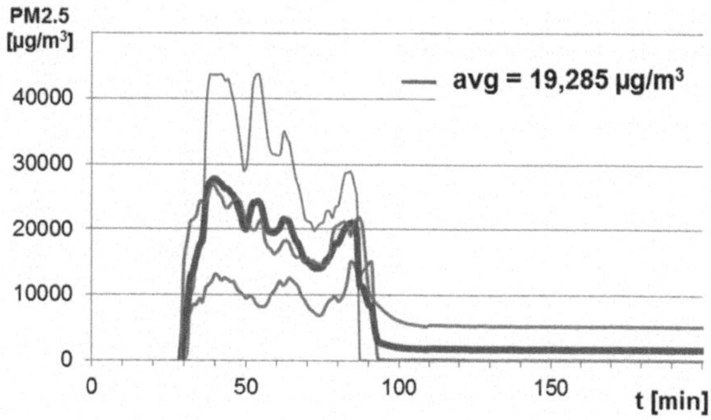

Fig. 22.7 Exposure levels of $PM_{2.5}$ for DFWL A on D7–D9 with a traditional cookstove

Levels for CO and CO_2 only with the traditional stove as a pollution source show higher levels than recommended for CO (379 ppm in average) and lower levels for CO_2 (1,464 ppm in average). The higher levels for CO are already known from the cookstoves research and those for CO_2 are lower as recommended even with the DFWL A as a second pollution source (664.8 ppm in average).

However the use of the DFWL A simultaneously with a traditional stove regarding $PM_{2.5}$ showed an average level of 19,285 $\mu g/m^3$ (with peaks over 43,000 $\mu g/m^3$) having an average increase of nearly 30 % from the values with a traditional cookstove as the only pollutant. The evolution of the exposure levels of $PM_{2.5}$ on D7–D9 with both burning sources can be seen in Fig. 22.7.

An unexpected result was observed in that a traditional cookstove, as the only source of pollution, reaches levels of sulfur dioxide SO_2 far exceeding the permissible exposure values from various organizations, such as the WHO. SO_2 is not a typical gas taken into account in typical cookstove emission tests.

The intensity of this emissions were so high that they even exceed the maximum possible measurement levels of the instruments (>15 ppm), hence the concentration levels during the full test couldn't be monitored properly for both cases (DFWL alone 14 ppm and in addition with a traditional cookstove 11.52 ppm in average respectively). The disrupted evolution of SO_2 on D7–D9 for the case of simultaneous pollution can be seen in Fig. 22.8.

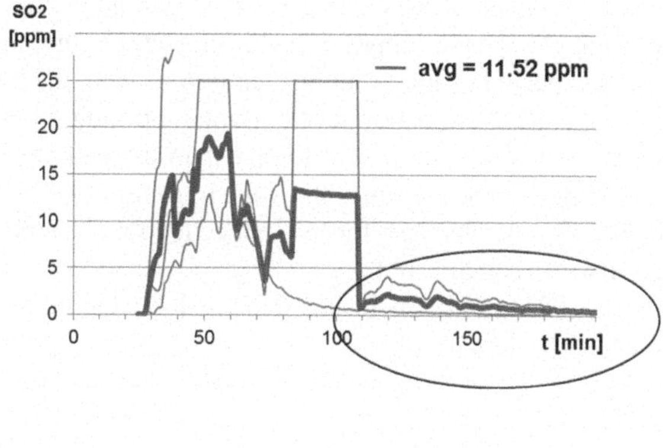

Fig. 22.8 Exposure levels of SO_2 for DFWL A on D7–D9 with a traditional cookstove

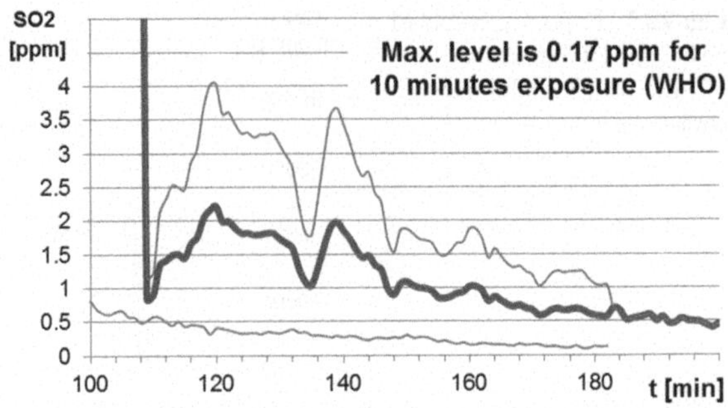

Fig. 22.9 Remaining exposure levels of SO_2 after turning off both pollution sources on D7–D9

SO_2 emission speeds of 0.5 and 1.5 ppm/min with DFWL A alone and with both polluting sources respectively could be observed. In addition one hour after having turned off both polluting sources, still remain in the room higher levels of SO_2 as recommended by the WHO. This can be seen in a zoomed picture of previous figure, that is shown on Fig. 22.9.

Conclusions

The evaluated DFWL are less polluting than traditional cookstoves taking into account the exposure levels measured for the pollution from the traditional stove alone in this study for $PM_{2.5}$ and CO. Carbon monoxide emissions do not exceed the recommended exposure levels. However these values for particulate matter represent around 60 % of the measure taken for the traditional cookstove.

This could be a serious problem, since both traditional devices exceed by far the exposure limits recommended by WHO and are commonly used simultaneously in rural households. This finding could set an additional challenge to improved cooking stove (ICS) programs, since the usage of ICS alone wouldn't be enough for those households using in addition DFWL.

In addition both tested DFWLs seem to exceed during the whole burning process the recommended exposure levels of SO_2 by the WHO for 10 min. There isn't a value for one hour exposure time, but it would be much lower than the 10 min one, making clearer the scope of health impacts by the normal usage of DFWL.

Unexpected results were SO_2 emissions from the traditional cookstove above the recommended exposure levels (when burning alone and simultaneous with a DFWL). These values were higher than the equipment measure capacity and representing at least ten times the values from the DFWL alone. Traditional ICS tests could open the research for SO_2, since it could be an invisible pollutant in the actual efforts for reducing IAP.

The CO_2 emission levels remained by far under the exposure limits recommended by OSHA.

Discussion

This study invites further research on indoor air pollution, taking into consideration sulfur dioxide SO_2 and its health implications, either through DFWL or traditional cooking stoves. It is recommended to test other typical diesel burning gases like nitrogen dioxide and evaluate its implications on health comparing results with WHO air quality guidelines.

Furthermore it is suggested that more traditional DFWL are collected from the field and tested with the same methodology so there can be enough evidenced data that serves as consistent basis for a baseline of the pollution from these devices, since they are all different among each other. When this is reached, it is suggested to carry out evaluations with one representative type of DFWL and play with other variables like different types of carburant agents (wicks) since it has been observed that different varieties of wicks emit larger or smaller amounts of fine particles.

The tests performed have also shown the existence of high levels of SO_2, emitted only by wood burning stoves. For this reason, it is suggested to consider the levels of SO_2 emitted during combustion when validating improved stoves.

In the absence of complete simultaneous measurement of pollutants during the tests on DFWL and traditional stoves, it is suggested to perform this test with equipment that allows for a wider range of measurement and records other pollutants, such as nitrogen oxides and sulfur. It is also advisable to measure on different days, following similar time schedules, in order to control environmental variables.

The testing in laboratory may be helpful to have first impressions of the potential inherent risk these devices have. However, the real risks for the people under real exposure levels can only be measured on the field under real weather and ventilation conditions in real houses, and furthermore, under manipulation from real users. Only then the potential and the real exposure levels can be linked and concrete consequences to health be inferred.

It is understood that the high pollutant exposure levels recorded (especially $PM_{2.5}$ and SO_2) imply a risk to people who use these traditional devices for lighting and cooking in their homes. So, being the most risky average exposure levels from SO_2 and $PM_{2.5}$ by breathing contaminated air from the DFWL or traditional stoves, people should be warned about the risk of carrying out activities within a closed environment due to the presence of these polluting sources.

References

Aprovecho Research Center. (2014). Disponible en: http://www.aprovecho.org/lab/index.php. Consultado el 09 de marzo del 2013.

Fullerton, D. G., Bruce, N., & Gordon, S. B. (2008). Indoor air pollution from biomass fuel smoke is a major health concern in the developing world. *Transactions of the Royal Society of Tropical Medicine and Hygiene, 102*(9), 843–851.

GACC. (2012). International Workshop Agreement, Global Alliance for Clean Cookstoves, an Initiative Led by the United Nations Foundation, Washington, DC, USA.

Jacobson, A., Bond, T. C., Lam, N. L., & Hultman, N. (2013). *Black carbon and kerosene lighting: An opportunity for rapid action on climate change and clean energy for development*. Washington, DC: Global Economy and Development, The Brookings Institution.

Lam, N. L., Smith, K. R., Gauthier, A., & Bates, M. N. (2012a). Kerosene: A review of household uses and their hazards in low- and middle-income countries. *Journal of Toxicology and Environmental Health, Part B: Critical Reviews, 15*(6), 396–432.

Lam, N. L., Chen, Y., Weyant, C., Venkataraman, C., Sadavarte, P., Johnson, M. A., et al. (2012b). Household light makes global heat: High black carbon emissions from kerosene wick lamps. *Environmental Science and Technology, 46*, 13531–13538.

MEM. (2011). *Dirección General De Electrificación Rural del Ministerio de Energía y Minas, Plan Nacional de Electrificación Rural* (PNER) 2012–2021. Lima. http://dger.minem.gob.pe/ArchivosDger/PNER_2012–2021/PNER-2012-2021%20Texto.pdf. 17 Apr 2014.

Mills, E. (2012). *Health impacts of fuel-based lighting*. Lumina Project Technical Report #10. http://light.lbl.gov/pubs/tr/Lumina-TR10-health-impacts.pdf. 13 Aug 2014.

MINAM. (2013). *Servicio Nacional de Meteorología e Hidrología del Perú—SENAMHI*: http://elcomercio.pe/lima/sucesos/contaminacion-lima-aumento-veces-anos_1-noticia-1674872. 17 Feb 2014.

Morawska, L., Moore, M. R., & Ristovski, Z. D. (2004). *Health impacts of ultrafine particles: Desktop literature review and analysis*. Report to the Australian Department of the Environment and Heritage.

OECD. (2010). *World energy outlook* 2010. http://www.worldenergyoutlook.org/media/weo2010.pdf. 8 Apr 2013.

Smith, K. R., Mehta, S., & Maeusezahl-Feuz, M. (2004). Indoor air pollution from household use of solid fuels. *Comparative Quantification of Health Risks: Global and Regional Burden of Disease Attributable To Selected Major Risk Factors, 2*, 1435–1493.

UNEP. (2012). *Map world sulphur*. Disponible en: http://www.unep.org/transport/pcfv/PDF/Maps_Matrices/world/sulphur/MapWorldSulphur_June2012.pdf. 17 Feb 2014.

WHO—Organización Panamericana de la Salud. (2013). http://www.iarc.fr/en/media-centre/pr/2012/pdfs/pr213_E.pdf. 15 July 2014.

Chapter 23
Analysis of Decision-Making for Off-Grid Rural Electrification in Colombia

Adriana M. Valencia J.

Abstract Due to their implicit costs, the technology options currently selected to expand national energy production and to provide energy to off-grid areas of Colombia lead to higher Government expenses, inflated cross-subsidization from higher-income electricity consumers to other consumers, and to limited electricity services for some. The technology option decision ultimately has an impact on energy access and duration of service. The overall cost of electricity provision through diesel, the fuel most frequently used, often increases in response to costs that are not normally taken into account. Therefore, it is suggested that decision making for off-grid rural electrification be more comprehensive, mindful of final costs of service for the end user, and attentive to the long term sustainability of the service.

Keywords Off-grid electricity · Technology choice · Diesel

Introduction

Colombia's contrasting geography, with forests comprising more than 50 % of the country, as well as the armed conflict, makes it very difficult and costly to provide electricity to people located in areas where there is no grid-connected electricity service (off-grid areas). More than 1.2 million people live in these often difficult to access and remote locations of Colombia (this is 2 % of the total population, located in 52 % of the country's territory (IPSE 2013).

The low population densities found in these locations (which can be as low as 2 people/km^2, compared to an average population density of 40 people/km^2—PAHO 2012) adds to the cost per connection, compared to grid-connected areas.

A.M. Valencia J. (✉)
Energy Specialist, Washington, DC, USA
e-mail: adrianam@berkeley.edu

S. Groh et al. (eds.), *Decentralized Solutions for Developing Economies*,
Springer Proceedings in Energy, DOI 10.1007/978-3-319-15964-5_23

In addition, there are about 7 million people who meet their energy needs using small (<500 kW) diesel or gasoline generators, which often provide limited hours of service.[1] The hours of service have been reported to be 6 h in 99 % of the localities [of which there are more than 1400 localities (IPSE 2013)], 12–24 h in less than 10 localities and 24 h in only two localities (Flórez et al. 2009).

With the above in mind, it can be said that about 17 % of the Colombian population (the population reached 48.3 million people in 2013) lacks adequate access to electricity.

Transmission costs can be especially high and therefore these locations are mostly served with decentralized small diesel systems. 97 % of electricity provided in off-grid areas is diesel based, 2 % is small hydro and the rest are mostly photovoltaic (Flórez et al. 2009; UPME Personal Communication 2012). While more recent data is needed, in 2001 it was reported that 44 % of the people living without electricity still used candles for lighting (AENE and Bailly 2001).

Research Objectives

The main question guiding the research for this paper was: What are the main drivers that affect the technology choice decision/decision-making process in electricity sector expansion for off-grid electricity projects in Colombia?

Methods

The author analyzed the existing literature on the topic. In addition, several interviews and conversations were held with more than 30 decision makers and relevant experts in Colombia. A survey was conducted to understand the decision-making process of the institution in charge of rural electrification in Colombia.[2]

57 responses were received for projects with different types of technologies including diesel, small hydropower, solar-diesel, and biomass. The projects surveyed were spread out in various areas of Colombia. The criteria used for selecting projects were as follows: 1. Projects in various geographical areas of the country; 2. Projects with various installed capacities to be able to compare costs; and 3. Projects with various types of technologies (including grid connections/grid repairs).

[1] There are communities which are located in areas denominated as 'grid-connected' that do not have electricity (UPME 2003).

[2] This article is written based largely on research conducted for the author's Ph.D. dissertation (Valencia 2008).

The survey was made available in electronic form, in which it was possible to answer some questions by selecting from a multiple-choice drop-down menu, and, in addition, for all the questions there was the possibility of providing open-ended responses. The surveys, along with the interviews and literature research corroborated and complemented the information presented in the section Results.

This article focuses on technology choice and does not explore barriers to successful implementation of renewable energy for rural electrification, the subject of a different article published by the author (Valencia and Caspary 2008). However, it should be noted that there are some factors that affect the quality of electricity provision, such as poor service management, the lack of a direct person responsible for the service, and low technical and administrative capacity by municipalities (UPME 2003).

Results

I. Main actors. Most off-grid electrification projects in Colombia have ultimately been the responsibility of two government institutions:

(i) **UPME**, Colombia's Energy and Mining Planning Unit, has the mission of planning mining and energy development, information management and support the formulation of sector policy, in coordination with the actors involved. In the past, project approval decision making was largely influenced by UPME because the financial evaluation was the task of this government entity. Now the financial evaluation is also the responsibility of the IPSE (see below).

(ii) **IPSE**, the Colombian Planning Institute for Off-grid Energy Solutions, is the government organization in charge of selecting technology options for off-grid regions of Colombia and the institute has an ultimate decision to determine the technology choice.[3,4]

[3] There are two main funds that are tapped into for off-grid electrification in Colombia: FAZNI and FNR. The FAZNI fund (designated by a Law No. 633 of 2000) provides Col$1 (US$0.0005)/kWh to offer financial assistance for off-grid electrification. The FNR (National Royalties Fund) (established by Law No. 756 of 2002) designates partial funds for the same purpose. The majority of projects have been financed using FAZNI funds and with the use of these funds the technology (technical decision) is in the hands of the IPSE.

[4] In cases when the IPSE joins universities or multilateral institutions for the implementation of innovative or pilot projects, then the technology type is a joint decision between these entities involved.

There are other actors involved in diesel electrification decisions in Colombia are as follows:

(i) **Estructuradores**. *Estructuradores* for electrification initiatives are normally electrical engineers who are hired by communities or by local or regional governments to formulate and design a project that is then presented to the FAZNI committee (Fund for the Electrification of Off-grid Regions) for approval and consequently for funding. The *estructuradores* normally present projects that are diesel based because: (a) this is the technology they know, (b) this technology has been known to work, and (c) it is easy and fast to implement and operate.

(ii) **Community members**. Community members sometimes also have a say on the type of technology that is installed in their communities.[5]

(iii) **Small companies**. *Estructuradores* also subcontract or consult with small companies to help them design electrification systems. Small companies are also sometimes directly contacted by communities to help bring electricity.

II. Factors influencing the decision. The institutions that have been in charge of off-grid electrification in Colombia (see above) have been influenced by various incentives or disincentives regarding the type of technology to be implemented in electrification projects. These incentives include: (a) regulations that have resulted in extensive diesel fuel subsidies—which have kept (in the past, because this is recently changing) diesel prices lower than they would otherwise be and previous incentives that provided tax exemptions and made it easy for entities to purchase diesel power plants; (b) the lack of available information on alternative energy solutions (i.e., on their ease of use, on resources available to use the technologies, on the functioning of the equipment)[6]; and (c) the apparent lack of any obligation for these government institutions to compare alternative technology options.[7]

Accordingly, diesel has most often been the fuel of choice, and the decision has been influenced by:

(i) **The information available for decision making**, such as renewable energy resource potential maps that are readily available in Colombia, is either mission or not sufficiently detailed for off-grid energy planners. Likewise, detailed demographic information is also inaccessible and often outdated. Collecting information in off-grid areas can also be difficult due to the armed conflict; and

[5] Communities are often aware of the difficulties that are encountered in acquiring diesel fuel and how costly this can be to them.

[6] Specifically, the survey conducted for the research revealed that, for 84 % of the projects surveyed, the availability of information was a factor influencing the technology choice decision.

[7] However, IPSE and UPME carefully compare grid versus diesel electrification solutions.

(ii) **The limited methodology (or lack thereof) for technology selection**, hampers technology specific comparisons. When asked during interviews how technologies were compared to one another, the respondents revealed that cost comparison was basically nonexistent. There is one methodology that has potential to promote technology comparisons, yet it has not even minimally achieved its potential. The methodology has a section in which developers are asked whether they want to compare alternative technology options. If the project developer answers negatively, the section is simply skipped. In short, there is no obligation to complete a technology comparison to apply for funds, nor is there an obligation to perform this task for purposes within the institution. Although there is a methodology that project developers need to follow to acquire subsidies for off-grid electrification projects (cross-subsidy from other customers, FAZNI), the methodology needs large improvement and better use.

III. Unaccounted costs. Important additional costs are incurred in electricity production through diesel in Colombia. These costs, which are not normally considered in cost calculations, include: high costs of fuel transport, high costs of diesel (due to illegal markets), and frequent diesel plant replacements (either because of involved high opportunity costs in transaction or because of the high costs of spare parts). These costs are not included in government planners' calculations, in part, because they are normally the responsibility of municipality or location being provided the service.

(i) **Cost of fuel transport**. As studies corroborate, the greatest factor affecting the cost of electricity production through diesel is the cost of fuel (AENE and Bailly 2001) and the fuel price is largely influenced by the cost of transport. Whereas in some places the cost of transport is only Col$450/gallon, there are other regions of the country where the transport of the fuel can be as costly as Col$10,290/gallon (around US$0.22/gallon vs. US$5.15/gallon, respectively). Cost differences are largely due to the geography and topography of the regions, and to whether fuel needs to be transported by airplane, boat, truck, mule, or a combination of these modes, for up to 6 h. These costs add great variability to the final cost per kWh produced depending on the locations; thus, when in some locations the *transport cost* per kWh can be only US$0.02/kWh, the cost in other locations can be as high as US$0.40/kWh, with the average cost being around US$0.05/kWh.[8,9]

(ii) **Cost of diesel and illegal markets**. Border regions have had special privileges that governments continue to support. Law 677 was created in 2001 in

[8] Calculated by author using data provided by IPSE, a fuel efficiency of 0.08 gal/kWh (thus assuming that 12.5 kWh can be produced from each gallon of fuel). Fuel efficiency improves slightly with larger plant sizes, however, the majority of plants in Colombia are small.

[9] In addition, the transport of diesel fuel in off-grid areas of Colombia is mostly done by one company, meaning that there has not been competition to transport the fuel and to thus keep the prices of fuel transport low (UPME Personal Communication 2008).

Colombia with the purpose of attracting and generating new investment to strengthen the process of national exports. The law creates the term "special economic zones of export", which benefits five cities in five departments. This law has various incentives, which include tariff exemptions and special customs privileges (República de Colombia Ministerio de Comercio 2001). This has included fossil fuel, and most diesel in Colombia is imported. However, according to interviews, this has led to black markets. Specifically, this has led to the resale of the fuel to those who do not benefit from the border incentives (i.e., the diesel that is destined for some locations for specific uses is resold to other customers for higher prices; or diesel is purchased in neighboring countries and then resold). Evidently, fuel sales have decreased in non-border regions, while they have increased in border regions with the implementation of the above mentioned Law (UPME 2004).

Unfortunately, diesel is a desirable input for illicit trade, e.g., for production of drugs—both for processing these and the transport of trade. This has resulted in high prices, less fuel availability, and the interruption of diesel delivery to intended destinations (and therefore fewer hours of service in some communities). Therefore, fuel price differences can be large, at less than US$2/gallon in some places compared to more than US$7 dollars/gallon in others. These characteristics have also led to fuel quality distortions, which modify the technical characteristics required for electricity production and which, in turn, lead to the deterioration of the diesel power equipment.

(iii) **Diesel plant replacements**. The cost of equipment (diesel plants) is normally not significant unless the equipment has to be replaced constantly. Unfortunately, diesel equipment replacement generally occurs every 4 years or less in Colombia and therefore the costs of electricity generation increase. In fact, during the interviews it was found that the replacement happens as often as every 2 years. Diesel plants become damaged for various reasons but this is often due to lack of or poor maintenance. These damages tend to be more frequent in coastal zones due to climate conditions. Regardless, when the plants become damaged, instead of the replacement of a part, etc., it is common practice in Colombia to purchase a new plant and replace the old one. These costs are not taken into consideration in cost calculations for off-grid electricity provision and they can be significant. Including the equipment breakdown at a periodicity of once every 4 years, in the end, the total generation costs can increase by US$0.03–US$0.10/kWh.

IV. Location of service prioritization. Associated with the decision on the technology choice for electricity provision is the decision on which zones to prioritize for electricity delivery. The research found that this decision, in terms of location, is based, in practice, on a number of criteria and factor combinations (in no specific order): (1) lack of electricity availability; (2) population density; (3) geopolitical importance[10]; (4) availability of resources for power generation; and (5) payment capacity (IPSE Personal Communication 2008). It is important to highlight the third criterion listed because, as found in the research (and corroborated by other authors), off-grid electrification in Colombia bends to regional political pressure and is highly influenced by political aspects (Aristizábal Ángel 2005). This is why there has been a trend to provide fast installation solutions, such as with diesel power plants.

Discussion and Conclusions Applicable Beyond Colombia

The technology choice decision in Colombia, and in other countries in Latin America, has largely been based on following the status quo.[11] As is the case in Colombia, diesel is the easiest technology alternative available. The analysis conducted in this article provides evidence that factors other than economic ones have been influencing decision making for off-grid electrification. This goes against the idea of other authors that energy policies are normally based on cost minimization.[12] The research found the contrary: for 95 % of the projects surveyed it was found that cost was not the main factor influencing the technology choice decision.

The limited use of methodology of planners and developers for selecting electricity technologies present a challenge to provide adequate electricity solutions in off-grid areas. It is imperative to have technology selection methods that are simple to use and that take into account all the aspects that need to be reviewed before deciding on an electrification technology (or a combination of technologies). Aspects to be included in the technology selection should especially take into account the ultimate cost of the solution to the country and to the end users. A challenge is added however, when resource availability and even demographic data are not immediately accessible to planners, to include as variables in methodologies to be used.

To conclude, the first step to improve the decision making process is to ensure there is sufficient technical capacity in the main institutions involved in making the

[10] Geopolitical importance means areas where there is priority to provide "State presence", which beyond military patrols, also means priority to provide other services such as electricity. These are normally areas that are located on or close to national borders and also areas where there are public order (e.g., violence) problems.

[11] Another clear example is Nicaragua, where on the Atlantic Coast diesel-powered generators are commonly used for microgrid generation due to their low capital costs, modularity, and the widespread availability of expertise in selling, operating, and maintaining the systems (Casillas 2008).

[12] Other authors have revealed that energy policies in other countries have in the past been normally based only on cost minimization (although more recently other criteria have also been recognized as being important and therefore there is a growing interest in multi-criteria decision analysis for rural electrification (Georgopoulou et al. 1997).

decision. These institutions need professionals trained in alternative technologies and economic analysis. Alternative technologies include solar photovoltaic panels, wind generators, micro and mini hydropower plants and a combination of these (hybrid systems).

A single decision-making tool will not be optimal for all rural electrification projects around the world, and therefore the idea in this article is not to prescribe a specific one. However, there are decision-analysis methodologies, even ones that have been developed in Colombia, which can be of valuable use for project planners.[13]

It is hoped that government entities and independent project planners will be more objective and careful in their selection of electrification technologies and long-term electrification expansion, justifying their selection with technology comparisons, with the inclusion of community impacts, and under long-term planning horizons. It is recommended that, for ease of use and consistency, the use of these tools be standardized in the planning institution(s) of the country in question.

Furthermore, all expenses for calculating the cost of a technology choice need to be taken into account, including the cost of fuel transport, the cost of operation and maintenance, and the final cost to the end user. Electricity provision decisions should weigh the costs-benefits of different levels of access (i.e. number of hours of service/day) related to the solutions to be provided.

Lastly, for the projects to be sustainable, more emphasis will have to be placed on project design. This means, again, including communities in the decision-making process, training locals to maintain and operate the equipment, and including the possibility of helping the communities develop further through new or more effective productive uses of energy.

Technology choice methodologies that include the above mentioned factors, will ideally lead to more reliable, cleaner, and less costly solutions to off-grid regions.

References

AENE Consultoría, & Hagler Bailly Services. (2001). *Establecimiento de un plan estructural, institucional y financiero, que permita el abastecimiento energético de las zonas no interconectadas con participación de las comunidades y el sector privado.* Santa Fe de Bogotá. Prepared for the Government of Colombia.

Aristizábal Ángel, J. A. (2005). *Políticas y estrategias en pro de un uso racional de los recursos para la energización de la zona no interconectada de Colombia.* Master thesis, Universidad Nacional, Medellín, Colombia.

[13] These tools include multi-criteria decision analyses such as the Renewable Energy for Sustainable Livelihoods, developed by the Imperial College of London (United Kingdom) and the National University of Colombia, and the HMO tool developed by the Universidad de Antioquia, Colciencias, ISA, and Carbono y Bosques. Other well more known tools include HOMER and RETScreen.

Casillas, C. (2008). *Wind generation: A step towards energy independence on the Atlantic Coast of Nicaragua*. Master's thesis, University of California, Berkeley.

Flórez, A., J, D. Tobón, & G. Castillo (2009). Ha sido efectiva la promoción de soluciones energéticas en zonas no interconectadas (ZNI) de Colombia?: Un análisis de la estructura institucional. *Cuadernos de Administración, 22*(38), 219–245 (enero-junio).

Georgopoulou, E., Lalas, D., & Papagiannakis, L. (1997). A multicriteria decision aid approach for energy planning problems: The case of renewable energy option. *European Journal of Operational Research, 103*, 483–500.

Instituto de Planeación de Soluciones Energéticas (IPSE). (2008). Personal communication.

IPSE. (2013). Institutional video. Available online at http://www.ipse.gov.co/ipse/

PAHO. (2012). Organización Panamericana de la Salud. Salud en las Américas, Edición de 2012: Volumen de países, Colombia.

República de Colombia, Ministerio de Comercio. (2001). Ley 677. DIARIO OFICIAL. AÑO CXXXVII. N. 44509. 4, AGOSTO, 2001. PAG. 11. Available online at http://www.mincomercio.gov.co/econtent/documentos/normatividad/leyes/ley_677_2001.pdf

Unidad de Planeación Minero Energética (UPME). (2003). *Plan Indicativo de Expansión de Cobertura del Servicio de Energía Eléctrica* (pp. 36, 51).

UPME. (2004). *Informe del Proceso de Asignación de Volumenes Máximos de Combustibles Líquidos Exentos del Iva y Global en las Zonas de Frontera.* Bogotá.

UPME. (2008). Personal Communication.

UPME. (2012). Personal Communication.

Valencia, A. M. (2008). *Missing links: Demystifying alternative energy use and improving decision making for increased off-grid electrification in Colombia*. Berkeley: University of California.

Valencia, A. M., & Caspary, G. (2008). *Expanding the benefits of globalization through electricity access*. New Haven, CT: Yale Center for the Study of Globalization. YaleGlobal.

Chapter 24
Barriers and Solutions to the Development of Renewable Energy Technologies in the Caribbean

Philipp Blechinger, Katharina Richter and Ortwin Renn

Abstract Despite large amounts of readily available renewable energy, most island states in the Caribbean are still heavily dependent on mostly imported fossil fuels for their energy production. Making use of empirical analyses, this paper explores the barriers to the development of RE for power generation in the Caribbean, and outlines a strategy of how to overcome these barriers. Semi-structured interviews with three "super-experts" serve to supplement the findings of a preceding literature review. Approximately 30 experts are consulted to confirm and rank the identified barriers to RE according to their importance. The end-product of this study is a ranking matrix that will serve as a strategy instrument for decision-makers, who are then able to prioritise barriers and initiate their removal.

Keywords Islands · Renewable energy · Barriers · Caribbean

Introduction

The Caribbean power generation sector depends on approximately 97 % of its energy production on imported fossil fuels (CIA 2014; IEA 2013; Byer et al. 2009). This causes not only locally harmful emissions of particular matter and nitric oxides

P. Blechinger (✉) · K. Richter
Reiner Lemoine Institut gGmbH, Ostendstraße 25, 12459 Berlin, Germany
e-mail: philipp.blechinger@rl-institut.de

K. Richter
e-mail: katharichter@hotmail.com

P. Blechinger
Department of Engineering, Berlin Institute of Technology, Fasanenstraße 89, 10623 Berlin, Germany

O. Renn
Department of Social Sciences V, University of Stuttgart, Seidenstr. 36, 70174 Stuttgart, Germany
e-mail: ortwin.renn@sowi.unistuttgart.de

S. Groh et al. (eds.), *Decentralized Solutions for Developing Economies*, Springer Proceedings in Energy, DOI 10.1007/978-3-319-15964-5_24

but also emissions of greenhouse gases causing global warming and climate change (IPCC 2014). One of the solutions to reduce the fossil fuel based power generation is the implementation of renewable energies.

Within the Caribbean area, abundant resources exist for solar and wind based power generation with annual irradiation ranging from 1700 to 2300 kWh/m^2 year and average wind velocities from 5 to 9 m/s (Stackhouse and Whitlock 2008; Xsitaaz and Clarke 2014). These resources are available on every island, and are supplemented by a huge potential for hydro power generation on larger islands with high mountains (IRENA 2012). In addition, geothermal potential can be found on all volcanic islands of the Eastern Caribbean island belt (Joseph 2008). Biomass could also be a renewable option to substitute fossil fuels based on high agricultural yields on Caribbean islands (IRENA 2012).

Even though the Caribbean is rich in clean and sustainable natural resources for economic competitive power generation, the implementation of renewable energy technology is rather slow (Shirley and Kammen 2013). As a result, the islands experience high electricity prices, energy poverty and grid connectivity deficits. In addition, the limited amount of imported fuel poses a serious challenge to a sustainable energy production due to the region's projected increase in population and thus energy demand (Insulza 2008). Furthermore, the, severe effects of climate change on the Caribbean island states are already showing up and act as a forceful reminder of the negative side effects of fossil fuel combustion. Despite recent RE promotion efforts throughout the region, more drastic measures are required to promote renewable energy and to remove existing barriers if CARICOM's set goal of a 20 % renewable electricity capacity share by 2017 is going to be reached (CARICOM 2013). This paper sets out to explore the diverse barriers and strategies to overcome those barriers that impede the use of RE for electricity production in the Caribbean today. The paper focuses on all Caribbean island states, excluding Cuba and Haiti due to their special economic and political situations. The analysis in this paper investigates not only technical and economic, but also political and social barriers.

While different geographical and political circumstances affect overarching regional analyses, this paper uses empirical research methods to identify and categorise barriers to RE into a framework that decision makers within Caribbean islands can apply. Whereas much work has been done on the barriers to RE in general (Painuly 2001; Verbruggen et al. 2010), only one academic study (Ince 2013) has focused on this specific region. Ince has set a baseline for the scientific understanding of barriers for implementing renewables in the Caribbean and pointed out that more research is needed, especially quantitative studies. While reports of different institutions have targeted the subject, they lack clear scientific methodology (CREDP 2010, 2011; IDB 2011).

Tackling both the currently slow uptake of renewable energies (RE) in the Caribbean, as well as the lack of evidence-based strategies to overcome barriers of implementation, our work aims to deliver a comprehensive overview on the most important barriers. The empirical analysis covers the views of all important stakeholders in the Caribbean power generation sector.

For this research the four identified main categories of barriers are technical, economic, political and social constraints (Blechinger 2013; Negro et al. 2012). The contribution of the present paper consists of an elaboration on these barriers, and the development of a rating matrix that includes a strategy on how to prioritise and initiate their removal.

Thus the central questions pursued are the following:

- What are the barriers to the development of RE in the Caribbean?
- Which are the most important barriers?
- What measures can be implemented to overcome these barriers?

These research questions are examined along the following structure of the paper. First the applied methods are explained and broad literature review on the methodologies is given. This is followed by the presentation of the results which are discussed in the next section. Within the discussion recommendations to remove the identified barriers are presented. Finally, the paper ends with a summary in the conclusion section.

Methods

In order to answer the three main research questions, a three-fold analysis is performed, consisting of a literature review which is complemented by a qualitative and quantitative investigation. Firstly, a literature review of peer-reviewed papers and reports extracted the existing expertise on barriers to RE and revealed the challenges to sustainable electricity production in the Caribbean. The consulted literature consisted of primary and secondary sources, as well as intergovernmental reports.

The material reviewed in the first step has been limited to literature that concerns itself with wider barriers to renewable energy in general on the one hand, and with limitations to renewable energy implementation on small island states and remote areas in low-income countries on the other hand. The search criteria followed the clustering of barriers in categories developed by Blechinger (2013) and Negro et al. (2012). The analysis sharpened the focus on market failure, so as to decide whether to include this component as an additional main barrier to the previously identified technical, political, economic and social categories. Furthermore, the literature review has been limited to cover renewable energy use for electricity production only.

In a second step, a qualitative survey closely elicits current difficulties in the implementation of RE. To this end, semi-structured interviews were conducted with three "super experts" who have diverse and extensive professional experience within the Caribbean energy sector. Since this form of data gathering provides in-depths results of discursive and explorative nature, it results in a holistic understanding of the expert's opinions on the interrelated barriers to renewable energy (Merriam 1988; Bogdan and Biklen 2003; Guba and Lincoln 1994; Magoon 1977; Patton 1980). Therefore, while the literature provides us with the barriers per se, the rich, deep data from the expert interviews contributes meaning to the factual developments and explores why the participants hold their respective views (Stainback and Stainback 1988; Joubish et al. 2011; Creswell 1994). Consequently, the questions are primarily attitudinal, seeking the expert's view and understanding of the topic, which clearly points out the nature of the Caribbean specific barriers and potential solutions of how to remove them. The data reflect subjective judgment, which implies that interpretation plays a major part in processing the results (Creswell 2003; Denzin 2011; Rossman and Rallis 2003).

Expert assessments were selected for an analysis of interaction and proposals for promoting renewable energy (Maxwell 2013). The interviewees approached are associated with the Caribbean Electric Utility Services Corporation (CARILEC), the Caribbean Community Secretariat (CARICOM) and the Deutsche Gesellschaft für Zusammenarbeit (GIZ). The latter organisation looks back on more than ten years of project experience in the Caribbean, and was heavily involved in the Caribbean Renewable Energy Development Programme (CREDP). The comprehensive process of an in-depth study allows only for a small number of interviewees. We limited the amount of interviews to three, however triangulated by a quantitative study thereafter.

The interview technique follows Witzel's conceptualisation of the problem-centred interview, but is also characterised by the methods of expert interview (Witzel 1989; Meuser and Nagel 1991). The experts have been selected based on their knowledge on RE, and represent different stakeholders, namely utilities, government and the private sector respectively. The results from the preceding literature review served as an interview guide, thereby thematically organising the insights from the literature review into a coherent and consistent approach to the analysis and comparative review of the outcomes (Witzel 1989). The previous data collection on barriers to RE via literature review thus serves as heuristic-analytical framework to generate ideas for the questions that are asked, however not dominating the talk (Witzel 2000). The previous categorisation of the barriers assisted the research team in identifying, coding and categorising patterns (Joubish et al. 2011), which resulted in an alteration of the list of barriers (Atteslander 2008). The final outcome of the interviews was thus organised as inductive/deductive mutual relationship (Witzel 2000), thereby avoiding a mere confirmation of the literature review.

The aggregation of the results of these two steps culminated in a list of 31 detailed barriers. In the third, quantitative research step, they have been subsequently weighted empirically through another round of questioning. Via email and/or telephone, over 100 experts from the private and public sector, utilities, international organisations (IOs) and academia were presented with a questionnaire containing the list of barriers, and were asked to rank them on a Likert scale from 0 to 5 (cf. Fig. 24.1). This psychometric, summated scale measures the importance of all the barriers through intensity of feelings and attitudes about each barrier (Bryman 2002; Likert 1932; McIver 1981; Spector 1992). The inclusion of a "don't know category" ensures that respondents with a non-attitude answer in a way that corresponds with another variable systematically (Paulhus and Reid 1991; Schnell et al. 2008; Schuman and Presser 1981; Converse and Presser 1986).

5	4	3	2	1	0	Z
Highest importance	High importance	Moderate importance	Low importance	Very low importance	Absolutely no import	Don't know

Fig. 24.1 Likert scale from highest importance (*5*) to absolutely no importance (*0*) and "don't know" category

To allow for more in-depth interpretation, the questionnaire contains a comment section for the participants to further elaborate. The analysis recognises the possible impact of the social desirability bias, whereby especially representatives of energy ministries might want to present their organisation in a favourable light, thereby distorting the answers (Paulhus 2002). However, also the private sector and utility representatives might deliberately exaggerate their basic agency and communion values such as personal agency and commonalty relating to their role in the advancement of renewable energy technologies (Paulhus 1984, 1991; Bardwell et al. 2001; Sullivan and Scandell 2003).

The methodological triangulation allows for a more holistic, complete and contextual understanding of barriers to RE, and provides a connection of different perspectives from different sectors (Banister et al. 1994; Denzin 1978; Flick 1995, 2011; Holtzhausen 2001; Jakob 2001). Through the combination with the qualitative analyses, the quantitative research step serves as confirmation and validation (Tashakkori and Teddlie 2003; cf. *Phasenmodell* p. 3 Jakob 2001), it increases the validity and credibility of findings (Patton 1990; De Vos 2002), and finally enhances the result's trustworthiness (Ralph 1999).

The summation of the weightings finally permitted a detailed clustering of the barriers, whereby the mean of the responses has been evaluated for both the separate stakeholder groups, as well as for the overall sample size. The end-product of this study has been a rating matrix of the identified and categorised barriers and sub-barriers. Since the ranking follows the importance and impact of the barriers, this matrix serves as a strategy instrument to allow for their removal by political and economic decision makers.

Results

The first step of the research, the literature review, produced a list of 32 barriers to renewable energies in the Caribbean, grouped into the aforementioned four broad categories. While the bulk of the analysed literature pointed in the general direction of each barrier and assisted the team in the formulation of the key and supporting questions of the interviews, it was the crucial information extracted from the responses of Mr. Williams (CARICOM), Mr. Homscheid (GIZ/CREDP) and Mrs. Jean (CARILEC) that allowed for the creation of a thorough list of Caribbean-relevant barriers to RE. Literature on barriers to renewables on small island states, for example, frequently mentioned natural barriers such as limited availability of natural resources or land as restriction to the implementation of RE (IRENA 2012; Ince 2013, Del Rio 2011). Since the former found no mentioning in the interviews, it was dropped from the list, while the latter was modified as barrier to be included as "Land use competition on islands". Homscheid (2014) illustrates this by saying "[l]and is available but it comes with certain problems. You can't put up a wind farm in the midst of a hotel development area." Williams highlighted two barriers: the risk aversion of commercial banks, as well as the lack of evidence-based assessments of RE potentials as barriers to their funding and implementation. Both aspects, were then included in the list. According to Homscheid (2014), there is no "study that was looking at the complex economics comparing one vs. the other [RE], looking at the scaling effect."

In the literature, efficiency constraints of RE technologies were given high priority as a barrier to their development (Ince 2013; Timilsina et al. 2012; Painuly 2001), yet this assumption could not be confirmed in the interviews, it was therefore deleted from the list. A significant social barrier frequently pointed to in the literature was the consumer resistance to RE, and their preference for the status quo (Reddy and Painuly 2004; Painuly 2001; Verbruggen et al. 2010; Sovacool 2009; Ince 2013). However, the interviews indicated that consumers were mostly concerned with high electricity prices (Jean 2014; Williams 2014), and possibly in favour of RE if they reduced the price level. The second step of the analysis thus altered the list, e.g. by incorporating "short terms of procurement contracts" (ECLAC and GTZ 2004) into other financial barriers, while adding "strong fossil fuel lobby" as social barrier.

Table 24.1 represents the barriers as listed in the questionnaire. The questionnaire is available for download from the Reiner-Lemoine Institute's website (2014), and contains a detailed description of the individual barriers.

Within the timeframe of this research, 30 participants coming from various scientific and practical backgrounds were asked to respond to a questionnaire developed by the research team. All experts selected were identified as highly prestigious experts for renewable energies in the Caribbean in a previous analysis. The sample includes seven respondents from the private sector (excluding utilities), six from utilities, seven from international organisations and five from governmental and five from academic institutions. Figure 24.2 and Table 24.2 show the overall ranking of all barriers by the five stakeholder groups who participated.

The six most important barriers (importance higher than 3.5 equals high importance) to all five groups are "lack of regulatory framework and legislation for private investors", "gap between policy targets and implementation", "high initial investments", "lack of regulatory framework for independent power producers (IPPs) and power purchase agreements (PPAs)", "diseconomy of scale", and "utility monopoly of production, transmission and distribution of electricity", respectively. For the five most important barriers the variance is relatively low, which indicates a shared perception of the barriers' importance among the respondents. The sixth most important barrier shows a larger variance which means more extreme values

Table 24.1 Unranked barriers to RE in the Caribbean

1.	*Technical barriers*
1.1.	Natural conditions
1.1.1.	Land use competition on islands
1.1.2.	RE impact on landscapes and ecosystems
1.1.3.	Natural disasters
1.1.4.	Lack of evidence-based assessment of RE potentials
1.2.	Technical constraints
1.2.1.	Lack of technical expertise and experience
1.2.2.	Low availability of RE technologies
1.3.	Infrastructure
1.3.1.	Inappropriate transport and installation facilities
1.3.2.	Unsuitable transmission system and grid stability issues with decentralised RE
2.	*Economic barriers*
2.1.	Price/cost
2.1.1.	High initial investments
2.1.2.	High transaction costs
2.1.3.	Diseconomy of scale
2.2.	Financial aspects
2.2.1.	Lack of access to low cost capital or credit
2.2.2.	Lack of understanding of project cash flows from financial institutions
2.2.3.	Lack of private capital
2.3.	Market failure/distortion
2.3.1.	Utility monopoly of production, transmission and distribution of electricity
2.3.2.	Small market sizes
2.3.3.	Lock-in dilemma (conventional energy supply structures block REs)
2.3.4.	Fossil fuel subsidies and fuel surcharge
3.	*Political barriers*
3.1.	Policy
3.1.1.	Gap between policy targets and implementation
3.1.2.	Lack of incentives or subsidies for RE
3.2.	Institutional capacity
3.2.1.	Lack of formal institutions
3.2.2.	Lack of RE experts on governmental level
3.3.	Regulatory
3.3.2.	Lack of legal framework for IPPs and PPAs
3.3.2.	Lack of regulatory framework and legislation for private investors
4.	*Social barriers*
4.1.	Consumer behaviour/awareness
4.1.1.	Lack of social norms and awareness
4.1.2.	Lack of educational institutions

(continued)

Table 24.1 (continued)

4.2.	Interaction networks
4.2.1.	Lack of RE initiatives
4.2.2.	Lack of local/national champions/entrepreneurs
4.2.3.	Strong fossil fuel lobby
4.3.	Cultural
4.3.1.	Dominance of cost over environmental issues
4.4.	Psychological/moral
4.4.1.	Preference for status quo

Sources ECLAC (2009), Arenas (2013), Weisser (2004a, b), Beck and Martinot (2004), ESMAP (2009), Boyle (1994), Unruh (2000), CREDP (2010), Union of Concerned Scientists (2002), Owen (2006), Timilsina et al. (2012), Quadir et al. (1995), IEA (2011), LCCC (2012)

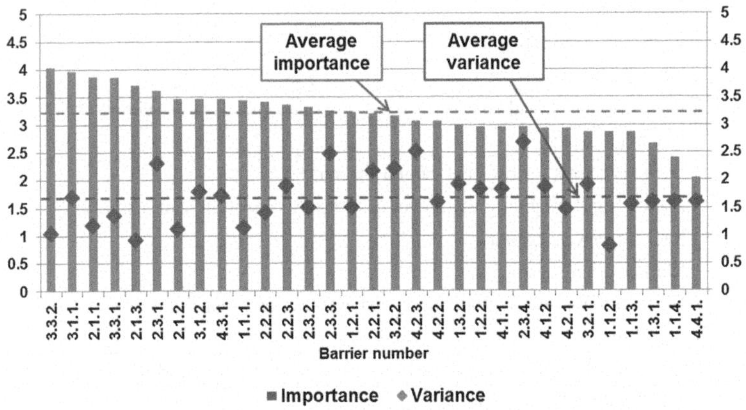

Fig. 24.2 Results of empirical weighting of barriers—overall importance and variance of each barrier

within the responses could be found. More details are explained in the discussion section.

The two barriers considered least important (importance lower than 2.5 equals to low importance) of all the groups were, beginning with the last one, "preference for status quo" and "lack of evidence-based assessments for RE potentials". Both barriers have a low variance in the importance weighting.

Overall it can be seen that the most important barriers are of economic and political nature. The most significant six items are followed by "high transaction costs" and "lack of incentives or subsidies for RE". The first social barrier is "dominance of cost over environmental issues" on rank nine with moderate to high importance. Almost the same importance is found for the item "land use competition on islands", the highest ranked technical barrier with an overall rank of ten. The remaining barriers are all ranked between importance values of 3.5 and 2.5.

Table 24.2 Results of assessment of each barrier's importance and variance by all stakeholders and importance by each single stakeholder group

Barrier no.	Barrier name	Imp. overall	Var. overall	Imp. government	Imp. organisation	Imp. private	Imp. research	Imp. utility
3.3.2.	Lack of regulatory framework and legislation for private investors	4.03	1.03	4.2	4.0	4.3	4.4	3.3
3.1.1.	Gap between policy targets and implementation	3.97	1.70	3.8	3.9	4.4	4	3.7
2.1.1.	High initial investments	3.87	1.18	4.4	3.9	3.7	4.2	3.3
3.3.1.	Lack of legal framework for IPPs and PPAs	3.86	1.36	4.2	4.3	4.0	4.2	2.7
2.1.3.	Diseconomy of scale	3.71	0.92	3.2	3.6	3.6	4.4	3.8
2.3.1.	Utility monopoly of production, transmission and distribution of electricity	3.62	2.30	3.8	4.1	4.2	4.2	1.8
3.1.2.	High transaction costs	3.47	1.12	3.4	3.7	2.9	4.2	3.3
4.3.1.	Lack of incentives or subsidies for RE	3.47	1.78	3.4	3.7	3.9	3.4	2.8
2.1.2.	Dominance of cost over environmental issues	3.47	1.72	3.6	2.9	3.7	3.4	3.8
1.1.1.	Land use competition on islands	3.45	1.14	3.4	3.1	3.2	3.8	3.8
2.2.2.	Lack of understanding of project cash flows from financial institutions	3.41	1.41	4	3.3	3.7	4	2.2
2.2.3.	Lack of private capital	3.37	1.90	3.6	3.6	3.3	4	2.5
2.3.2.	Small market sizes	3.32	1.50	3.4	3.3	3.8	3.8	2.3

(continued)

Table 24.2 (continued)

Barrier no.	Barrier name	Imp. overall	Var. overall	Imp. government	Imp. organisation	Imp. private	Imp. research	Imp. utility
2.3.3.	Lock-in dilemma (conventional energy supply structures block REs)	3.25	2.47	3	4.0	3.7	4.2	1.5
1.2.1.	Lack of technical expertise and experience	3.23	1.51	3.8	3.4	3.0	4	2.2
2.2.1.	Lack of access to low cost capital or credit	3.21	2.16	2.6	3.8	3.6	3.2	2.7
3.2.2.	Lack of RE experts on governmental level	3.17	2.21	4.4	3.7	3.1	3	1.7
4.2.3.	Strong fossil fuel lobby	3.07	2.51	2.8	4.2	3.7	3.5	1.3
4.2.2.	Lack of local/national champions/entrepreneurs	3.07	1.60	3.8	3.6	2.9	3	2.2
1.3.2.	Unsuitable transmission system and grid stability issues with decentralised RE	3.00	1.93	3.2	2.2	2.3	4.4	3.2
1.2.2.	Low availability of RE technologies	2.97	1.83	3.2	3.3	2.6	4	2.0
4.1.1.	Lack of social norms and awareness	2.97	1.83	3.8	3.6	3.3	2.4	1.7
2.3.4.	Fossil fuel subsidies and fuel surcharge	2.96	2.68	2.2	3.7	3.2	4.25	1.7
4.1.2.	Lack of educational institutions	2.93	1.86	3.6	3.7	3.1	2.4	1.7
4.2.1.	Lack of RE initiatives	2.93	1.46	3.4	3.3	3.0	2.6	2.3

(continued)

Table 24.2 (continued)

Barrier no.	Barrier name	Imp. overall	Var. overall	Imp. government	Imp. organisation	Imp. private	Imp. research	Imp. utility
3.2.1.	Lack of formal institutions	2.87	1.92	3.6	3.0	2.6	3.8	1.7
1.1.3.	RE impact on landscapes and ecosystems	2.86	0.81	2.8	2.7	3.3	2.4	3.0
1.1.2.	Natural disasters	2.86	1.57	3.2	2.4	2.8	3.2	2.8
1.3.1.	Inappropriate transport and installation facilities	2.66	1.61	3.2	2.6	1.7	3	3.0
1.1.4.	Lack of evidence-based assessment of RE potentials	2.39	1.60	2.4	2.2	2.9	2.6	1.8
4.4.1.	Preference for status quo	2.04	1.61	1.8	2.7	3.0	1.6	1.0

The variances are higher than average showing that the consensus on the ranking has been relatively low and more extreme evaluation values on both sides of the distribution curve were the rule.

Looking at the results for the five stakeholder groups, it becomes evident that there are differences in the number of barriers perceived as highly important. Researchers rank fourteen barriers with a score value of 4 or higher, thus considering these to be of high or highest importance. The participating representatives of governments and organisations ranked a total of five barriers with 4 or higher, while representatives of the private sector merely ranked four barriers within the high and highest importance. In contrast, representatives of the Caribbean utilities consider no barrier of an importance as high as 4 or above.

Another striking aspect is the discrepancy in perception of the barriers between the utilities and the rest of the stakeholders as it can be observed in Table 24.2. The items, "lack of regulatory framework and legislation for private investors" and "lack of legal framework for IPPs and PPAs" received high importance from all stakeholders with the remarkable exception of the utilities. "Lack of legal framework for IPPs and PPAs" is also seen as a highly significant barrier by the representatives of organisations. Private and academic representatives assign an importance score of four and higher to the overall second most important barrier "gap between policy targets and implementation". In comparison, political and academic representatives give an importance of four or higher to the third most important barrier "high initial investments", with the political group considering these most important. For researchers the most important barriers are "unsuitable transmission system and grid stability issues with decentralised RE" with an importance of 4.4, together with "diseconomy of scale" and "lack of regulatory framework and legislation for private investors". The first two have an overall importance of 3 (20th rank) and 3.7 (5th rank). "Land use competition on islands" is considered as most important barrier among utilities with an importance of 3.8, as well as "diseconomy of scale" and "dominance of cost over environmental issues". The first listed barrier by utility representatives is on rank 10 (importance of 3.45), the second listed is on rank 5 (importance of 3.7) and the third listed is on rank 9 (importance of 3.47) according to the overall ranking. From the governmental perspective the barriers "high initial investments" and "lack of RE experts on governmental level" are most important, with an average value of 4.4. As a comparison, overall "high initial investments" is ranked on number 3 (importance of 3.87), while "lack of RE experts on governmental level" is only ranked on 17 with an importance of 3.17.

Discussion

Three of the six most important constraints ranked alike by all stakeholder groups referred to political barriers, pointing to the leading role governments and regulatory agencies must play if they are to achieve CARICOM's 20 % renewable electricity capacity share by 2017. While two of these refer to the lack of legal and regulatory framework for renewable energy projects, one refers to the gap between policy targets and implementation. The responsibility of overcoming these three barriers lies with the governments, who must provide clear legislation and regulatory rules to secure and attract investments. Decisive government action can fill the gap between good intent and actual implementation. Ultimately, a clear regulatory framework laid out by the respective governments will allow for the removal of three of the six major barriers for implementation.

The other three most important barriers are of economic nature. Diseconomies of scale lead to high specific investment and power generation costs. In combination with the other economic barrier "high initial investments", the need for cost or price reduction to enhance the implementation of renewables is self-evident. Means of implementation may be direct or in-direct subsidies, such as tax reduction for investments or secured feed-in tariffs. By adjusting market sizes so as to become more attractive to investors could provide a significant solution that especially targets the issues of diseconomy of scale. Alternatively, the creation of a single Caribbean-wide market would allow for a similar attraction of investors in RE. The monopolistic structure of the power generation sectors in most Caribbean island states is nominated as sixth most important barrier. On smaller islands, reaping benefits of a market liberalisation is impeded by the small overall size and a lack of capacities to create proper competition without a decrease in service quality. To overcome the negative effects of monopolies two of the aforementioned measurements can serve as solution: the creation of a Caribbean wide market would generate more competition on the one hand, while on the other hand sufficient regulation could direct utilities on smaller island states towards the implementation of renewables.

Most of the solutions for the most important barriers can be provided under governmental guidance. However, local governments need support to implement these solutions. Political representatives see the lack of renewable energy experts on governmental level as one of the most important barriers. This call for support from the inside perspective of the governments clearly shows that tasks such as the creation of a regulatory framework cannot be met by current governments alone. Support of local and national organisations may strengthen the administrative capacities and renewable energy expertise that is required.

While all stakeholders of the power generation sector should contribute to overcoming the aforementioned barriers, the results expose some discrepancies in the perception of the barriers, reflected in the variance of the results. Researchers are much more sceptical about the success of renewables and rate about three times as many barriers of high importance as the private sector, governments and organisations do. In contrast, utilities express confidence in developing renewables by rating relatively few barriers of importance, which might be a reflection of their monopolistic position in the market. Yet, the private sector seems to confirm this issue by including the utilities' monopoly market distortion in their top three barriers. For this special barrier the high variance shows the different perception of the importance indicator. By virtue of their position, the utilities have given their monopolistic position in the market a low importance in impeding the development of RE. Indeed, utilities have rated their market position between low and very low importance and might even be seen as chance to implement renewables due to the power of a monopolistic utility, whereas private and academic stakeholders blame the monopolies for holding back renewables.

Other interesting barriers that received different ratings (highest variances) among the respondents are "fossil fuel subsidies and fuel surcharge", "strong fossil fuel lobby" and "lock-in dilemma (conventional energy supply structures block REs)". These barriers all relate to the power of the prevailing energy system and its decision makers, pointing to their unwillingness to change the power generation structure. This is mostly considered as barrier by new players in the market, as well as by renewable favouring organisations or research institutions. With utilities being representatives of the current energy system, they have ranked these barriers very low, as did the governmental representatives. This disagreement is a big hurdle for RE implementation as the more powerful players (utilities and governments) would not target these barriers, however considered highly important by all the other players. Utilities and governments must issue clear and binding statements with regards to the implementation of renewables, thereby removing suspicion that they would prefer the fossil based status quo.

Moreover this example emphasizes the need for improving the dialogue and communication between utilities, private sector, governments, researchers and international organisations in the Caribbean. Like their counterparts from the IOs, the representatives of academia have demonstrated a similar trend in rating many of the barriers as important or very important. Moreover, they have prioritised barriers that were not perceived to be important by the other groups at all, which points to severe communication problems between those who are implementing renewable energy projects on the ground and academia. For this group, a lack of mutual understanding impedes the development of renewable energies.

Overall, the empirical results show clear trends that match the experience of the experts interviewed in the second research step and the insights from the literature. Even though the experts included in this study are regarded as high level renewable energy experts, their assessment is not necessarily valid for all Caribbean island states and their related renewable energy projects. It may be biased with respect to the location of the expert or its disciplinary background. More case studies from each island would be necessary to test the importance of the barriers along real implemented projects.

Conclusion

For this research, qualitative and quantitative empirical methods were applied in parallel. Both have been useful to provide answers to the research questions. A total of 32 barriers (8 technical, 11 economic, 6 political and 7 social) could be identified in response to the first research question. "Lack of regulatory framework and legislation for private investors", "gap between policy targets and implementation" and "high initial investments" topped the list as the most significant barriers. They were immediately followed by the items: "Lack of legal framework for IPPs and PPAs", "diseconomy of scale" and "utility monopoly of production, transmission and distribution of electricity". To overcome these barriers the experts interviewed suggested to install a sound and effective regulatory framework for private investors and more oversight over monopolistic utilities. In addition, a Caribbean wide market was also seen as one possibility to resolve some of the economic barriers.

Apart from identifying the key barriers to the development of RE in the Caribbean, the paper addressed the systemic, overarching lack of communication and mutual understanding between the RE key players. Most of the experts were convinced that cheaper electricity prices and an environmentally sustainable and independent energy supply would be feasible if the actors were cooperating. By these recommendations, this study will advance the implementation of RE in the Caribbean and thus contribute to the region's energy security, access and sovereignty, as well as the diversification and decarbonisation of its energy production.

The findings reflect the specific character of the Caribbean region, but they can also be applied to other small developing island states with similar framework conditions. In general the methodology can easily be applied to other regions for identifying and evaluating region-specific ranking of barriers. For future research it would be interesting to compare the results of the Caribbean area to barriers for implementing renewable energies on other islands states worldwide.

Acknowledgments One author (P. Blechinger) gratefully acknowledges the Reiner-Lemoine Foundation for financing his research work. In addition we would like to thank the RLI off-grid team for valuable discussions and support during this research work.

References

Arenas, D. (2013). Transition to renewable energy in the small island developing states of the Caribbean. Master thesis, Faculty of Technology, Policy and Management, DELFT University of Technology, Netherlands.

Atteslander, P. (2008). *Methoden der empirischen Sozialforschung*. Berlin: Erich Schmidt Verlag.

Banister, P., Bruman, E., Parker, I., Taylor, M. & Tindall, C. (Eds). (1994). *Qualitative methods in psychology: A research guide*. Buckingham, UK: Open University Press.

Bardwell, W. A., Ancoli, I. S., & Dimsdale, J. E. (2001). Response bias influences mental health symptom reporting in patients with obstructive sleep apnea. *Annals of Behavioral Medicine, 23* (4), 313–317.

Beck, F., & Martinot, E. (2004). Renewable energy policies and barriers. *Encyclopedia of Energy*, 1–22.

Blechinger, P. (2013). Regional and structural differences of barriers to implement renewable energies: Implications for less or least developed countries. In *Proceedings of the International Conference Micro Perspectives for Decentralised Energy Supply* (pp. 56–59). Berlin, Germany: Technische Universität Berlin, Zentrum Technik und Gesellschaft, Promotionskolleg Mikroenergie-Systeme.

Bogdan, R. C., & Biklen, S. K. (2003). *Qualitative research for education: An introduction to theories and methods* (3rd ed.). Needham Heights, MA: Allyn & Bacon.
Boyle, S. (1994). Making a renewable energy future a reality: Case studies in successful renewable energy development. *Renewable Energy, 5*(2), 1322–1333.
Bryman, A. (2002). *Social research methods* (2nd ed.). Oxford: University Press.
Byer, T., Crousillat, E., & Dussan, M. (2009). Latin America and the Caribbean region energy sector—retrospective review and challenges. Technical paper 123/09. Washington, D.C.: The International Bank for Reconstruction and Development/The World Bank, U.S.A.
CARICOM. (2013). Caribbean sustainable energy roadmap (C-SERMS), phase 1: summary and recommendation for policymakers. http://www.caricom.org/
CARILEC. http://www.carilec.com/
CIA. (2014). The world factbook. www.cia.gov/library/publications/the-world-factbook/
Converse, J. M., & Presser, S. (1986). *Survey questions: Handcrafting the standardized questionnaire*. Beverly Hills, CA: Sage Publications.
CREDP. (2010). Analysis of the potential solar energy market in the Caribbean. http://credp.org/
CREDP. (2011). Final evaluation of CREDP programme. http://credp.org/
Creswell, J. W. (1994). *Research design: Qualitative and quantitative approaches*. Thousand Oaks, CA: Sage.
Creswell, J. W. (2003). *Research design: Qualitative, quantitative, and mixed methods approaches* (2nd ed.). Thou-sand Oaks, CA: Sage.
De Vos, A. S. (2002). Combined quantitative and qualitative approach. In A. S. De Vos, H. Strydom, C. B. Fouche, & C. S. L. Delport (Eds.), *Research at grass roots: For the social sciences and human service provisions* (2nd ed.). Pretoria: Van Schaik Publishers.
Del Rio, P. (2011). Analysing future trends of renewable electricity in the EU in a low-carbon context. *Renewable and Sustainable Energy Reviews, 15*, 2520–2533.
Denzin, N. K. (1978). *The research act: A theoretical introduction to sociological research methods* (2nd ed.). New York: McGraw-Hill.
Denzin, N. C., & Lincoln, Y. S. (Eds.). (2011). *Handbook of qualitative research* (4th ed.). Thousand Oaks, CA: Sage.
ECLAC. (2009). A study on energy issues in the Caribbean: Potential for mitigating climate change. http://www.cepal.org/default.asp?idioma=IN
ECLAC & GTZ. (2004). Renewable energy sources in Latin America and the Caribbean: Situation and policy proposals. http://www.cepal.org/default.asp?idioma=IN
ESMAP. (2009). Latin America and the Caribbean region energy sector: Retrospective review and challenges. Technical Paper 123/09. www.esmap.org
Flick, U. (1995). Triangulation. In U. Flick, E. von Kardorff, H. Keupp, L. von Rosenstiel, S. Wolff (Eds)., *Handbuch Qualitative Sozialforschung: Grundlagen, Konzepte, Methoden und Anwendungen* (2nd ed., pp. 432–434), Wein-heim: Beltz, Psychologie Verlags Union.
Flick, U. (2011). Triangulation. In G. Oelerich & H.-U. Otto (Eds.), *Empirische Forschung und Soziale Arbeit* (pp. 232–328). Wiesbaden: Verlag für Sozialwissenschaften.
Guba, E. G., & Lincoln, Y. S. (1994). Competing paradigms in qualitative research. In N. C. Denzin & Y. S. Lincoln (Eds.), *Handbook of qualitative research* (pp. 105–117). Thousand Oaks, CA: Sage.
Holtzhausen, S. (2001). Triangulation as a powerful tool to strengthen the qualitative research design: The resource-based learning career preparation programme (RBLCPP) as a case study. Higher Education Close Up Conference 2, 16–18 July 2001, Lancaster University. Available at http://www.leeds.ac.uk/educol/documents/00001759.htm
Homscheid, S. (2014). Personal communication, February 3rd, 2014.
IDB. (2011). Renewable energy best practices in promotion and use for Latin America and the Caribbean. Capital Markets and Financial Institutions Division, Discussion Paper No. IDB-DP-190. www.iadb.org
IEA. (2011). Renewable energy: Policy considerations for deploying renewables. November 2011, www.iea.org
IEA. (2013). World energy outlook 2013, 1–708.

Ince, P. D. M. (2013). Drivers and barriers to the development of renewable energy industries in the Caribbean. Doctoral Dissertation, Department of Graduate Studies, University of Calgary, Canada.

Insulza, J. M. (2008). Energy and development in South America. In C. J. Arnson, C. Fuentes, & F. R. Aravena (Eds.), *Energy and development in South America: Conflict and cooperation*. Woodrow Wilson International Centre for Scholars: Washington, D.C.

IPCC. (2014). Climate change 2014: Mitigation of climate change. Contribution of Working Group III to the Fifth Assessment Report of the Intergovernmental Panel on Climate Change. United Kingdom, New York, NY, USA: Cambridge University Press, Cambridge

IRENA. (2012). Renewable energy profiles Caribbean. September 2012, www.irena.org

Jakob, A. (2001). On the triangulation of quantitative and qualitative data in typological social research: Reflections on a typology of conceptualizing "uncertainty" in the context of employment biographies. *Forum Qualitative Social Research, 2*(1), 1–29.

Jean, A. A. (2014). Personal communication February 17, 2014.

Joseph, E. P. (2008). Geothermal energy potential in the Caribbean region seismic research unit, University of the West Indies, 2008.

Joubish, M. F., Khurram, M. A., Ahmed, A., Fatima, S. T., & Haider, K. (2011). Paradigms and characteristics of a good qualitative research. *World Applied Sciences Journal, 12*(1), 2082–2087.

LCCC. (2012). Energy policy and sector analysis in the Caribbean 2010–2011. Available at http://www.ecpamericas.org/data/files/Initiatives/lccc_caribbean/LCCC_Report_Final_May2012

Likert, R. (1932). A Technique for the Measurement of Attitudes. *Archives of Psychology, 22* (140), 1–55.

Magoon, A. J. (1977). Constructivist approaches in educational research. *Review of Educational Research, 47*, 651–693.

Maxwell, J. A. (2013). *Qualitative research design: An interactive approach* (3rd ed.). Thousand Oaks, CA: Sage.

McIver, J. P. (1981). Criminal mobility: A review of empirical studies. In S. Hakim & G. F. Rengert (Eds.), *Crime spillovers* (pp. 20–47). Beverly Hills, CA: Sage Publications.

Merriam, S. B. (1988). *Case study research in education: A qualitative approach*. San Francisco, CA: Jossey-Bass.

Meuser, M., & Nagel, U. (1991). ExpertInneninterviews—vielfach erprobt, wenig bedacht. Ein Beitrag zur qualitati-ven Methodendiskussion. In D. Garz & K. Kraimer (Eds.), *Qualitativ-empirisch Sozialforschung* (pp. 441–468). Opla-den: Westdeutscher Verlag.

Negro, S. O., Alkemade, F., & Hekkert, M. P. (2012). Why does renewable energy diffuse so slowly? A review of innovation system problems. *Renewable and Sustainable Energy Reviews, 16*(6), 3836–3846.

Owen, A. D. (2006). Renewable energy: Externality costs as market barriers. *Energy Policy, 34*(5), 632–642.

Painuly, J. P. (2001). Barriers to renewable energy penetration; a framework for analysis. *Renewable Energy, 24*(1), 73–89.

Patton, M. O. (1980). *Qualitative evaluation methods*. Beverly Hills, CA: Sage.

Patton, M. Q. (1990). *Qualitative evaluation and research methods*. (2nd ed.). Newbury Park, CA: Sage Publications.

Paulhus, D. L. (1984). Two-component models of socially desirable responding. *Journal of Personality and Social Psychology, 46*(3), 598–609.

Paulhus, D. L. (1991). Measurement and control of response bias. In J. P. Robinson, P. R. Shaver, & L. S. Wrightsman (Eds.), *Measures of personality and social psychological attitudes* (pp. 17–59). San Diego, CA: Academic Press.

Paulhus, D. L. (2002). Socially desirable responding: The evolution of a construct. In H. Braun, D. N. Jackson, & D. E. Wiley (Eds.), *The role of constructs in psychological and educational measurement* (pp. 67–88). Hillsdale, NJ: Erlbaum.

Paulhus, D. L., & Reid, D. (1991). Enhancement and denial in socially desirable responding. *Journal of Personality and Social Psychology, 60*(2), 307–317.

Quadir, S. A., Mathur, S. S., & Kandpal, T. C. (1995). Barriers to dissemination of renewable energy technologies for cooking. *Energy Conservation Management, 36*(12), 1129–1132.

Ralph, E. G. (1999). Oral-questioning skills of novice teachers: ... Any questions? *Journal of Instructional Psychology, 26*(4), 286–297.

Reddy, S., & Painuly, J. (2004). Diffusion of renewable energy technologies: Barriers and stakeholders' perspectives. *Renewable Energy, 29*(9), 1431–1447.

Reiner-Lemoine Institute. (2014). Publications, February. http://reiner-lemoine-institut.de/sites/default/files/140218_re_caribbean_questionnaire_blechinger_richter.pdf

Rossman, G. B., & Rallis, F. S. (2003). *Learning in the field: An introduction to qualitative research* (2nd ed.). Thousand Oaks, CA: Sage.

Schnell, R., Hill, P. B., & Esser, E. (2008). *Methoden der empirischen Sozialforschung* (8th ed.). München: Olden-bourg Wissenschaftsverlag.

Schuman, H., & Presser, S. (1981). *Questions and answers in attitude surveys*. Thousand Oaks: Sage Publications.

Shirley R., & Kammen D. (2013). Renewable energy sector development in the Caribbean: Current trends and lessons from history. *Energy Policy, 57*, 244–252, ISSN 0301-4215. http://dx.doi.org/10.1016/j.enpol.2013.01.049

Sovacool, B. (2009). Rejecting renewables: The socio-technical impediments to renewable electricity in the United States. *Energy Policy, 37*, 4500–4513.

Spector, P. E. (1992). *Summated rating scale construction: An introduction*. Newbury Park: Sage Publications.

Stackhouse, P. W., & Whitlock, C. H. (2008). Surface meteorology and solar energy (SSE) release 6.0 NASA SSE 6.0. Earth Science Enterprise Program, National Aeronautic and Space Administration (NASA).

Stainback, S., & Stainback, W. (1988). *Understanding & conducting qualitative research*. Reston, VA, Dubu-que, Iowa: Council for Exceptional Children.

Sullivan, B. F., & Scandell, D. J. (2003). Psychological needs and response bias: An examination of Paulhus and John's reformulation. *North American Journal of Psychology, 5*(2), 279–287.

Tashakkori, A., & Teddlie, C. (2003). *Handbook of mixed methods in social & behavioral research*. Thousand Oaks, CA: Sage Publications.

Timilsina, G. R., Kurdgelashvili, L., & Narbel, P. A. (2012). Solar energy: Markets, economics and policies. *Renewable and Sustainable Energy Reviews, 16*, 449–465.

Union of Concerned Scientists. (2002). Barriers to renewable energy technologies (pp. 1–11). Cambridge, MA. Retrieved from http://www.ucsusa.org/clean_energy/smart-energy-solutions/increase-renewables/barriers-to-renewable-energy.html?print=t

Unruh, G. C. (2000). Understanding carbon lock-in. *Energy Policy, 28*(March), 817–830.

Verbruggen, A., Fischedick, M., Moomaw, W., Weir, T., Nadaï, A., Nilsson, L. J., et al. (2010). Renewable energy costs, potentials, barriers: Conceptual issues. *Energy Policy, 38*(2), 850–861.

Weisser, D. (2004a). On the economics of electricity consumption in small island developing states: A role for renewable energy technologies? *Energy Policy, 32*(1), 127–140.

Weisser, D. (2004b). Costing electricity supply scenarios: A case study of promoting renewable energy technologies on Rodriguez, Mauritius. *Renewable Energy, 8*, 1319–1347.

Williams, J. (2014). Personal communication January 27, 2014.

Witzel, A. (1989). Das problemzentrierte Interview. In G. Jüttemann (Hrsg.) (Ed.), *Qualitative Forschung in der Psycho-logie. Grundfragen, Verfahrensweisen, Anwendungsfelder* (pp. 227–256). Heidelberg: Asanger.

Witzel, A. (2000). The problem-centered interview. In *Forum: Qualitative Social Research* (Vol. 1 (1)), Art. 22. http://nbn-resolving.de/urn:nbn:de:0114-fqs0001228

Xsitaaz, T. C., & Clarke, R. M. (2014). Large-scale wind energy potential of the Caribbean region using near-surface reanalysis data. *Renewable and Sustainable Energy Reviews, 30*, 45–58, ISSN 1364-0321. http://dx.doi.org/10.1016/j.rser.2013.09.018

Chapter 25
The Role of Gender Concerns in the Planning of Small-Scale Energy Projects in Developing Countries

Julia Terrapon-Pfaff, Carmen Dienst and Willington Ortiz

Abstract Energy poverty affects women in developing countries more severely than it affects men; at the same time, women have less control over household resources and are often not involved in decision-making processes on energy matters. In order for transition processes of energy systems to be sustainable, these gender-related concerns need to be addressed. Although this link is widely recognized, gender aspects are still not well perceived in the planning of energy projects. To better understand the role of gender concerns in project planning, the research presented in this paper evaluates concepts of small-scale sustainable energy projects with regard to their gender sensitivity. The data originates from an expert evaluation process and was analyzed with focus on gender-related aspects. The results show that even in sustainable energy projects the issue is still not high on the agenda.

Keywords Gender · Sustainable energy transitions · Project concepts · Developing countries

Introduction

More than 2.6 billion people worldwide still lack access to an affordable and dependable energy supply to meet their fundamental energy needs (IEA and OECD 2012). Without access to energy, it is unlikely to reduce the poverty of these people and to further their development. Although both men and women are affected by

J. Terrapon-Pfaff (✉) · C. Dienst · W. Ortiz
Wuppertal Institute for Climate, Environment and Energy GmbH,
P.O. Box 100480, 42004 Wuppertal, Germany
e-mail: julia.pfaff@wupperinst.org

C. Dienst
e-mail: carmen.dienst@wupperinst.org

W. Ortiz
e-mail: willington.ortiz@wupperinst.org

S. Groh et al. (eds.), *Decentralized Solutions for Developing Economies*,
Springer Proceedings in Energy, DOI 10.1007/978-3-319-15964-5_25

energy poverty, the negative implications associated with the lack of available, affordable and reliable energy supply disproportionally affect women and girls due to their traditional socio-cultural roles. In these traditional roles women are usually responsible for household tasks such as cooking, food processing, heating, water supply and washing. Providing and managing the energy sources necessary to fulfill these household tasks is also generally the woman's responsibility. Accordingly, it is mostly women who shoulder the burden of domestic energy supply in developing countries. This fact is validated in numerous studies that provide empirical evidence to support the assertion that women spend more time and travel longer distances than men to collect biomass fuels to meet domestic energy needs (e.g. Pachauri and Rao 2013; Action 2012; Oparaocha and Dutta 2011; Parikh 2011; Wickramasinghe 2003). Due to their traditional tasks in the home, women and girls are also more severely affected by the high levels of indoor smoke pollution arising from the direct combustion of biomass. Consequently, since the UN Conference on Women in Beijing in 1995, the international development community has emphasized the potential welfare improvements that access to energy services can bring to the lives of women in developing countries (Clancy 2009).

As well the fact that women have more to gain than men in terms of health and welfare, the implementation of sustainable energy technologies is also associated with the potential empowerment of women. Freeing women's time by providing reliable energy services potentially allows women to engage in income-generating activities and education (Clancy et al. 2012; Misana and Karlsson 2001). However, although this would appear to be a realistic and reasonable outcome of improved access to energy for women, the empirical evidence to support these assumptions is sparse (Dienst et al. 2013; UNDP 2011; World Bank 2008).

Despite the anticipated gender differentiated effects that sustainable energy access can have, the gender dimension itself influences energy projects in multiple ways. Gender can have substantial effects on technology preferences and on how these technologies and the generated energy are used (Pachauri and Rao 2013). Due to their gender roles, poor women and men also have different attitudes to paying for certain energy services (Skutsch 2005). In addition, gender can also play a significant role for the efficiency and long-term sustainability of energy projects (Oparaocha and Dutta 2011; Cecelski 2000).

All of these realities substantiate the need to integrate the gender dimension into development interventions addressing energy issues. In line with this necessity the role of women in development interventions has already been discussed since the early 1970s. In this first period the efforts were concentrated on women, who where mainly seen as passive beneficiaries of development aid. Particularly the potential welfare that access to energy services can bring to the lives of women in developing countries has been emphasized. In the late 1970s the focus shifted from women as recipients of aid to women as an idle economic asset. It was acknowledged that women could play an essential role for the economic development, if their access to credit, land and employment would be increased (Cecelski 2004).

This acknowledgment resulted in a series of development programs that were primarily focused on increasing the economic activities of women. But over time it became clear that interventions focusing on women in isolation did not achieve the expected success. In the course of the growing recognition, that not the sex itself but the socially attributed roles in the households and in society make women more vulnerable to poverty, the focus changed from women to gender. The emphasis on gender aligns to the fact that the change of attributes and relationships within social and cultural systems can only be achieved if the complex connections between women and men are taken into consideration (Lambrou and Piana 2006; Cecelski 2004; Cecelski 2000). Since the UN Conference on Women in Beijing in 1995 the gender dimension has been widely adopted by the international development community (Clancy 2009). Later in the year 2000 the United Nations Millennium Declaration further highlighted the importance of gender for development by naming gender equality one of the eight MDGs. Therewith gender equality was postulated as a key factor to achieve sustainable development.

However, despite theoretical knowledge and the growing number of publications on the relationship between sustainable energy and gender, little progress has been made in terms of gender equality and there is still much to learn about better integrating gender aspects into transition processes for more sustainable energy systems (Fig. 25.1). This gap, between gender mainstreaming rhetoric in concepts, guidelines, management tools etc. and the practice, where gender-sensitive energy policies, programs and projects are still the exception, has been described by numerous authors (e.g. Alston 2014; Cecelski 2004; Skutsch 1998); however, little information exists on why this discrepancy persists.

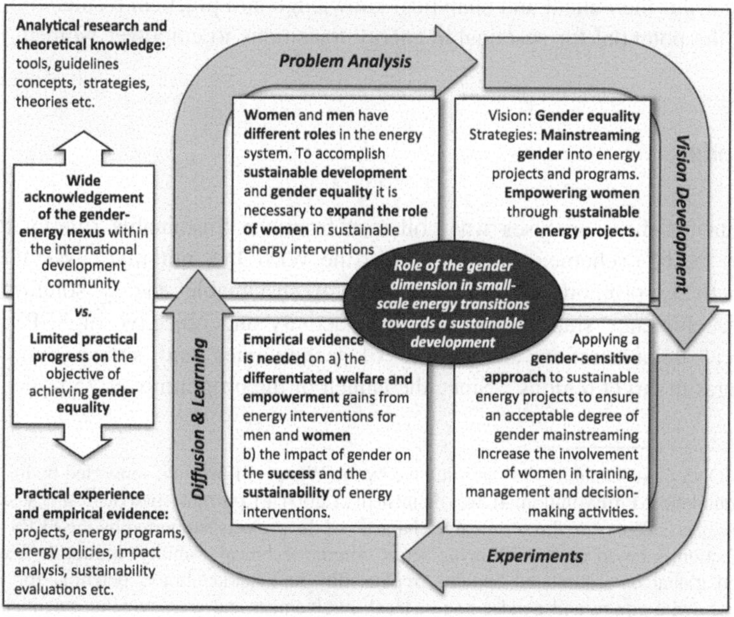

Fig. 25.1 The gender dimension within the transition cycle towards sustainable energy systems

Research Objectives

In view of these observations, by analyzing the role of gender in sustainable energy projects, this paper aims to address the question of why gender implications have been widely conceived in theory but have not been transferred into practice. The study concentrates on small-scale and local projects that are designed to meet the needs of energy-poor households and communities and should, therefore, be sensitive to gender concerns.

Within the transition cycle these projects can be seen as experiments (Fig. 25.1), where the theoretical and analytical knowledge is converted into practice in the form of subsidized technology innovations, demonstration projects or market development activities. These experiments are central to a sustainable transition process because they provide insights into the underlying developments and are often the basis for systemic change (Geels 2011). Analyzing the concepts of these project-level experiments can help to better understand and accompany this stage of the transition process.

The objective is to provide empirical evidence on (a) how gender concerns are perceived in the design of small-scale sustainable energy projects in developing countries and (b) if technology choices or geographical factors play a role in the extent to which gender is integrated into these experiments. These analyses are the first step in more comprehensive research effort that aims to analyze the role of gender and of women's empowerment in local sustainable energy projects. The overall objective of this research is to increase the understanding about how to better transfer theoretical and analytical knowledge into practical results in order to unlock the potential for sustainable energy transitions to empower women.

Methods

The data for these analyses was collected from the Sustainable Energy Project Support (SEPS) scheme, which is part of the WISIONS initiative.[1] WISION has been actively promoting the introduction of sustainable energy solutions and resource efficiency since 2004. The main activity and objective of SEPS is the "practical demonstration and testing of technologies and support for active exchanges in target regions". Since the launch of the program over 75 projects and

[1] "WISIONS of sustainability" is an initiative by the Wuppertal Institute supported by the Swiss-based foundation ProEvolution. It was launched in 2004 to promote practical and sustainable energy projects. To ensure the sustainable character of the projects supported by the SEPS scheme their selection is based on the following set of criteria: technical viability, economic feasibility, local and global environmental benefits, replicability and marketability, potential for poverty reduction, social equity and gender issues, local involvement and employment, potential, sound implementation strategy and dissemination concept. For more detailed information on the program, please visit the website www.wisions.net.

knowledge exchanges with innovative approaches that respond to energy needs at local level have been implemented in 41 countries.

To better understand the role gender concerns play in these small-scale projects, this study starts at the beginning of the project cycle by examining and evaluating the gender-sensitivity of the project concepts. In total, 192 project proposals submitted between 2007 and 2012 (of which only a small share were realized through SEPS funding) were evaluated.

The proposals that were taken into account passed the pre-validation process and were then evaluated by at least two experts against a set of well-established sustainability criteria. For the detailed examination of the proposals against these criteria, the experts provide marks on a four-point scale (0 = the proposal fails to address the issue or cannot be judged against the criterion due to missing or incomplete information; 1 = poor; 2 = fair; 3 = good; 4 = very good). For this study, the data collected was revisited and analyzed according to whether the project contributed to gender equity or addressed gender-related issues.

Results

The first analysis provides empirical evidence about the extent to which gender concerns were incorporated into the design of small-scale sustainable energy projects in developing countries (Fig. 25.2).

Looking at the distribution across the entire sample of the 192 evaluated proposals, about 16 % of the project proposals scored less than 1, meaning that gender issues were not addressed at all, 21 % of the proposals only reached 1 point, indicating that gender issues were only poorly addressed and a further 10 % attained an average score of 1.5, positioning these project concepts between 'poor' and 'fair' in terms of their gender-sensitivity. Overall, 47 % of the projects were considered to address gender issues less than fairly (Fig. 25.3). Project concepts that were rated as

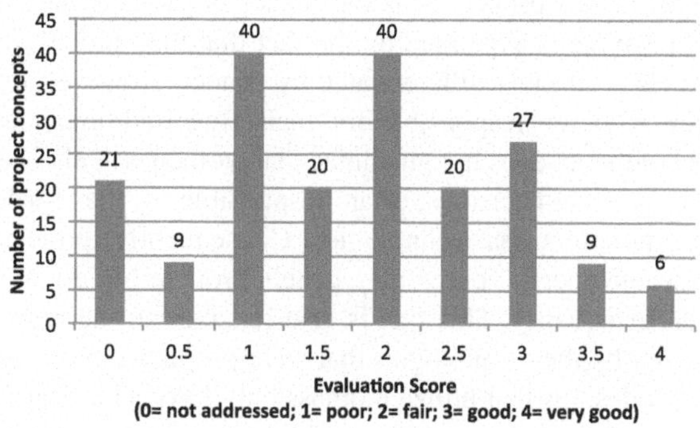

Fig. 25.2 Distribution of scores corresponding to the question of whether the project concepts contributed to gender equity or addressed gender-related issues

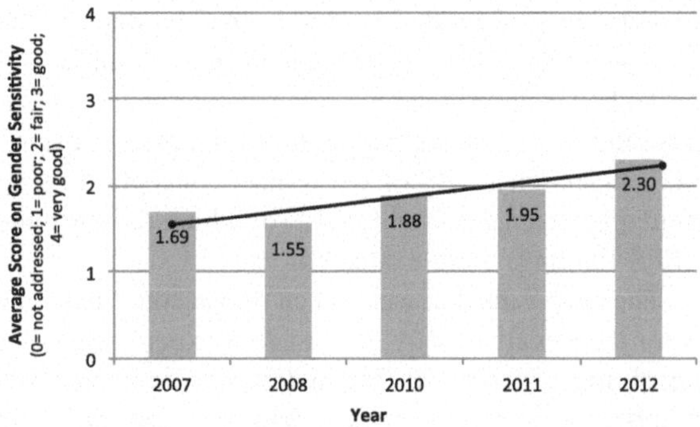

Fig. 25.3 Average score per project concept for the project cycles 2007–2012

'good' with regards to their contribution to gender-related issues on the other hand represented only 18 % of the sample. Only 3 % of the evaluated sample achieved the maximum score of 4, meaning that these project concepts fully addressed gender-related issues or contributed strongly to gender equality. These results provide empirical confirmation of the observation that despite the wide uptake of gender mainstreaming rhetoric in research literature and the availability of numerous guides on how to integrate gender into project concepts in the development cooperation literature, gender still does not play a prominent role in energy project planning.

However, when examining the data in relation to the development over time of gender awareness in project design, it can be observed that over the last 6 years the average score increased by over 0.6 points (Fig. 25.3). This trend in the data allows for the assumption that in recent years gender issues have been better incorporated into small-scale energy project planning. This might indicate that the theoretical knowledge about the importance of gender in sustainable energy transitions and lessons learned from previous project-level experiments is slowly becoming more widely adopted by practitioners and project planners.

With regards to renewable energy technologies many publications point out that gender issues are most widely addressed in the context of biomass fuels (Skutsch 2001; Dutta 2003). The stronger focus on gender in projects that use biomass as renewable energy source is explained by the fact that the burden of providing and using biomass fuels is strongly differentiated by gender. However, the fact that it is generally women who are responsible for managing traditional biomass energy sources at household level does not guarantee that women will also be the ones who are trained, who make the decisions, or are responsible for the management and/or ownership of modern biomass technologies. Consequently, projects focusing on biomass as renewable energy source may benefit women but do not automatically contribute to gender equality. This fact is also reflected in the results of the evaluation characterized by the technologies that were proposed in the project concepts (Fig. 25.5). Project designs that utilize biomass only have an average score of 1.84, similar to solar (1.82) and hydro technologies (1.86). Only project concepts that focus on wind power score higher with an average of 2.14, meaning that these,

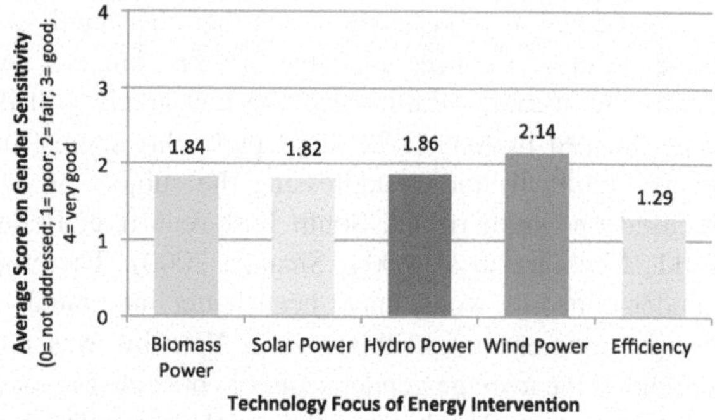

Fig. 25.4 Average score according to technology focus

at least, fairly address gender-related issues. However, wind projects represent only about 4 % of the evaluation sample, meaning that the number may not be considered as statistically significant. Taken as a whole, Fig. 25.4 shows that, based on the analyzed sample, the choice of technology has no significant influence on how far gender-related concerns are considered as part of the project idea. Project concepts that focus on energy efficiency measures contribute even less to gender equality, scoring on average 1.29. This low rating can be explained by the fact that energy efficiency measures in the evaluated concepts mainly addressed the optimization of energy appliances in public buildings. Often, the benefits of these types of improvements cannot be differentiated by gender, which explains the low scores for the contribution to gender equality.

As well as differentiation in terms of technology, the project designs were analyzed in terms of the effect of geographical location on gender concerns. The 192 reviewed project concepts were supposed to be implemented in over 40 different countries; nearly half (47 %) were located across Asia, 32 % were located in Africa and 14 % in Latin America, while only 5 project concepts were supposed to be implemented in the Middle East. The average gender-sensitivity scores of the project ideas in different regions are presented in Fig. 25.5. The low results for projects concepts in the Middle East are particularly noticeable. However, as the number of submitted project concepts for the Middle East is very small, these low scores cannot

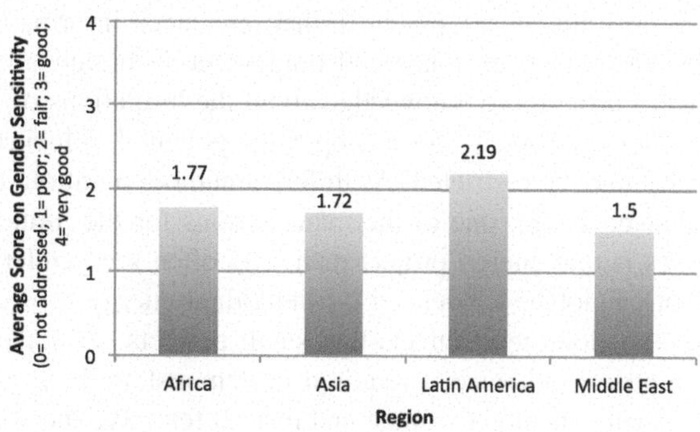

Fig. 25.5 Average score according to geographic region

be considered as representative. On the other hand, the better perception of gender issues in project concepts located in Latin America can be considered to be statistically significant. Additionally, the low gender-sensitivity rating of projects concepts located in Asia (1.72) is of particular interest because a lot of research and many empirical studies addressing the subject of gender and energy have been focused on Southern and South East Asia (e.g. Reddy and Nathan 2013; Parikh 2011; Malhotra et al. 2004; Shailaja 2000). Therefore, the expectation was that gender concerns would have been better integrated into concepts for sustainable energy interventions in this region. Yet, this was not the case in the evaluation sample as the average gender score of concepts for Asia (1.72) and Africa (1.77) did not even score 2, which is the level designated for 'fairly' addressing gender concerns.

Discussion

The main constraint of the presented study lies in the fact that it is limited to analyzing project concepts, which restrict the ability to predict in what way projects would actually contribute to gender equality in practice. While project concepts might aim to integrate gender concerns in various forms during different project stages the process of implementing these approaches into practice might proof to be more difficult than expected. This results in less gender-sensitive implementation models compared to the proposed project concept. Another restriction of the study is that the evaluation is based on expert judgments. There is a risk that these assessments can be biased based on expert's background and experience. Therefore, the extend to which gender issues are incorporated might be evaluated differently by different experts.

Notwithstanding these limitations, the results suggest that nearly a decade after Cecelski (2004) and Skutsch (2005) criticized the lack of perception of gender concerns in the planning processes of energy projects and programs, surprisingly little progress has been made. In the evaluation sample gender-related concerns are still only incorporated to a limited extent into project concepts across technologies and regions. However, there are indications that this situation is gradually changing. To push this development further, enhanced information and knowledge exchange between science, donor and implementing organizations and practitioners is needed. Furthermore, knowledge about the limitations of the gender approach must be further explored as, for example, the potential influence of small projects on gender roles may be restricted. As different authors point out (e.g. Pachauri and Rao 2013; Skutsch 1998), one of the main reasons for the limited uptake of gender perspectives is that energy project managers often see gender equality as a cultural matter that cannot be influenced by individual energy projects. To really understand the type of long-term effects that small projects can have, further research is needed. Empirical evidence is required to demonstrate how, or even if, local project-level experiments affect women and men differently, and whether the role of women can really be changed in the long term or if, eventually, men reassert their privileges. Therefore, in order for gender to become a positive driving force, instead of a barrier, in the transition process towards sustainable local energy systems, it is necessary to generate more transformation knowledge by closer analyzing the impacts and learning from the conducted experiments.

Conclusion

The results of this study show that even in concepts for small-scale sustainable energy projects gender, as a subject is still not on the forefront of the agenda. Gen der related issues were on average only addressed fairly. Merely 3 % of the sample was categorized as fully incorporating gender issues. The additional analyses that distinguished the integration of gender concerns based on the renewable energy resources and geographical location produced similar results. Although small differences existed between renewable energy sources as well as geographic regions, these proved not to be significant. That gender related issues are not well perceived in the conceptualization of sustainable energy projects suggests that the discourse around the gender dimension of energy has not yet reached the practical as well as the empirical level. Thus more efforts are necessary to intensify the discussion.

References

Action, P. (2012). *Poor people's energy outlook 2012*. UK: Warwickshire.

Alston, M. (2014). Gender mainstreaming and climate change. *Women's Studies International Forum*. Vol. 47, Part B, pp. 287–294.

Cecelski, E. (2000). *The role of women in sustainable energy development*. Colorado: National Renewable Energy Laboratory Golden.

Cecelski, E. (2004). *Re-thinking gender and energy: Old and new directions*. ENERGIA/EASE Discussion Paper.

Clancy, J. (2009). Late developers: Gender mainstreaming in the energy sector. http://www.devstud.org.uk/aqadmin/media/uploads/4ab8efeb3f827_SA3-clancy-dsa09.pdf. Accessed 10 Jan 2014.

Clancy, J., Winther, T., Matinga, M., & Oparaocha, S. (2012). Gender equity in access to and benefits from modern energy and improved energy technologies. In *World Development Report: Gender Equality and Development*. World Bank.

Dienst, C., König, J., Ortiz, W., & Terrapon-Pfaff, J. (2013). Small is beautiful!? Evaluating the sustainability of small-scale renewable energy projects in developing countries. In M. Schäfer, N. Kebir, & D. Philipp (Eds.), *Micro perspectives for decentralized energy supply. Proceedings of the international conference*, Berlin, 27 Feb–1 Mar 2013.

Dutta, S. (2003). *Mainstreaming gender in energy planning and policies, UNESCAP project on capacity building on integration of energy and rural development planning*. Background paper for expert group meeting, New York.

Geels, F. W. (2011). The multi-level perspective on sustainability transitions: Responses to seven criticisms. *Environmental Innovation and Societal Transitions, 1*, 24–40.

International energy agency (IEA)/Organization for Economic Co-operation and Development (OECD). (2012). *World energy outlook 2012: WEO2012 measuring progress towards energy for all.* Paris: International Energy Agency.

Lambrou, Y., & Piana, G. (2006). *Energy and gender in rural sustainable development.* Rome: Food and Agriculture Organization of the United Nations.

Malhotra, P. R., Neudoerffer, C. & Dutta, S. (2004). A participatory process for designing cooking energy programmes with women. *Biomass and Bioenergy, 26,* 147–169.

Misana, S., & Karlsson, G. (2001). *Generating opportunities: Case studies on energy and women.* New York: United Nations Development Programme.

Oparaocha, S., & Dutta, S. (2011). Gender and energy for sustainable development. *Current Opinion in Environmental Sustainability, 2011*(3), 265–271.

Pachauri, S., & Rao, N. D. (2013). Gender impacts and determinants of energy poverty: Are we asking the right questions? *Current Opinion in Environmental Sustainability, 5,* 205–215.

Parikh, J. (2011). Hardships and health impacts on women due to traditional cooking fuels: A case study of Himachal Pradesh, India. *Energy Policy, 39*(2011), 7587–7594.

Reddy, B., & Nathan, H. (2013). Energy in the development strategy of Indian households—the missing half. *Renewable and Sustainable Energy Reviews, 18*(2013), 203–210.

Shailaja, R. (2000). Women, energy and sustainable development. *Energy for Sustainable Development, Energy for Sustainable Development, 4*(1), 45–64.

Skutsch, M. M. (1998). The gender issue in energy project planning. Welfare, empowerment or efficiency? *Energy Policy, 26*(12), 945–955.

Skutsch, M. M. (2001). Why special attention to gender in electrification programmes? www.energia.org/pubs/papers/skutsch_electrification.pdf. Accessed 10 Jan 2014.

Skutsch, M. M. (2005). Gender analysis for energy projects and programmes. *Energy for Sustainable Development, 9*(1), 37–52.

United Nations Development Programme (UNDP). (2011). *Towards an "energy plus" approach for the poor. A review of good practices and lessons learned from Asia and the Pacific.* UNDP, Bangkok.

Wickramasinghe, A. (2003). Gender and health issues in the biomass energy cycle: impediments to sustainable development. *Energy for Sustainable Development, 7*(3), 51–61.

World Bank. (2008). *The welfare impact of rural electrification: A reassessment of the costs and benefits. An IEG impact evaluation.* Washington: The World Bank.

The manufacturer's authorised representative in the EU is Springer Nature Customer Service Centre GmbH, Europaplatz 3, 69115 Heidelberg, Germany. If you have any concerns regarding our products, please contact ProductSafety@springernature.com

Printed and bound by CPI Group (UK) Ltd, Croydon, CR0 4YY
25/09/2026
02222836-0011